AF614719

METHODS IN MOLECULAR BIOLOGY™

For other titles published in this series, go to
www.springer.com/series/7651

Systems Biology in Drug Discovery and Development

Methods and Protocols

Edited by

Qing Yan

PharmTao, Santa Clara, CA, USA

Editor
Qing Yan
PharmTao
Santa Clara, CA
USA
qyan@pharmtao.com

ISSN 1064-3745 e-ISSN 1940-6029
ISBN 978-1-60761-799-0 e-ISBN 978-1-60761-800-3
DOI 10.1007/978-1-60761-800-3
Springer New York Dordrecht Heidelberg London

Library of Congress Control Number: 2010934107

Cover illustration: See Figure 1 of Chapter 13 for more information.

Printed on acid-free paper

Humana Press is part of Springer Science+Business Media (www.springer.com)

Preface

Currently, the drug discovery industry has reached the bottleneck. Adverse drug reactions (ADRs) are one of the leading causes of death and illness in the US. In the mean time, although the industry is spending a tremendous amount of money and time, high-profile drug withdrawals are increasing, with fewer FDA approvals of new drugs. The "one-drug-fits-all" model has not been successful. There is an urgent need to change the current drug discovery and development process that has high cost, low efficacy, and high ADRs. It is necessary to develop personalized medicine that treats whole systems and brings the right drug to the right patient with the right dosages.

Systems biology emerged with the realization that genes, molecules, tissues, and organs do not work alone but interact with each other in a whole system. Combined with pharmacogenomics studies, systems biology would provide a holistic and thorough understanding of health and medicine. Such understanding would change the emphasis of medicine from diseases to humans and enable the transformation from disease treatment to prevention and health promotion.

This book has several features that readers may find helpful to their work. First of all, it focuses on translational methods through applying systems biology approaches directly in drug development and clinical practice. One of the major challenges that needs to be resolved in current bioscience is the translation of basic studies into better clinical outcomes. This book is written in response to this challenge through highlighting the development of translational medicine based on systems biology.

We hope that these approaches may help make some breakthroughs and advancement toward the realization of personalized medicine, which is also the second feature of the book. That is, most of the methods and protocols described in the book are geared toward the development of individualized therapeutics.

The third feature is that this book provides both practical methods and comprehensive resources that can be used for solving complex problems in medicine. A wide range of approaches are introduced with problem-solving objectives, from theoretical and computational analyses to experimental steps.

The fourth feature is that this book integrates the advancement of science with innovative technologies. While the first part of the book describes cutting-edge technologies and methods in the field, the second part illustrates how the technologies can be applied in science for disease understanding and therapeutic discovery.

The first part of the book introduces basic and novel concepts, as well as advanced technologies in systems biology for efficient drug discovery and development. Such concepts include proteomics, cell behavior, interactomes, and multi-drug targets. The technologies include computational modeling, Bayesian networks, translational bioinformatics, quantitative proteomics methods, microarrays, and RNA interference (RNAi). These technologies can help us with the identification of biomarker genes and pathways and understanding the interactions among genes, drugs, and diseases.

One of the potential results from systems biology studies is the identification of novel drugs tailored for individuals. Concepts such as proteomics, toxicoproteomics, epigenetics, and their roles in systematic drug target discovery and clinical trial design are introduced in the first chapter of this book. For instance, current proteomics technologies include two-dimensional gel electrophoresis (2D-GE), mass spectrometry (MS), and protein arrays.

Based on proteomic studies, multiple-target approaches are novel ways for the design of drugs against atherosclerosis, cancer, depression, psychosis, and neurodegenerative diseases (*see* Chapter 2). Novel computational and mathematical modeling are the essential methods for dealing with complex proteomic data and understanding of genetic interaction networks involved in these processes. Quantification approaches are important for the identification of protein biomarker signatures and the study of interactomes.

For example, stable isotope labeling by amino acids in cell culture (SILAC) is a quantitative proteomics method (*see* Chapter 3). It can be combined with high-resolution MS as a potent tool for functional analyses. In combination with RNAi, SILAC can address many of the systems-wide approaches that were previously impossible.

Better understanding of intracellular and cellular circuits would contribute to systems biology approaches to drug discovery. Mathematical modeling of molecular modules in a cell can link intracellular molecular machinery and cellular activity to enable the understanding of cell behavior. The cell behavior includes avoidance reaction, escape reaction, conjugation, chemotaxis, cell division, stochastic ball movement, search reaction, and Ca^{2+} concentration-dependent movement (*see* Chapter 4).

The elucidation of regulatory networks and pathways from proteomic data reveals how proteins regulate each other, which is important for drug design. Computational methods to the understanding of the functional roles of cellular networks include "static" models, as well as dynamical and stochastic simulations (*see* Chapter 5).

These methods are useful for interpretation of high-throughput interaction data, finding gene expression patterns, and building predictive models. For instance, Gaussian Bayesian network methodology is useful for the analyses of static and dynamic time series data (*see* Chapter 6). Bayesian network inference methods in the analysis of flow cytometry data can be used to evaluate regulatory network topology.

Furthermore, analyses of the large-scale self-regulatory behavior of the cell may help establish comprehensive models of the cell and genes action (*see* Chapter 7). Such methods for analyzing microarray data include principal component analysis (PCA), clustering, tree building, self-organizing map (SOM), and bootstrapping.

Based on these data mining approaches, translational bioinformatics is a powerful method to bridge the gap between systems biology research and clinical practice. Translational bioinformatics would bring novel insights in the identification of biomarkers and systemic interactions. Methods of data integration and data mining can provide decision support for both researchers and clinicians (*see* Chapter 8).

Part II of this book focuses on the application of these methods in the translation of systems biology into understanding of disease states and development of personalized therapeutics. These diseases include cardiovascular disease, cancer, lupus erythematosus, influenza, drug abuse, and brain injury. Most of these diseases have close associations with certain responses such as inflammation and immunological reactions.

For example, inflammation is a complex response involved in many diseases including rheumatoid arthritis, asthma, cancer, diabetes, atherosclerosis, Alzheimer's, and obesity. Translational applications of computational simulations applied to inflammation are reviewed in this book, such as agent-based modeling (ABM) and equation-based modeling (EBM) (*see* Chapter 9). Translational systems biology modeling efforts at various levels, from the systemic level to the cellular level are described.

The immune system plays an important role in the development of personalized medicine for a variety of diseases including cancer, autoimmune diseases, and infectious diseases. The integration of immunoinformatics with systems biology may lead to a better understanding of immune-related diseases at various levels for the development of individualized vaccines and drugs. Basic concepts and various methods are explained in detail with examples in different diseases (*see* Chapter 10).

In the development of cardiovascular disorders such as atherosclerosis, chronic inflammation also plays an important role. Systems biology tools and comprehensive bioinformatics packages can be used for the development of cardiovascular drugs and the detection of beneficial and adverse effects. Detailed protocols and methods are described in this book, from experimental design and tissue collection, to bioinformatics analysis of functional processes and pathways (*see* Chapter 11).

Cancer is a complex disease at tissue, organism, and population levels. Various computational and mathematical approaches are necessary for analyzing different types of cancer data. Comprehensive approaches are discussed in this book, from data- and process-driven theoretical methods to experimental methodologies in cancer research, as well as applications in the clinical context (*see* Chapter 12).

Systemic lupus erythematosus (SLE) is a disease with inappropriate response to self-antigens. Different levels of mechanisms and causes are reviewed in this book, including genetic, environmental, and hormonal factors (*see* Chapter 13).

Influenza virus infection is a public health threat worldwide. It is urgent to develop effective methods and tools for the prevention and treatment of influenza. New personalized vaccines, adjuvants, and drugs may result from the understanding of interactions of host genetic, environmental, and other factors. A comprehensive overview is provided in this book about systems biology studies for the development of the optimal prevention and treatment strategies for influenza (*see* Chapter 14).

Drug abuse, such as the abusive use of methamphetamine (METH), is a growing problem that may cause harmful effects to the human brain. This book describes the current protocols for analyzing the drug abuse problem, from the in vitro cell culture to in vivo rat models, as well as bioinformatics methods for establishing interaction maps to study altered functions (*see* Chapter 15).

Traumatic brain injury affects millions of patients every year in the USA, but currently no FDA-approved drugs are available to treat the problem. Traditional single drug-to-target approach has failed to work. Systems biology-based approaches may provide promising solutions through focusing on converging pathways and downstream biomarkers as potential therapeutic intervention and targeting points (*see* Chapter 16).

By covering topics from fundamental concepts to advanced technologies, this book can be used by biomedical students and professionals at all levels who are interested in integrative studies in molecular biology, genetics, bioinformatics, bioengineering, biochemistry, physiology, pathology, microbiology, immunology, pharmacology, toxicology, drug discovery, and clinical medicine. Written by leading experts in the field, this book intends to provide a practical, state-of-the-art, and holistic view of the translation of systems biology into better drug discovery and personalized medical practice.

I would like to thank all of the authors for their innovative input and sophisticated contributions to this exciting new field. I also thank the series editor, Dr. John Walker, for his help with the editing.

Qing Yan, MD, PhD
Santa Clara, CA

Contents

Contributors

GARY AN • *Department of Surgery, Center for Inflammation and Regenerative Modeling, Northwestern University, Chicago, IL, USA; McGowan Institute of Regenerative Medicine, University of Pittsburgh, Pittsburgh, PA, USA*
DECLAN G. BATES • *Systems Biology Lab, Department of Engineering, University of Leicester, Leicester, UK*
TIZIANA BONALDI • *Department of Molecular Oncology, European Institute of Oncology, Milano, Italy*
JAMES L. BOWN • *University of Edinburgh, Edinburgh, UK*
MELINDA BUCHANAN • *Department of Psychiatry, Center for Neuroproteomics and Biomarkers Research, McKnight Brain Institute of the University of Florida, Gainesville, FL, USA*
BULBUL CHAKRAVARTI • *Department of Biology and Department of Chemistry, York College, City University of New York, CA, USA*
DEB N. CHAKRAVARTI • *Department of Biology and Department of Chemistry, York College, City University of New York, CA, USA*
ALESSANDRO CUOMO • *Department of Molecular Oncology, European Institute of Oncology, Milano, Italy*
BENJAMIN L.DE BIVORT • *Rowland Institute at Harvard, Cambridge, MA, USA*
DANA FARATIAN • *Breakthrough Research Unit, Centre for Research in Informatics and Systems Pathology (CRISP), University of Edinburgh, Edinburgh, UK*
ANTONINA N. GIZATULLINA • *Biophysics & Bionics Lab, Department of Physics, Kazan State University, Kazan, Russia*
MARK S. GOLD • *Department of Psychiatry, Department of Community Health and Family Medicine, Center for Neuroproteomics and Biomarkers Research, McKnight Brain Institute of the University of Florida, Gainesville, FL, USA*
KARLFRIED GROEBE • *ProteoSys AG, Mainz, Germany*
MARCO GRZEGORCZYK • *Department of Statistics, TU Dortmund University, Dortmund, Germany*
DAVID J. HARRISON • *University of St. Andrews, St. Andrews, UK*
RONALD L. HAYES • *Banyan Biomarkers, Inc., Alachua, FL, USA*
ROBERT KLEEMANN • *BioSciences, Department of Vascular and Metabolic Diseases, TNO-Quality of Life, Leiden, The Netherlands*
FIRAS H. KOBEISSY • *Department of Psychiatry, Center for Neuroproteomics and Biomarkers Research, McKnight Brain Institute of the University of Florida, Gainesville, FL, USA*
NIKOLAY V. KOTOV • *Biophysics & Bionics Lab, Department of Physics, Kazan State University, Kazan, Russia*
VASILEIOS C. KYTTARIS • *Division of Rheumatology, Beth Israel Deaconess Medical Center, Harvard Medical School, Boston, MA, USA*

SIMON P. LANGDON • *University of St. Andrews, St. Andrews, UK*
STEPHEN F. LARNER • *Banyan Biomarkers, Inc., Alachua, FL, USA*
BUDDHADEB MALLIK • *Keck Graduate Institute of Applied Life Sciences, Claremont, CA, USA*
YURI NIKOLSKY • *GeneGo, Inc., St.Joseph, MI, USA*
ALPAN RAVAL • *Keck Graduate Institute of Applied Life Sciences, Claremont, CA, USA; School of Mathematical Sciences, Claremont Graduate University, Claremont, CA, USA*
SHANKAR SADASIVAN • *Department of Psychiatry, Center for Neuroproteomics and Biomarkers Research, McKnight Brain Institute of the University of Florida, Gainesville, FL, USA*
ANDRÉ SCHRATTENHOLZ • *ProteoSys AG, Mainz, Germany*
V. ANNE SMITH • *University of Abertay Dundee, Dundee, UK*
VUKIC SOSKIC • *ProteoSys AG, Mainz, Germany*
YOSHINORI UMEZAWA • *Kings College London, St John's Institute of Dermatology, London, UK*
NAJL V. VALEYEV • *Systems Biology Lab, Department of Engineering, University of Leicester, Leicester, UK; Kings College London, St John's Institute of Dermatology, London, UK*
RAVISHANKAR R. VALLABHAJOSYULA • *Keck Graduate Institute of Applied Life Sciences, Claremont, CA, USA*
YORAM VODOVOTZ • *Center for Inflammation and Regenerative Modeling, McGowan Institute of Regenerative Medicine and Department of Surgery, University of Pittsburgh, Pittsburgh, PA, USA*
KEVIN K.W. WANG • *Department of Psychiatry, Center for Neuroproteomics and Biomarkers Research, McKnight Brain Institute of the University of Florida, Gainesville, FL, USA; Center of Innovative Research, Banyan Biomarkers, Inc., Alachua, FL, USA*
QING YAN • *PharmTao, Santa Clara, CA, USA*
ZHIQUN ZHANG • *Center of Innovative Research, Banyan Biomarkers, Inc., Alachua, FL, USA*

Part I

Methodologies of Systems Biology in Drug Discovery and Development

Chapter 1

Proteomics and Systems Biology: Application in Drug Discovery and Development

Bulbul Chakravarti, Buddhadeb Mallik, and Deb N. Chakravarti

Abstract

Studies of complex biological systems aimed at understanding their functions at a global level are the goals of systems biology. Proteomics, generally regarded as the comprehensive study of the expression of all the proteins at a particular time in different organs, tissues, and cell types is a key enabling technology for the systems biology approach. Rapid advances in this regard have been made following the success of the human genome project as well as those of various animals and microorganisms. Possibly, one of the most promising outcomes from studies on the human genome and proteome is the identification of potential new drugs for the treatment of different diseases and tailoring the drugs for individualized patient therapy. Following the identification of a new drug candidate, knowledge on organ and system-level responses helps prioritize the drug targets and design clinical trials based on their efficacy and safety. Toxicoproteomics is playing an important role in that respect. In essence, over the past decade, proteomics has played a major role in drug discovery and development. In this review article, we explain systems biology, discuss the current proteomic technologies, and highlight some important applications of proteomics and systems biology approaches in drug discovery and development.

Key words: Proteomics, Systems biology, Drug development, Drug discovery, Mass spectrometry, Two-dimensional gel electrophoresis, 2D-GE, Protein arrays

1. Introduction

Systems biology refers to the study of the interrelationships of all the different elements in a biological system and not studying them in an isolated manner one at a time. As explained by Weston and Hood (1), biological information moves from the genome of organisms to ecologies, the relationship between organism and their environment, in a hierarchical manner (DNA → RNA → protein → biomodules, protein interactions, protein or gene regulatory networks → cells → organs → individuals → populations

Qing Yan (ed.), *Systems Biology in Drug Discovery and Development: Methods and Protocols*, Methods in Molecular Biology, vol. 662, DOI 10.1007/978-1-60761-800-3_1, © Springer Science+Business Media, LLC 2010

of individuals → ecologies). As evident from the above hierarchy, changes in environmental signals can change the biological information at each level. Thus, for the systems biology approach, it is important to gather information at as many levels as possible followed by the integration and meaningful interpretation of such data in the context of the biology of the specific organism. By integrating the biological knowledge, the ultimate goal of systems biology is to understand how the molecules act together within the network of interactions that makes up life. One of the major challenges of systems biology is to determine the architecture of protein and gene regulatory networks and to understand how their behaviors are integrated to carry out biological functions.

Progress in systems biology research has been possible due to advancements as well as cross-disciplinary research among different branches of science and technology such as biology, chemistry, computer science, bioinformatics, engineering, mathematics, and physics. In this regard, specific mention should be made of the human genome project describing the human genes and *cis*-control elements; the availability of high throughput platforms for genomics, proteomics, and metabolomics, which makes possible the rapid acquisition of global data sets; high speed DNA sequencers; DNA arrays; rapid methods for genotyping; high throughput proteomics capability including protein chip array and particularly mass spectrometry (MS); availability of the internet for acquiring and disseminating the knowledge-base of large global data sets on DNA, RNA, proteins as well as their interactions.

For diagnosis as well as prognosis of any disease, it is fundamentally important to understand the underlying molecular basis associated with the disease. Until very recently, the detection of diagnostic biomarkers as well as the development of a drug target was focused mostly on single molecules. Due to rapid technical advancement particularly in omics biology, such as genomics, proteomics, and metabolomics, it is becoming increasingly evident that a cellular behavior associated with normal physiology or a disease pathology is the outcome of interaction at various levels which take place among different cellular components. One of the major goals of systems biology is to understand the role of protein biomodules (groups of proteins that perform a particular function such as galactose metabolism, protein synthesis, etc.) as well as their interconnections which give rise to networks in physiology and pathology. It is thus important to conduct basic research on a systems biology-based approach to understand the normal biological systems as well as the pathological states. The successful treatment of many diseases will depend on overcoming genetic or protein defects such as genetic mutation, aberrant protein–protein interaction, protein misfolding, protein mislocalization, etc.

Knowledge of the physiologically healthy system will help us pinpoint any such defect present in a diseased system. With the help of continuously growing technology platforms, particularly due to rapid advancement of the proteomic technology, it is becoming increasingly feasible to carry out a systems biology approach to profile global cellular and molecular changes associated with healthy and diseased state. This level of information is helpful for a more rational and effective drug design to overcome the malfunctioning of the network and come up with a better treatment strategy.

2. Proteomics and Systems Biology

Proteome includes all proteins present in a cell, tissue, or organ at a specific point of time. Analogous to the term genomics, which involves the studies of an organism's entire genome, the studies on the entire set of proteins is called proteomics. Proteomics, the parallel separation, detection, quantification, and identification of all proteins present in a cell, tissue, or organ as well as the analysis of their properties such as posttranslational modifications and interactions provide a more detailed information about any particular biological system compared to that obtained from the genome or mRNA expression profiling of that system. A vast amount of DNA sequence information is available for both eukaryotes and prokaryotes whose entire genome has been sequenced. The genome is rather static and mRNA profiling gives a snapshot of gene expression level at a certain time and condition. Although nucleotide microarray for mRNA profiling is useful for multiplexed comparative analysis of gene expression levels which provide important insights into various cellular mechanisms under different conditions (2), proteins are in general far more directly involved in performing almost all of the cellular activities. Interestingly, as it is turning out, the number of proteins in a cell is much higher than the number of genes due to differential splicing and different posttranslational modifications. Apart from this, sometimes there is a lack of correlation between mRNA expression and protein expression levels due to varying stabilities and life times (3–5). In fact, rather than just providing quantitative measurement of different proteins, detailed proteomic analysis provides lot more information such as posttranslational modification, subcellular localization, etc. As a result, the analysis of the proteome is far more informative than mRNA expression profiling and offers information with different type of value. Proteomics is becoming increasingly important to enhance our knowledge in biological as well as medical research. However, studies on protein expression face several analytical challenges since the proteins are expressed over a wide dynamic range, 1–10^6 in cells, and greater than 1–10^9 in serum. While the most abundant proteins in serum such as albumin is present in mg/mL level (30–50 mg/mL), low abundant proteins such as IL-6 is present in pg/mL level (0–5 pg/mL) (reviewed by Anderson and Anderson (6)). However, no single method can measure protein expression levels over such a wide dynamic range. Also, there is no amplification method for proteins analogous to the polymerase chain reaction (PCR) for amplification of genes. In addition, there is a change in protein expression pattern under various developmental, physiological, pathological, and environmental conditions which requires a way of taking global snapshots of patterns of protein expression. The combination of different proteomic technologies is required for such purpose. The goal of proteomics is multifaceted and involves a study of proteins with reference to their expression, posttranslational modification, protein–protein interaction, protein–DNA interaction, protein–lipid interaction, protein processing and turn over, cellular and subcellular localization, at a global level. In fact, significant progress has been made toward achieving these goals since the term "Proteomics" was coined (7). However, it is necessary to integrate proteomic data with other information such as gene, mRNA, and metabolite profiles to fully understand how the system works.

3. Current Proteomic Technologies

As explained above, there is currently a vast collection of DNA sequence information as well as gene expression data at the mRNA level. However, these data do not provide direct information on the levels of proteins or their states of modification. Due to the recent advancement in the field of biological MS in conjunction with protein/DNA-sequence database search algorithms for the identification of proteins from MS data, it is possible to identify proteins with unprecedented speed and accuracy. However, it still remains a challenge to obtain quantitative information regarding the levels of identified proteins as well as site-specific modifications and their extent within individual protein molecules. In the sections below, we will describe different methods used in proteomics for the quantitative comparison of individual proteins present in cell pools that differ in some respect from one another (such as normal vs. disease) and for accurately determining changes in the levels of posttranslational modifications (such as phosphorylation, glycosylation) at specific sites of the individual proteins. Some of these methods can be applied directly to mixture of proteins and do not require a complete separation of individual protein components.

The widely used methodologies for proteomics technology in general depends on (a) protein and/or peptide separation using two-dimensional-gel electrophoresis (2D-GE) or liquid chromatography, and (b) protein identification using MS. All these procedures are highly dependent on computer technology such as software for the image analysis of 2D gels, automated spot picking from the 2D gels, DNA or protein sequence database searches, proteome databases, and protein interaction maps (8). A third method using protein microarrays is also becoming a widely used proteomic technology for quantitative proteomics. In addition, surface-enhanced laser desorption/ionization time-of-flight mass spectrometry (SELDI-TOF-MS)-based ProteinChip® System has also been used for proteomic studies.

3.1. Two-Dimensional Gel Electrophoresis and Mass Spectrometry

3.1.1. Two-Dimensional Gel Electrophoresis

2D-GE has become a key technology in proteomics because of its high resolution as well as the ability to detect proteins with different kinds of posttranslational modification. The technique of 2D-GE was first introduced by Margolis and Kenrick (9) and later modified by O'Farrell (10) and Klose (11). In 2D-GE, proteins are separated by two orthogonal parameters, isoelectric point (pI) and molecular weight. The initial step of separation, isoelectric focusing (IEF) separates proteins based on differences in pI (the pH at which the net charge of any protein molecule is zero). This step uses precast immobilized pH gradient (IPG) strips and is available from vendors such as GE-Health Care Life Sciences and Bio-Rad Laboratories. The second dimension is sodium dodecyl sulfate-polyacrylamide gel electrophoresis (SDS-PAGE), which separates proteins based on their molecular weight; precast polyacrylamide gels with a variety of chemistries and gradients are commercially available from different vendors such as Bio-Rad Laboratories.

There are some limitations of 2D-GE, such as poor reproducibility due to gel-to-gel variation, difficulty in automation, as well as poor ability to detect the low abundant proteins, hydrophobic proteins and proteins with very high or very low molecular weight and pI. The problem encountered in detecting low abundant proteins can be overcome by choosing an appropriate technique that enriches low abundant proteins prior to separation.

The method of 2D-GE is widely used to separate proteins from two (or more) different samples followed by the visualization of separated protein spots with an appropriate stain. Following staining, the 2D gel images are captured using a charge couple device-based system or a laser scanner. The identification of protein spots which have undergone differential expression or post-translational modification is carried out using computer-assisted appropriate image analysis software (such as Image Master Platinum 2D, GE Healthcare Life Sciences; PD Quest, Bio-Rad Laboratories; Progenesis SameSpots, Nonlinear dynamics, etc). For expression proteomics, different fluorescent stains such as SYPRO

Ruby (Bio-Rad Laboratories), Deep Purple (GE Healthcare Life Sciences), Flamingo (Bio-Rad Laboratories); as well as silver stain (Bio-Rad laboratories); and Biosafe coomassie (Bio-Rad Laboratories) are used. Fluorescent stains are also available for the detection of posttranslational modification present in glycoproteins (Pro-Q Emerald 300 glycoprotein stain, Molecular Probes) as well as phosphoproteins (Pro-Q Diamond Phosphoprotein gel stain, Molecular Probes). The stains used for 2D-GE should be compatible with the mass spectrometer since the proteins of interest are typically identified by MS following their separation by 2D-GE. Although a number of improvements have resulted in an increased reproducibility of proteome pattern between different laboratories using the 2D-GE technique, some major concerns still remain, for example, about the inability to resolve all the proteins of interest present in a biological sample.

Two-dimensional difference gel electrophoresis (2D-DIGE) (12) is a relatively new technique in 2D-GE for a multiplex quantitative analysis of the component proteins of related but different protein samples (see review by Chakravarti et al. (13)). Commercially available CyDye DIGE Fluor minimal dyes (CyDye DIGE Fluor Cy2 minimal dye, CyDye DIGE Fluor Cy3 minimal dye and CyDye DIGE Fluor Cy5 minimal dye), with similar structures but different spectral characteristics, are widely used for 2D-DIGE. All of the CyDye DIGE fluor minimal dyes contain *N*-hydroxy succinimidyl ester group which forms covalent bonds with the ε-amino groups of the lysine residues of a protein through an amide linkage. In the DIGE technology, prior to electrophoretic separation, different protein samples are labeled with different fluorescent dyes, pooled together, and the proteins present in the pooled samples are separated by IEF followed by SDS-PAGE. Thus, it is possible to coseparate and codetect different but usually related protein samples on the same gel (12, 14). The technique of 2D-DIGE is carried out in four different steps: Step (1) – usually, two different but related sets of protein samples (such as normal and diseased tissue extract) are labeled with Cy3 and Cy5 separately. A pooled protein sample consisting of the same amount of all individual samples for one particular set of experiment is labeled separately with Cy2 and this is used as the pooled internal standard. Step (2) – separation using 2D-GE of the mixture of one set of samples, that is, one protein sample labeled with Cy3, one protein sample labeled with Cy5, and one pooled internal standard labeled with Cy2 is carried out on the same gel. Step (3) – the same gel is scanned three times with fluorophore-specific (Cy2, Cy3, and Cy5) excitation and emission wavelength and images are captured separately. Step (4) – image analysis and intragel quantitative comparison of all spots are carried out using the DeCyder Differential Analysis Software or Image Master Platinum 2D (GE-Health Care) or any other appropriate software such as PDQuest (Bio-Rad Laboratories) and Progenesis SameSpots (Nonlinear Dynamics). However, the DeCyder software is most widely used for DIGE image analysis. The use of the internal standard containing all the test samples ensures that all proteins present in the samples are represented, thereby allowing intragel (within the same gel) as well as intergel (within a number of gels) matching. Since the pooled internal standard is labeled with Cy2, intragel differences in protein abundances are determined by calculating the average ratio (Cy3:Cy2):(Cy5:Cy2). Inter gel comparison is carried out to find out the statistical significance of variation of any particular protein spot between different samples. Contrary to the traditional 2D-GE, the application of 2D-DIGE is restricted to protein expression and not for posttranslational modification of the proteins.

For DIGE analysis, in addition to CyDye DIGE Fluor minimal dyes used for minimal labeling, CyDye DIGE Fluor saturation dyes are available which consist of two different dyes namely CyDye DIGE Fluor saturation dye Cy3 and CyDye DIGE Fluor saturation dye Cy5. These dyes have no intrinsic charge, contain thiol reactive maleimide groups and label all cysteine sulfhydryl groups in a protein sample. It is particularly useful for labeling protein samples available in limited quantity. Saturation labeling is much more sensitive than minimal labeling, as more fluorophore is incorporated into each protein species. However, proteins that do not contain cysteine cannot be labeled using saturation dyes.

Since in saturation labeling, a Cy2 fluor is not available, the internal standard is labeled with one of the CyDyes and samples are labeled with the other CyDye. As a result of this, for similar experiments, saturation labeling requires twice as many gels as minimal labeling. For saturation labeling of a new sample, it is important that labeling conditions must be optimized to ensure the complete reduction of disulfide bonds and stoichiometric labeling of cysteine residues.

3.1.2. Mass Spectrometry

Several review articles on MS-based proteomics are available (15–19). Protein identification by MS can be carried out by the analysis of the whole protein ("top-down" proteomics) or that of peptides obtained from enzymatic or chemical cleavage of the protein ("bottom-up" proteomics). For either the "top-down" or the "bottom-up" approach, it is crucial to separate the proteins and/or peptides, as applicable, using different methods of separation, such as reverse phase, ion-exchange, size-exclusion, IEF, etc. Many of these techniques can be carried out either off-line or online with the mass spectrometer.

Mass spectrometer measures the mass/charge (m/z) ratio of the gas phase ions. The mass spectrometer consists of an ion source, a mass analyzer and a detector. The ion source converts the analyte molecules into gas-phase ions, the mass analyzer separates the analyte ions according to their *m/z* ratios, and the detector detects and records the number of ions at each *m/z* value. In proteomics research, proteins and/or peptides are ionized by any one of two different soft ionization techniques: (a) electrospray ionization (ESI) (20), and (b) matrix-assisted laser desorption/ionization (MALDI) (21). There are four different types of mass analyzers which are commonly used for proteomics research. They vary in their physical principles and analytical performance – quadrupole ion trap (QIT), linear ion trap (LIT) or linear trap quadrupole (LTQ), time of flight (TOF), and Fourier transform ion cyclotron resonance (FTICR). There are several hybrid instruments which combine the capabilities of different mass analyzers such as Q (quadrupole)-TOF, TOF–TOF, and LTQ-FTICR. Orbitrap is a new type of Fourier transform mass analyzer with high resolution, mass accuracy and dynamic range which is available in hybrid instruments such as LTQ Orbitrap (Thermo Fisher Scientific).

Tandem mass spectrometry (MS/MS) is a widely used technique for protein identification in which mass spectra of peptide fragment ions are obtained to determine amino acid sequence information and posttranslational modification. Among the different fragmentation techniques, collision-induced dissociation (CID) is used most widely. More recent fragmentation techniques such as electron capture dissociation (ECD) (22, 23) and electron transfer dissociation (ETD) are used as well (24, 25).

A widely used application of MS in quantitative proteomics is the identification of differentially expressed proteins which uses stable isotope tags to distinctly label proteins from two different conditions. In this method, the proteins present in two sets of similar samples are first separately labeled with different isotopes, combined together and then digested to yield labeled peptides. The peptides containing two different labels from the two sets of samples are separated by multidimensional liquid chromatography followed by their analysis using MS/MS. Proteins are identified by automated database searches of corresponding peptide MS/MS data and relative protein abundances are obtained from the intensity of the ion current in the mass spectra. Isotope coded affinity tag (ICAT) reagents (26) are the most commonly used isotope tags for this purpose and consist of three main components (a) a reactive group (iodoacetic acid) with specificity for cysteine residues; (b) a linker labeled with either light hydrogen (d0) or heavy hydrogen (such as eight deuterium: d8) isotope; and (c) an affinity tag (biotin) for the solid-phase capture and isolation of labeled peptides. Two different samples containing similar protein mixtures obtained from two different conditions are labeled separately with either the hydrogen-containing reagent or the deuterium-containing reagent. The two samples are then mixed together, digested with trypsin, and isotopically labeled peptides are purified by avidin affinity chromatography. The purification step eliminates noncysteine-containing peptides and thus reduces the complexity of the sample to be analyzed. The disadvantage of this method is that cysteine residues might be absent in the proteins of interest or might not be available for alkylation using cysteine-specific ICAT reagent due to posttranslational modifications of these residues. However, by using alternative chemistries, other amino acid residues can also be labeled with a stable isotope and analyzed (27–29). While this leads to different coverage if used alone, additional coverage can be obtained if the data obtained are combined with that from the ICAT approach. For example, Goodlett et al. (27) used permethyl esterification of peptides for the relative quantification of two different protein extracts and automated de novo sequence derivation of the same dataset. The method did not require the presence of cysteine-containing proteins in the extract. Each of the protein mixture was digested with trypsin separately and then methylated with d0- or d3-methanol. By this process, carboxylic acid residues present on the side chain of glutamic acid, aspartic acid, and the carboxyl terminus were converted to their corresponding methyl esters. The separate mixture of d0- and d3-containing peptides were combined and analyzed by liquid chromatography tandem mass spectrometry (LC-MS/MS). Following correlative database searching of peptide tandem mass spectra, the parent proteins of the methylated peptides were identified. It is possible to obtain ratios of the proteins in the two original mixtures by normalization of the area under the ion current intensity curve for identical charge state of d0- to d3-methylated peptides. These investigators also developed an algorithm to derive peptide sequence de novo by the comparison of tandem mass spectra of the d0- and d3-peptide methyl esters. In another approach (28), using 2-methoxy-4,5-dihydro-1H-imidazole, a lysine specific labeling reagent, it was possible to convert the lysine residues at the C-terminus as well as tryptic digests of the proteins to 4,5-dihydro-1H-imidazole-2-yl derivatives. Since the mass spectra of the derivatized peptides show greater number of more intense features than their underivatized counter parts, more information could be obtained from peptide-mapping experiments. Also, this label can be produced as stable isotopic forms containing four deuterium atoms which can be used for differential quantitative studies. Several improvements made in isotope tagging and data analysis have made stable isotope tagging a useful technique for quantitative proteomics studies.

Interestingly, some other chemical reactions have also been used to introduce tags into specific sites of peptides or proteins to probe specific functions of proteins. For example, phosphorylated peptides have been isolated using isotope labeling and selective chemistries to enrich phosphoproteins from a complex mixture (30–34). Since level of phosphorylation plays an important role in several disease conditions, the quantitative comparison of this posttranslational modification in control and disease cells is helpful to monitor disease progression, cure as well as the identification of putative drug candidates. In fact, phosphoproteomics has been used as a new research front leading to drug discovery. It is estimated that approximately one-third of the total cellular proteins undergoes posttranslational reversible phosphorylation and play important roles in the regulation of cellular signaling network in response to external stimuli and signal transduction mechanisms (35–37). Dysregulation in the processing of these signals has been implicated to be associated with human pathologies, including many forms of cancers (36, 37). Thus, it is of no surprise that a new research front is being developed for high-throughput structure-function profiling of phosphoproteomes, and the information is used to find effective drug targets as well as for the development of early detection of disease biomarkers. In eukaryotes, serine residues are the most common sites of phosphorylation (~90%), followed by threonine (~10%) whereas tyrosine residues are rarely phosphorylated (~<0.05%) (37). Phosphoproteomics involves a high-throughput identification and quantitative analysis of the phosphorylation states of proteins in a given biological sample and relating them to the signaling events as well as understanding of the underlying biological implications. A database has been developed to maintain the information of all possible phosphorylated sites in all possible proteins (http://www.phosphosite.org). It already lists >30,000 phosphorylation sites on >17,000 proteins. As the analysis of new samples continues, the database will grow larger with time. We will briefly discuss the sample preparation prior to MS based high-throughput profiling of phosphoproteomes.

Sample enrichment prior to MS is necessary particularly for the analysis of phosphoproteins. The dynamic range of proteomes and phosphoproteomes are in the same order (~10^9) (37), but the analysis of phosphoproteomes is difficult due to the substoichiometric and transient phosphorylation of these proteins. The goal of studying phosphoproteomes is to identify sites of phosphorylation, the quantification of such site stoichiometry, and monitoring the link between cellular perturbations and the temporal changes in phosphorylation levels. Many proteins that are involved in signaling networks are only sparingly expressed. Together with this, the transient phosphorylation/dephosphorylation dynamics generate a pool of extremely low levels of phosphopeptides after proteolytic digestion in comparison to large number of total peptides in cell or tissue extract. Consequently, the MS signal intensities of phosphopeptides of interest are greatly suppressed and thus the efficacy of such method is questionable. Therefore, the enrichment of the phosphorylated proteins and peptides in the sample before MS is necessary. Currently a number of phosphopeptide enrichment strategies are followed based on the principle of immunoprecipitation and some specialized chromatographic techniques. A number of recent review articles are available that discuss in detail about these methods including the choices of one method over the other depending on the final objective (33, 37, 38). For example, immunoprecipitation using antiphosphotyrosine antibodies is generally applied to enrich phosphotyrosine proteins and peptides since compared to phosphorylated serine and threonine residues, the occurrence of tyrosine phosphorylation is very low. On the other hand, antibodies specific to phosphoserine and phosphothreonine are not very effective for their enrichment and therefore are not used routinely. For global qualitative phosphoproteome profiling, the most common technique used for the enrichment process is the immobilized metal affinity chromatography (IMAC). It depends on the high affinity of phosphate groups for metal ions such as Fe^{3+}, Zn^{2+}, and Ga^{3+}. Although the highly negative environment around phosphate groups have higher preference to bind with strong metal cations, weakly acidic peptides may also bind with the metal ions leading to the enrichment of nonphosphorylated acidic peptides. Consequently, during IMAC, nonphosphoproteins and peptides will be eluted along with phosphorylated substrates as well. However, a multistep separation and/or mixture of multiple metal ions could be employed to reduce the heterogeneity of the enriched proteins and peptides. Also, the conversion of carboxylate groups to ester groups can eliminate this nonspecific interaction (34). The titanium dioxide-based metal oxide affinity chromatography (MOAC) is another form of IMAC with some added advantages, such as shorter preparation time and increased capacity compared to the IMAC resin. However, nonspecific retention of nonphosphorylated peptides is also a problem for this technique. Other methods such as strong cation exchange (SCX), a blend of anion and cation exchangers (ACE), are also used for the selective enrichment of phosphopeptides. In practice, generally multistep enrichment strategy is followed using a combination of the above techniques. The development of technically less complicated and less expensive enrichment strategies is expected in future.

The quantification of phosphoproteins has important application in clinical research. Changes in phosphorylation level before and after a cellular perturbation are linked with the specific biological response. Generally, phosphorylated samples are labeled with nonradioactive stable isotopes through chemical modifications or through the addition of labeled amino acids in cell culture. Label-free methods are also used. However, recent developments in this field have mostly been devoted to the effective enrichment strategies and qualitative identification of global or specific phosphorylation sites rather than the quantitation aspects. One of the most used quantitation methods in the multiple chemical modification to incorporate stable isotopes is "isobaric tag for relative and absolute quantitation" (iTRAQ). The isobaric tags are a group of designed chemical reagents, each one consisting of a "reporter region," a "balance region," and an amine-specific "peptide reactive group region" (iTRAQ™ Reagents, Applied Biosystem, http://www.appliedbiosystems.com). In a specially designed and controlled MS/MS fragmentation method, the bonds between the reporter region and the balance region as well as that between the balance region and the reactive group region are broken along with the usual peptide fragmentation. In a four-plex version (currently eight-plex is also available) of the iTRAQ reagents, the individual masses of the reporter region and the balance region are different, but the combined mass is maintained constant. Therefore, even if a digested peptide fragment is tagged with four such reagents, the MS signal will not split. However, during MS/MS, one can locate the distinct reporter signals which are designed in a way such that these signals do not interfere with peptide fragment ions. The peak area determined for the reporter ions can quantify the corresponding peptides originating from different samples. In addition, the MS/MS ion pattern or signature will allow the identification of the protein from sequence database using suitable algorithm. Thus, iTRAQ is a method for the simultaneous identification and quantitation of proteome including the phosphoproteome.

Another powerful quantitative technique is "stable isotope labeling with amino acids in cell culture" (SILAC). A specific cell type is grown in two different media, one containing natural amino acids and the other containing an isotopic form. All proteins expressed in two different media will be expected to incorporate either natural or isotope-containing amino acids. Following enzymatic digestion and mixing of the samples, each peptide is expected to generate a pair of two peaks with difference in mass based on the isotope type in MS run. The different peak intensities for such pairs can give an estimate of the relative quantities of a particular protein compared to others. Significant isotope incorporation requires multiple cell divisions and because of this reason SILAC is not a suitable method of quantitative proteome analysis for the comparison of primary cells or tumor tissues. Different advantages and disadvantages and precautionary measures required at different stages have been discussed in detail by Nita-Lazar et al. (37). For absolute quantitative estimation, it is necessary to add an internal peptide to the sample (39).

3.2. Protein Arrays

DNA microarrays enable the measurement of mRNA expression level of thousands of genes in a single experiment. Similarly, protein arrays can be used to achieve high throughput protein analysis in parallel and is becoming a key proteomics technology. Protein arrays are solid-phase ligand binding assay systems using immobilized protein spots on surfaces such as glass, membranes, microtiter wells, etc. Bioinformatics support is important for data handling. The protein spots can be homogenous or heterogeneous and may be antibody, a cell or phage lysate, nucleic acid, drug, recombinant protein or peptide, etc. Protein arrays can provide valuable information to detect proteins, monitor their expression levels, posttranslational modification, and functions as well as investigate protein–protein interactions. Using protein arrays, it is possible to carry out parallel multiplex screening of multiple interactions, such as protein–antibody, protein–protein, protein–ligand or protein–drug, enzyme–substrate screening, etc. Usually a very small amount of sample is required. Although used widely, 2D-GE and MS-based proteomics techniques may not be able to detect low abundant proteins. This demands other sensitive and easily accessible high throughput technologies for the detection and differential expression of proteins in health and disease. Protein array technology is useful for such purposes and can be applied to protein expression profiling ranging from limited numbers of proteins to global proteomic analysis. It can be used in target identification as well as in the validation process. In general, 2D-GE and MS-based techniques can be used for the initial identification of biomarkers and subsequently specific antibodies can be used in protein microarrays for rapid high throughput screening of a large number of samples. These techniques complement each other.

Protein microarray formats can be divided into two major classes – forward phase arrays and reverse phase arrays (RPA). In the forward phase array format, the analyte(s) of interest is present in solution phase and the capture molecules (usually antibodies) are immobilized on the solid surface. Usually each array measures one test sample such as a cellular lysate or serum sample representing a control or particular disease condition and multiple analytes are measured simultaneously. On the contrary, in the RPA format, the array may contain samples from a number of patients or from the same patient at different stages of disease or treatment and each array is incubated with one detection protein (e.g., antibody). The single analyte concentration can be compared among different patients or from same patients with different state of disease progression or different time points of drug treatment. The detection of the array is carried out using a tagged antibody, ligand or cell lysate or serum. The signal intensity is proportional to the binding of the tagged molecule bound to the analyte molecule (40). The multiplex analysis of the same spot for the detection of multiple analytes is possible using dual color infrared dye-labeled antibodies as well as quantum dots (41, 42).

For yeast (*Saccharomyces cerevisiae*), the whole proteome array which consisted of 5,000 purified proteins were deposited onto a glass microscope slide. This was used for global studies on protein–protein and protein–lipid interactions (43). In spite of such success, it is important to remember that protein chips arrays are much more challenging than DNA arrays due to the following reasons – the solubility of the protein can vary widely, their expression levels have a broad dynamic range, and proteins are less stable than DNA. In addition, it is generally difficult to preserve the native structures of proteins on the glass slides. In contradistinction, for DNA microarrays, only the nucleotide sequence needs to be maintained.

While used in combination with laser capture microdissection (LCM) (44) technology, RPA can be particularly useful for gaining information, particularly about cell signaling molecules and posttranslational modification from tissue sections obtained from normal and diseased individuals. In fact, the RPA platform has been used to explore a variety of signaling pathways involved in malignant progression and tumor biology (4). For example, preliminary data indicate that protein kinase C is downregulated during the progression of prostate cancer in a study of microdissected cells from normal, stroma, and prostate tumors (45). Prostate tumors, like most tumors, are solid tumors that grow as solid masses of tissue and consist of two distinct but interdependent compartments: the parenchyma containing neoplastic cells and the stroma that the neoplastic cells induce and in which they are dispersed. If validated, this finding on protein kinase C and prostate cancer could have profound effects on the rationale behind some current therapies (46). Also, by applying RPA in normal prostate epithelium, prostate intraepithelial neoplasia, and invasive prostate cancer tissues (45, 47), it has been possible to identify specific molecular changes like changes in signaling molecules, such as Akt, GSK3B, PKC-R, and p38, that are implicated with cancer progression between different tissues. For these studies, the cancerous tissues were isolated by LCM, arrayed on nitrocellulose membranes, and probed with appropriate antibodies. These examples illustrate the importance of proteomics technology coupled to signal pathway profiling in providing novel insights into the cellular processes. In fact, this technique can be applied to study normal physiology and pathology of any tissue at the level of the proteome by comparing diseased and healthy tissues within the same patient which will be helpful for the development of individualized diagnosis and treatment strategies.

Extreme sensitivity for the detection of an analyte is an advantage of this technology – in fact, attomole level of a protein can be detected. Using this technology, it is possible to analyze signaling pathways using small numbers of cultured cells or cells isolated by

LCM from human tissue which were available during clinical trials (45, 47–49).

As pointed out by Sheehan et al. (50), different probes, such as antibodies, aptamers (oligonucleotides or peptides that bind to specific molecular targets such as proteins), ligands and drugs, required for protein microarrays, often cannot be obtained with predictable affinity or specificity. In many instances, it is difficult to obtain highly specific antibodies or suitable protein-binding ligands. Since the turning on or off of signaling pathways depend on post-translational modifications as well as protein–protein interactions, it is important to have specific antibodies or probes that are specific for the detection of such modifications or interactions.

It is strongly recommended that the specificity of the antibodies should be thoroughly assessed and validated by Western blot analysis before using those antibodies in protein array format. Such validation may include the demonstration of single band of appropriate molecular weight in Western blot analysis of a complex biological mixture similar to the one to be analyzed by microarray format. For a phosphospecific antibody, it may be validated by differential detection of protein bands from lysates prepared from control and activated cells in which the signaling pathway is known to be activated. It will be helpful to have a real-time validation of the antibodies by printing out negative and positive controls in each array. A web posting of a set of validated antibodies can be found at the www.home.ccr.cancer.gov/ncifdaproteomics/. Cooperation among different funding agencies and international consortia will aid in the generation of large comprehensive libraries of fully characterized specific antibodies, ligands, and probes. Interestingly, such initiatives are already in progress by the Human Proteome Organization (www.hupo.org) and the Human Proteome Resource in Sweden (www.proteinatlas.org/intro.php). Individual investigators are also beginning efforts to provide the scientific community with critical antibody resources (51–53). A reagent resource to identify proteins and peptides of interest for the cancer community has been the topic of a workshop arranged by NCI (54).

We would like to point out that the global ICAT technology has certain advantages over protein chip array when it is necessary to quantify a large number of proteins. As one can understand, the generation of antibodies, proteins or protein capture agents as well as their deposition on the chip can be time-consuming, expensive, and technically challenging. Also, the capacity for accurate quantification can be problematic.

3.3. SELDI-TOF-MS-Based ProteinChip® System

Protein quantification can also be carried out using ProteinChip® array which depends on selective binding of proteins according to their charge or hydrophobicity from a complex biological mixture such as plasma followed by accurate and sensitive detection by

SELDI-TOF-MS (55). Bio-Rad Laboratories, Hercules, CA (www.bio-rad.com) is currently the source for SELDI technology products for the life science market, and Vermillion, Inc.(www.vermillion.com) is the source of these products for the molecular diagnostics market (www.vermillion.com/alliances_and_partnerships.cfm). SELDI allows the capture of proteins of interest on a chip and their subsequent analysis essentially by MALDI-TOF MS. This technology enables the comparative analysis of virtually any given protein-containing solution in a fast and simple process and requires only a minute amount of samples. A complex mixture of proteins can be applied to the spots of ProteinChip Arrays, which have been derivatized with specific chromatographic chemistries. The proteins interact and bind with the chromatographic array surface according to their interaction and binding potential. Subsequent on-spot washing removes salts or other unbound molecules. This on-chip retentate chromatography step thus leads to fewer background peaks that might interfere with the detection of low abundance species during MALDI-TOF MS. In fact, protein interaction studies or enzymatic reactions may be carried out directly on-spot under physiological conditions. The chromatographic surfaces provide a convenient support for the cocrystallization of matrix and target proteins, resulting in the formation of a homogenous layer on the spot which makes them ideal for subsequent MALDI-TOF analysis. It has various applications, including biomarker discovery and assay development (56), protein interaction studies (57), monitoring of enzymatic reactions (58) and for process proteomics approaches (59). Different kinds of quantification experiments have demonstrated the high sensitivity and reproducibility of the SELDI System. For example, the quantitative detection of myoglobin from 1% blood serum, the simultaneous capture and detection of IL-8 and epidermal growth factor from 20% blood serum, and the quantitative monitoring of phosphatase activity over time.

4. Applications of Phosphoproteomics in Drug Development

Quantitative phosphoproteomics provides useful tools for the development of target-oriented drug discovery and early diagnostic biomarkers. Multiple biochemical pathways leading to cancer initiation, development, and progression are in general due to irregular cellular signaling and thus involve an altered role of protein kinases (38). The identification and drug-induced alteration of the defects in any one or multiple such pathways may slow down the growth of cancer cells. Moreover, specific kinases are assumed to be the signatures for certain types of cancers.

This would help disease diagnosis as well as grouping patients that might require personalized therapy. Therefore, in recent years, much attention has been paid to the development of kinase-specific drug targets to treat cancers. Quantitative phosphoproteomics is considered to be an appropriate and potential method to study kinases and their roles in cancer therapy and diagnosis and a number of kinase-based approved anticancer therapeutics have been discussed by Yu et al. (38). The phosphorylation status of multiple proteins can also be effectively used to map the response of host cells to infection and thus can be useful to develop drugs for infectious diseases.

5. Toxicoproteomics and Its Application in Drug Development and Clinical Management

In the postgenomic era, proteomic techniques have been used to study the changes in protein profile following exposure to environmental toxins and drugs. This area of research is known as toxicoproteomics. The application of proteomics in toxicology can be divided into two broad overlapping areas: (a) mechanism of toxicity, and (b) screening and predictive toxicology. If it is possible to establish a relationship between toxic effects and protein markers and develop a database of such markers, it will be useful to screen a new drug for its potential toxic effects against a panel of predictive toxicity markers and will be an early indication of toxicity for further investigation. In the past, toxicity associated with any experimental drug was identified in animal studies according to the recommendation of the International Conference on Harmonization of Technical Requirements for Registration of Pharmaceuticals for Human Use (ICH; www.ich.org) and was dependent mostly on histopathological and biochemical parameters. The early detection of potential toxicity of potential drug candidates may cause significant reduction in time and cost for a new drug to become available in the market.

5.1. Hepatotoxicity

Liver is the major site of metabolism and detoxification in the body and often the target of drug-induced toxicities. Consequently, hepatotoxicity is used to predict adverse drug reactions. Drug-induced liver toxicity can be metabolism based (predictable) or idiosyncratic (unpredictable). Traditionally histopathological parameters such as hepatocellular vacuolation (steatosis) and periductual inflammation were used to measure the liver toxicity potential of investigational drugs in preclinical studies (60). Additionally, biochemical parameters such as elevated plasma level of secreted liver enzymes, such as alanine aminotransferase and aspartate aminotransferase were considered to be indicative of liver injury (61). But these methods cannot be applied for in vitro studies and fail to predict idiosyncratic toxicities. Recently, the use of proteomic techniques in predicting hepatotoxicity has been used by different laboratories. Examples include the development of rodent liver proteomic toxicity database based on studies on the effects of a range of xenobiotics on protein expression in the liver (62). Methylpyrilene, a widely used antihistamine was discontinued due to its carcinogenic property and was replaced by structurally similar pyrilamine which also has the antihistamine property but is not a carcinogen. Interestingly, using 2D-GE, while the former compound was found to change liver proteome profile, pyrilamine did not lead to such change (63). Such observation points toward the potential role of proteomics in the selection of lead candidates. Studies have also been carried out on the hypoglycaemic compound SDZPGU693 which stimulates glucose utilization in peripheral tissues (64). Using 2D-GE, liver samples of treated and untreated rats identified change in expression levels of several proteins which gave insights into the molecular mechanisms of both its pharmacological action and a toxic response. Similar studies by other investigators (65, 66) were able to identify proteins which are part of pathways known to be disrupted in drug-induced liver steatosis and hence their possible role in hepatotoxicity. The downregulation of some of the secretory proteins suggested the disruption of secretory pathway as a mechanism of liver toxicity. Interestingly, such downregulation could be detected as early as 6 h of treatment with the experimental compound suggesting a window of time that occurs before the manifestation of clinical symptoms and that time period can be analyzed by proteomic technology to predict the hepatotoxicity. Using two-dimensional biphasic liquid chromatography and tandem mass spectrometry (LC/LC/MS/MS), immunochemical techniques, and immortalized normal human hepatocytes, Gao et al. (67) tested hepatotoxicity of 20 different drugs. They observed increased levels of secretory proteins BMS-PTX-265 and BMS-PTX-837. For all 20 drugs, the elevations of BMS-PTX-265 correlated exactly with the known safety profile; whereas changes in BMS-PTX-837 correctly predicted the safety profile in 19 of 20 drugs (one false negative). In summary, the data support the preclinical in vitro method as a means to identify new biomarkers of liver toxicity, as well as the validity of the biomarkers themselves.

5.2. Nephrotoxicity

Apart from hepatotoxicity, nephrotoxicity is also used in the evaluation of drug toxicity using proteomic technology. Renal tissue is a sensitive organ for drug-induced toxicity for various reasons such as large blood flow and the presence of a variety of xenobiotics transporters and metabolizing enzymes. In addition, drugs and metabolites become concentrated in the tubules during urine production. Traditional evaluation of nephrotoxicity includes the measurement of serum metabolites such as blood urea nitrogen and serum creatinine, (68) as well as the identification of specific urine markers such as gamma glutamyl transferase or *N*-acetyl glucosamine (69) but are of inadequate sensitivity. Bandara et al. (70) identified a correlation between proteomic evaluation and conventional measurements in the assessment of renal proximal tubular toxicity. Using current proteomics technology of 2D-GE and MS, they identified the level of T-kininogen (a cysteine protease inhibitor expressed in kidney) to be elevated in rats following the treatment with cisplatin (and other known nephrotoxins) at early time points returning to basal level after 3 days of treatment suggesting T-kininogen may be required to counteract apoptosis in proximal tubular cells in order to minimize tissue damage following a toxic insult. Cycosporin A is an immunosuppressive drug whose use is restricted by adverse effects. Following the proteomic analysis of kidney homogenate of CsA-treated rats, vitamin D-dependent calcium-binding protein calbindin-D (28K) was identified as a novel biomarker for CsA-induced renal toxicity (71). Interestingly, while rats and humans displayed CsA-induced renal cytotoxicity, monkeys and dogs did not. In accordance with this phenomenon, calbindin-D was also regulated in a species-specific manner (72).

Gupta et al. (73) postulated a toxicoproteomics-based new drug development paradigm following appropriate target identification. It was suggested that the information obtained on drug toxicities from preclinical studies should be used throughout the development process as well as after regulatory approval as part of postmarketing surveillance. Also, at each stage of the process, the requirement for multiplexing decreases while that for specificity and sensitivity increases. In essence, high-throughput proteomics tools are necessary in the preclinical and early clinical trial stages for screening the proteome for biomarkers. However, once the relevant markers have been identified, then accurate and precise tests are necessary for clinical use. The suggested importance of toxicoproteomics at each stage of drug development is schematically represented in Fig. 1 and is described below:

5.3. Lead Optimization

This involves the chemical modification of a biologically active compound to fulfill all stereoelectronic, physicochemical, pharmacokinetic, and toxicologic properties required for clinical usefulness. In vitro assays using different cell lines, proteomics techniques such as 2D-GE-MS/MS or LC-MS/MS can be applied to find out adverse cellular changes associated with each drug candidate.

5.4. Preclinical Studies

Such studies are carried out in animal models to find out the in vivo safety, efficacy, and maximal tolerated dose. Using proteomics techniques such as 2D-GE-MS/MS or LC-MS/MS,

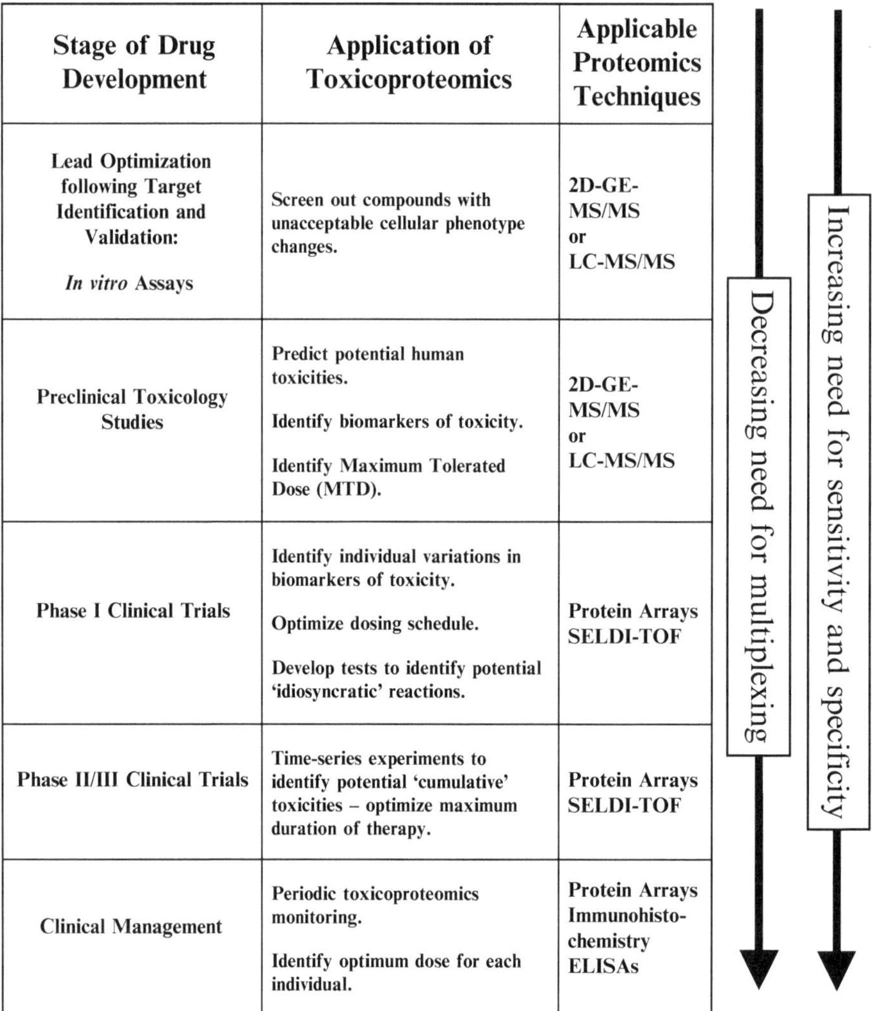

Stage of Drug Development	Application of Toxicoproteomics	Applicable Proteomics Techniques
Lead Optimization following Target Identification and Validation: *In vitro* Assays	Screen out compounds with unacceptable cellular phenotype changes.	2D-GE-MS/MS or LC-MS/MS
Preclinical Toxicology Studies	Predict potential human toxicities. Identify biomarkers of toxicity. Identify Maximum Tolerated Dose (MTD).	2D-GE-MS/MS or LC-MS/MS
Phase I Clinical Trials	Identify individual variations in biomarkers of toxicity. Optimize dosing schedule. Develop tests to identify potential 'idiosyncratic' reactions.	Protein Arrays SELDI-TOF
Phase II/III Clinical Trials	Time-series experiments to identify potential 'cumulative' toxicities – optimize maximum duration of therapy.	Protein Arrays SELDI-TOF
Clinical Management	Periodic toxicoproteomics monitoring. Identify optimum dose for each individual.	Protein Arrays Immunohisto-chemistry ELISAs

Fig. 1. New drug development paradigm in the context of toxicoproteomi. Drug development is moving toward a personalized approach with the identification of biomarkers for screening patients likely to have dose-related or idiosyncratic toxicities from an investigational compound. In the new paradigm, proteomic tools will be used extensively in preclinical and clinical studies to identify these predictive biomarkers. Diagnostic tests based on these markers may then be used in clinical medicine to predict drug-related toxicities. Reprinted from ref. (73) with permission from Bentham Science Publisher Ltd.

it may be possible to find out biomarkers as well as the mechanism of tissue toxicity and predict the possible toxic effect of a potential drug candidate in humans.

5.5. Phase I Clinical Trials

Following successful preclinical studies, Phase I trial is carried out in human populations to find out the safety of a potential drug candidate in humans as well as to find out one or more appropriate dosing schedule. Using proteomics technology (protein arrays and SELDI-TOF), it may be possible to identify markers indicative of drug-related toxicity. Such markers can be identified in serum, urine, leukocytes, or easily available tissues such as oral mucosa (74, 75). This can be helpful to personalize individual dose based on the level of these marker molecules in different individuals. It can also be helpful to find out the idiosyncratic reaction observed in any individual.

5.6. Phase II/III Clinical Trials

Following a successful Phase I trial, a Phase II or Phase III clinical trial would be helpful for the identification of markers of long-term or cumulative toxicity. This, in turn, helps further optimization of the dose of the drug, dosing frequency as well as maximum duration of the therapy that may be applicable to the patient.

5.7. Clinical Management

Use of toxicoproteomics will be particularly helpful for personalized medicine. Before treatment, the patient can be monitored for various biomarkers for dose-related or idiosyncratic toxicities. This will help in the choice of therapy as well as dose and the frequency of the drug to be given to the patient. Even during treatment, periodic monitoring for the toxic biomarker will be helpful to change the type of therapy or the dose or frequency of the drug.

6. Conclusion

Due to the tremendous rapid progress in proteomics and genomics technologies, it is possible to carry out detailed in-depth characterization of the molecular basis of different cellular functions contributing to normal or pathological cellular activities. In the past, drug discovery was mostly focused on a particular protein which could be connected to a particular disease. In recent years, there has been a shift toward strategies for the application of systems-oriented approaches. Such approaches could allow more effective therapeutic intervention based on a systems-oriented understanding of the disease biology and drug response. In essence, the emerging fields of systems biology and proteomics are offering new ways of diagnosis and prognosis of various disease as well as personalized medicine. Significant progress has been made in cancer research using systems biology and the proteomics platform. However, there still remains a lot of work to be done for the proper unraveling and understanding of systems biology. The understanding of protein and gene regulatory networks of biological systems will improve drug development efforts and prevention of diseases. As pointed out by Weston and Hood (1), targeting the key nodal points of networks will help circumvent the disease potentials emerging from defective genes (somatic or inherited) or pathological environmental stimuli. These nodes may therefore be more effective targets for therapeutic interventions. Bioinformatics play a major role since a large volume of different types of data need to be collected and integrated for the proper understanding of biological systems for accurate diagnosis and prognosis during drug therapy. Information technology and computational support are an essential part of such efforts.

While the expression and activities of several disease-associated molecules can provide important information, a full understanding of the participation and interaction of these molecules within the cellular control networks is equally important. It is important to understand the correlation of a drug target with complex cellular networks involved with the disease process in order to obtain high efficacy at low drug dosage and reducing the probability of development of drug-resistance. Inhibition of the activity of a signaling protein molecules without understanding its role in the overall cellular mechanism, can ultimately lead to an overall adverse effect. The recent advancement in the proteomics platform would help enormously to achieve this goal.

It is worth mentioning that although resources and technologies are increasingly available to identify and evaluate the efficacy and safety of different chemical compounds, such improvement has not been reflected in the approval of drugs against new targets – that is cellular molecules not perturbed or targeted by previous drugs. In spite of significant increase in research and development expenditure, there is a disproportionately low number of new drug targets. One of the examples of major success in this area includes the small molecule ABL kinase inhibitor imatinib mesylate (Gleevec, Novartis) in the treatment of chronic myelogenous leukemia.

Finally we would like to mention that ideal proteomics techniques for drug discovery require separation and characterization of the whole proteome of an organism under study. Other requirements include the identification of protein activity independent of protein abundance, and protein–protein and protein–small-molecule interactions. Much of the current proteomics research that is focused on target-protein identification would be impossible to carry out without the advancements in biological MS. Unfortunately, membrane proteins as well as low abundant proteins cannot be analyzed efficiently using some of the currently available separation and analytical techniques, such as 2D-GE. It is also necessary that these techniques should be user-friendly, automated, high-throughput, and cost-effective. Although enormous progress has been made in recent years, there still remains space for improvement in this area of proteomics research.

Acknowledgments

The authors would like to acknowledge the generous support provided by the Keck Graduate Institute of Applied Life Sciences, Claremont, CA, The Arnold and Mabel Beckman Foundation, the Ralph M. Parsons Foundation, and the National Science Foundation grant FIBR 0527023.

References

1. Weston AD, Hood L (2004) Systems biology, proteomics, and the future of health care: toward predictive, preventative, and personalized medicine. J Proteome Res 3:179–196
2. Heller MJ (2002) DNA microarray technology: devices, systems, and applications. Annu Rev Biomed Eng 4:129–153
3. Griffin TJ, Gygi SP, Ideker T, Rist B, Eng J, Hood L, Aebersold R (2002) Complementary profiling of gene expression at the transcriptome and proteome levels in *Saccharomyces cerevisiae*. Mol Cell Proteomics 1:323–333
4. Szkanderová S, Port M, Stulík J, Hernychová L, Kasalová I, Van Beuningen D, Abend M (2003) Comparison of the abundance of 10 radiation-induced proteins with their differential gene expression in L929 cells. Int J Radiat Biol 79:623–633
5. Greenbaum D, Colangelo C, Williams K, Gerstein M (2003) Comparing protein abundance and mRNA expression levels on a genomic scale. Genome Biol 4:117
6. Anderson NL, Anderson NG (2002) The human plasma proteome: history, character, and diagnostic prospects. Mol Cell Proteomics 1:845–67, Erratum in: Mol Cell Proteomics. 2, 50
7. Wilkins MR, Pasquali C, Appel RD, Ou K, Jean-Charles Sanchez OG, Yan JX, Gooley AA, Hughes G, Humphery-Smith I, Williams KL, Hochstrasser DF (1996) From proteins to proteomes: large scale protein identification by two-dimensional electrophoresis and amino acid analysis. Nat Biotechnol 14:61–65
8. Chakravarti DN, Chakravarti B, Moutsatsos I (2002) Informatic tools for proteome profiling. *BioTechniques* Suppl:4–10, 12–5
9. Margolis J, Kenrick KG (1969) 2 dimensional resolution of plasma proteins by combination of polyacrylamide disc and gradient gel electrophoresis. Nature 221:1056–1057
10. O'Farrell PH (1975) High resolution two dimensional electrophoresis of proteins. J Biol Chem 250:4007–4021
11. Klose J (1975) Protein mapping by combined isoelectric focusing electrophoresis of mouse tissues. A novel approach to testing for induced point mutations in mammals. Humangenetik 26:231–243
12. Unlu M, Morgan ME, Minden JS (1997) Difference gel electrophoresis: a single gel method for detecting changes in protein extracts. Electrophoresis 18:2071–2077
13. Chakravarti DN, Gallagher S, Chakravarti B (2004) Difference gel electrophoresis: application in quantitative proteomics research. Curr Proteomics 1:261–271
14. Tonge R, Shaw J, Middleton B, Rowlinson R, Rayner S, Young J, Pognan F, Hawkins E, Currie I, Davison M (2001) Validation and development of fluorescence two-dimensional differential gel electrophoresis proteomics technology. Proteomics 1:117–124
15. Han X, Aslanian A, Yates JR (2008) Mass spectrometry for proteomics. Curr Opin Chem Biol 12:483–490
16. Guerrera IC, Kleiner O (2005) Analysis of mass spectrometry in proteomics. Biosci Rep 25:71–93
17. Domon B, Alving K, He T, Ryan TE, Patterson SD (2002) Enabling parallel protein analysis through mass spectrometry. Curr Opin Mol Ther 4:577–586
18. Yates JR (2004) Mass spectral analysis in proteomics. Annu Rev Biophys Biomol Struct 312:212–217
19. Domon B, Aebersold R (2006) Mass spectrometry and protein analysis. Science 31:212–217
20. Fenn JB, Mann M, Meng CK, Wong SF, Whitehouse CM (1989) Electrospray ionization for mass spectrometry of large biomolecules. Science 246:64–71
21. Karas M, Hillenkamp F (1988) Laser desorption ionization of proteins with molecular masses exceeding 10,000 daltons. Anal Chem 60:2299–2301

22. Zubarev RA, Kelleher NL, McLafferty FW (1998) Electron capture dissociation of multiply charged protein cations. A nonergodic process. J Am Chem Soc 120:3265–3266
23. Zubarev R (2006) Protein primary structure using orthogonal fragmentation techniques in Fourier transform mass spectrometry. Expert Rev Proteomics 3:251–261
24. Syka JEP, Coon JJ, Schroeder MJ, Shabanowitz J, Hunt DF (2004) Peptide and protein sequence analysis by electron transfer dissociation mass spectrometry. Proc Natl Acad Sci USA 101:9528–9533
25. Pitteri SJ, Chrisman PA, Hogan JM, McLuckey SA (2005) Electron transfer ion/ion reactions in a three-dimensional quadrupole ion trap: reactions of doubly and triply protonated peptides with SO_2^-. Anal Chem 77:1831–1839
26. Gygi SP, Rist B, Gerber SA, Turecek F, Gelb MH, Aebersold R (1999) Quantitative analysis of complex protein mixtures using isotope-coded affinity tags. Nat Biotechnol 17:994–999
27. Goodlett DR, Keller A, Watts JD, Newitt R, Yi EC, Purvine S, Eng JK, von Haller P, Peters EC, Horn DM, Tully DC, Brock A (2001) A novel multifunctional labeling reagent for enhanced protein characterization with mass spectrometry. Rapid Commun Mass Spectrom 15:2387–2392
28. Peters EC, Horn DM, Tully DC, Brock A (2001) A novel multifunctional labeling reagent for enhanced protein characterization with mass spectrometry. Rapid Commun Mass Spectrom 15:2387–2392
29. Munchbach M, Quadroni M, Miotto G, James P (2000) Quantitation and facilitated de novo sequencing of proteins by isotopic N-terminal labeling of peptides with a fragmentation-directing moiety. Anal Chem 72:4047–4057
30. Oda Y, Huang K, Cross FR, Cowburn D, Chait BT (1999) Accurate quantitation of protein expression and site-specific phosphorylation. Proc Natl Acad Sci USA 96:6591–6596
31. Zhou H, Watts JD, Aebersold R (2001) A systematic approach to the analysis of protein phosphorylation. Nat Biotechnol 19:375–378
32. Goshe MB, Conrads TP, Panisko EA, Angell NH, Veenstra TD, Smith RD (2001) Phosphoprotein isotope-coded affinity tag approach for isolating and quantitating phosphopeptides in proteome-wide analyses. Anal Chem 73:2578–2586
33. Goshe MB, Veenstra TD, Panisko EA, Conrads TP, Angell NH, Smith RD (2002) Phosphoprotein isotope-coded affinity tags: application to the enrichment and identification of low-abundance phosphoproteins. Anal Chem 74:607–616
34. Ficarro SB, McCleland ML, Stukenberg PT, Burke DJ, Ross MM, Shabanowitz J, Hunt DF, White FM (2002) Phosphoproteome analysis by mass spectrometry and its application to *Saccharomyces cerevisiae*. Nat Biotechnol 20:301–305
35. Cohen P (2001) The role of protein phosphorylation in human health and disease. The Sir Hans Krebs Medal Lecture. Eur J Biochem 268:5001–5010
36. Olsen JV, Blagoev B, Gnad F, Macek B, Kumar C, Mortensen P, Mann M (2006) Global, in vivo, and site-specific phosphorylation dynamics in signaling networks. Cell 127:635–648
37. Nita-Lazar A, Saito-Benz H, White FM (2008) Quantitative phosphoproteomics by mass spectrometry: past, present, and future. Proteomics 8:4433–4443
38. Yu Li-R, Issaq HJ, Veenstra TD (2007) Phosphoproteomics for the discovery of kinases as cancer biomarkers and drug targets. Proteomics Clin Appl 1:1042–1057
39. Mirzaei H, Mcbee JK, Watts J, Aebersold R (2007) Comparative evaluation of current peptide production platforms used in absolute quantification in proteomics. Mol Cell Proteomics 7:813–823
40. Wilson DS, Nock S (2003) Recent developments in protein microarray technology. Angew Chem Int Ed Engl 42:494–500
41. Calvert VS, Tang Y, Boveia V, Wulfkuhle J, Schutz-Geschwender A, Olive DM, Liotta LA, Petricoin EF (2004) Development of multiplexed protein profiling and detection using near infrared detection of reverse phase protein microarrays. Clin Proteomics J 1:81–89
42. Geho DH, Lahar N, Ferrari M, Petricoin EF, Liotta LA (2004) Opportunities for nanotechnology-based innovation in tissue proteomics. Biomed Microdevices 6:2319
43. Zhu H, Bilgin M, Bangham R, Hall D, Casamayor A, Bertone P, Lan N, Jansen R, Bidlingmaier S, Houfek T, Mitchell T, Miller P, Dean RA, Gerstein M, Snyder M (2001) Global analysis of protein activities using proteome chips. Science 293:2101–2105
44. Espina V, Wulfkuhle JD, Calvert VS, VanMeter A, Zhou W, Coukos G, Geho DH, Petricoin EF, Liotta LA (2006) Laser capture microdissection. Nat Protoc 1:586–603
45. Grubb RL, Calvert VS, Wulfkuhle JD, Paweletz CP, Linehan WM, Phillips JL, Chuaqui R, Valasco A, Gillespie J, Emmert-Buck M, Liotta LA, Petricoin EF (2003) Signal

pathway profiling of prostate cancer using reverse phase protein microarrays. Proteomics 3:2142–2146

46. Tolcher AW, Reyno L, Venner PM, Ernst SD, Moore M, Geary RS, Chi K, Hall S, Walsh W, Dorr A, Eisenhauer E (2002) A randomized phase II and pharmacokinetic study of the antisense oligonucleotides ISIS 3521 and ISIS 5132 inpatients with hormone-refractory prostate cancer. Clin Cancer Res 8: 2530–2535
47. Paweletz CP, Charboneau L, Bichsel VE, Simone NL, Chen T, Gillespie JW, Emmert-Buck MR, Roth MJ, Petricoin EF III, Liotta LA (2001) Reverse phase protein microarrays which capture disease progression show activation of pro-survival pathways at the cancer invasion front. Oncogene 20:1981–1989
48. Wulfkuhle JD, Aquino JA, Calvert VS, Fishman DA, Coukos G, Liotta LA, Petricoin EF (2003) Signal pathway profiling of ovarian cancer from human tissue specimens using reverse-phase protein microarrays. Proteomics 3:2085–2090
49. Zha H, Raffeld M, Charboneau L, Pittaluga S, Kwak LW, Petricoin E III, Liotta LA, Jaffe ES (2004) Similarities of prosurvival signals in Bcl-2-positive and Bcl-2-negative follicular lymphomas identified by reverse phase protein microarray. Lab Invest 84:235–244
50. Sheehan KM, Calvert VS, Kay EW, Lu Y, Fishman D, Espina V, Aquino J, Speer R, Araujo R, Mills GB, Liotta LA, Petricoin EF, Wulfkuhle JD (2005) Use of reverse phase protein microarrays and reference standard development for molecular network analysis of metastatic ovarian carcinoma. Mol Cell Proteomics 4:346–355
51. Hanash S (2003) Disease proteomics. Nature 422:226–232
52. Tyers M, Mann M (2003) From genomics to proteomics. Nature 422:193–197
53. Agaton C, Galli J, Hoiden Guthenberg I, Janzon L, Hansson M, Asplund A, Brundell E, Lindberg S, Ruthberg I, Wester K, Wurtz D, Hoog C, Lundeberg J, Stahl S, Ponten F, Uhlen M (2003) Affinity proteomics for systematic protein profiling of chromosome 21 gene products in human tissues. Mol Cell Proteomics 2:405–414
54. Haab BB, Paulovich AG, Anderson NL, Clark AM, Downing GJ, Hermjakob H, Labaer J, Uhlen M (2006) A reagent resource to identify proteins and peptides of interest for the cancer community: a workshop report. Mol Cell Proteomics 5:1996–2007
55. Vorderwulbecke S, Cleverly S, Weinberger SR, Wiesner A (2005) Protein quantification by SELDI-TOF-MS based ProteinChip® system. Nat Methods 2:393–395
56. Wiesner A (2004) Detection of tumor markers with ProteinChip technology. Curr Pharm Biotechnol 5:45–67
57. Amaar YG, Thompson GR, Linkhart TA, Chen ST, Baylink DJ, Mohan S (2002) Insulin-like growth factor-binding protein 5 (IGFBP-5) interacts with a four and a half LIM protein 2 (FHL2). J Biol Chem 277:12053–12060
58. Boyle MDP, Romer TG, Meeker AK, Sledjeski DD (2001) Use of surface-enhanced laser desorption ionization protein chip system to analyze streptococcal exotoxin B activity secreted by *Streptococcus pyogenes*. J Microbiol Methods 46:87–97
59. Weinberger SR, Boschetti E, Santambien P, Brenac V (2002) Surface-enhanced laser desorption-ionization retentate chromatography mass spectrometry (SELDI-RC-MS): a new method for rapid development of process chromatography conditions. J Chromatogr B Analyt Technol Biomed Life Sci 782:307–316
60. Grunhage F, Fischer HP, Dauerbruch T, Reichel C (2003) Drug- and toxin-induced hepatotoxicity. Z Gastroenterol 41:565–578
61. Rej R (1989) Aminotransferases in disease. Clin Lab Med 9:667–687
62. Anderson NL, Taylor J, Hofmann JP, Esquer-Blasco R, Swift S, Anderson NG (1996) Simultaneous measurement of hundreds of liver proteins: application in assessment of liver function. Toxicol Pathol 24:72–76
63. Cunningham ML, Pippin LL, Anderson NL, Wenk ML (1995) The hepatocarcinogen methapyrilene but not the analog pyrilamine induces sustained hepatocellular replication and protein alterations in F344 rats in a 13-week feed study. Toxicol Appl Pharmacol 131:216–223
64. Arce A, Aicher L, Wahl D, Anderson NL, Meheus L, Raymackers J, Cordier A, Steiner S (1998) Changes in the liver protein pattern of female Wistar rats treated with the hypoglycemic agent SDZ PGU 693. Life Sci 63:2243–2250
65. Meneses-Lorente G, Guest PC, Lawrence J, Muniappa N, Knowles MR, Skynner HA, Salim K, Cristea I, Mortishire-Smith R, Gaskell SJ, Watt A (2004) A proteomic investigation of drug-induced steatosis in rat liver. Chem Res Toxicol 17:605–612
66. Meneses-Lorente G, Watt A, Salim K, Gaskell SJ, Muniappa N, Lawrence J, Guest PC (2006) Identification of early proteomic markers for hepatic steatosis. Chem Res Toxicol 19:986–998

67. Gao J, Ann Garulacan L, Storm SM, Hefta SA, Opiteck GJ, Lin JH, Moulin F, Dambach DM (2004) Identification of in vitro protein biomarkers of idiosyncratic liver toxicity. Toxicol In Vitro 18:533–541
68. Goodsaid FM (2004) Identification and measurement of genomic biomarkers of nephrotoxicity. J Pharmacol Toxicol Methods 49:183–186
69. Han WK, Bailly V, Abichandani R, Thadhani R, Bonventre JV (2002) Kidney Injury Molecule-1 (KIM-1): a novel biomarker for human renal proximal tubule injury. Kidney Int 62:237–244
70. Bandara LR, Kelly MD, Lock EA, Kennedy S (2003) A correlation between a proteomic evaluation and conventional measurements in the assessment of renal proximal tubular toxicity. Toxicol Sci 73:195–206
71. Steiner S, Aicher L, Raymackers J, Meheus L, Esquer-Blasco R, Anderson NL, Cordier A (1996) Cyclosporine A decreases the protein level of the calcium-binding protein calbindin-D 28kDa in rat kidney. Biochem Pharmacol 51:253–258
72. Aicher L, Wahl D, Arce A, Grenet O, Steiner S (1998) New insights into cyclosporine A nephrotoxicity by proteome analysis. Electrophoresis 19:1998–2003
73. Gupta N, Law A, Poddar R, Louie M, Ray A, Chakravarti DN (2005) Toxicoproteomics: applications in drug development. Curr Proteomics 2:97–101
74. Amacher DE (2010) The discovery and development of proteomic safety biomarkers for the detection of drug induced liver toxicity. Toxicol Appl Pharmacol 245:134–142
75. George J, Singh R, Mahmood Z, Shukla Y (2010) Toxicoproteomics: new paradigms in toxicology research. Toxicol Mech Methods [Epub ahead of print]

Chapter 2

Systems Biology Approaches and Tools for Analysis of Interactomes and Multi-target Drugs

André Schrattenholz, Karlfried Groebe, and Vukic Soskic

Abstract

Systems biology is essentially a proteomic and epigenetic exercise because the relatively condensed information of genomes unfolds on the level of proteins. The flexibility of cellular architectures is not only mediated by a dazzling number of proteinaceous species but moreover by the kinetics of their molecular changes: The time scales of posttranslational modifications range from milliseconds to years. The genetic framework of an organism only provides the blue print of protein embodiments which are constantly shaped by external input. Indeed, posttranslational modifications of proteins represent the scope and velocity of these inputs and fulfil the requirements of integration of external spatiotemporal signal transduction inside an organism. The optimization of biochemical networks for this type of information processing and storage results in chemically extremely fine tuned molecular entities. The huge dynamic range of concentrations, the chemical diversity and the necessity of synchronisation of complex protein expression patterns pose the major challenge of systemic analysis of biological models.

One further message is that many of the key reactions in living systems are essentially based on interactions of moderate affinities and moderate selectivities. This principle is responsible for the enormous flexibility and redundancy of cellular circuitries. In complex disorders such as cancer or neurodegenerative diseases, which initially appear to be rooted in relatively subtle dysfunctions of multimodal physiologic pathways, drug discovery programs based on the concept of high affinity/high specificity compounds ("one-target, one-disease"), which has been dominating the pharmaceutical industry for a long time, increasingly turn out to be unsuccessful. Despite improvements in rational drug design and high throughput screening methods, the number of novel, single-target drugs fell much behind expectations during the past decade, and the treatment of "complex diseases" remains a most pressing medical need. Currently, a change of paradigm can be observed with regard to a new interest in agents that modulate multiple targets simultaneously, essentially "dirty drugs." Targeting cellular function as a system rather than on the level of the single target, significantly increases the size of the drugable proteome and is expected to introduce novel classes of multi-target drugs with fewer adverse effects and toxicity. Multiple target approaches have recently been used to design medications against atherosclerosis, cancer, depression, psychosis and neurodegenerative diseases. A focussed approach towards "systemic" drugs will certainly require the development of novel computational and mathematical concepts for appropriate modelling of complex data. But the key is the extraction of relevant molecular information from biological systems by implementing rigid statistical procedures to differential proteomic analytics.

Key words: Network, Interactome, Multi-target drugs, Proteomics, Genomics, Systems biology, Posttranslational modification

Qing Yan (ed.), *Systems Biology in Drug Discovery and Development: Methods and Protocols*, Methods in Molecular Biology, vol. 662, DOI 10.1007/978-1-60761-800-3_2, © Springer Science+Business Media, LLC 2010

1. Genetic Interaction Network Analysis

Large-scale genome sequencing efforts have indicated an enormous scope of interactions, albeit only on the level of nucleic acids (1–3). Most of the insights into genetic interactions and networks come from studies using the yeast *Saccharomyces cerevisiae* (4, 5) but also more recently from multicellular organisms such as *Caenorhabditis elegans* (6) and mammalian cancer cell lines (7–9). The first genetic interaction maps in yeast were based on "synthetic lethal" interactions causing cell death by a combination of two gene deletions which by themselves have no deleterious effects. This type of phenomenon is also named "epistasis," and is commonly used to define "genetic interactions" in statistical terms (10). In eukaryotic organisms and cell lines, the recent development of RNAi libraries has enabled systematic genetic studies by using arrayed and pooled screens (8, 11).

A general understanding of the topology of genetic interaction networks in yeast has a wider importance, because similar principles are expected to underlie the relationship between genotype and phenotype in higher eukaryotic species. In terms of human disease, numerous modifiers and enhancers contribute to the complexity of phenotypes on a background of genetic, epigenetic and posttranslational stochasticity, but the structural and temporal determinants of the underlying networks remain unknown. Thus, mapping genetic networks in model organisms, such as yeast, supplies a framework for more complex systems, but next to structural interactions, temporal and spatial resolution of cellular sub compartments on the level of fast posttranslational events of proteins constitutes an important next level of modelling attempts.

Physical-interaction maps, generated by large-scale two-hybrid screening (12, 13) or affinity purification schemes followed by mass spectrometry (14–16), thus only provide an indirect link between genes and functional protein complexes functioning on time scales completely different as compared to nucleic acid turn over. So, next to valuable, but limited physical information (mostly taking into account structures related to the mere amino acid backbone of proteins, devoid of structural and kinetic consequences of posttranslational modifications), the genetic interaction map provides only limited functional information, largely identifying gene products that operate in functionally related pathways. Bluntly, the genetic information cannot tell when, where and how long protein partners will interact and what consequences result from very fast chemical changes, for example, oxidation, phosphorylation or proteolysis.

However, one significant finding made by systematic deletion studies in yeast showed that most eukaryotic genes are dispensable for viability, that is, there is an overwhelming degree of redundancy. Only about 20% of *Saccharomyces cerevisiae* genes are essential for haploid cells grown under laboratory conditions (3, 17). On the other hand, topological analysis of the yeast genetic interaction network predicts roughly 200,000 synthetic lethal interactions in the global yeast genetic network. Extrapolated to humans, this translates to a huge number of interactions with a specific phenotype potentially influenced by hundreds of different gene combinations (18). Only in a minority of cases, interactions can be deduced from genes or the directly related amino acid backbone information of proteins. In the vast majority of cases, low affinity interactions in highly redundant signalling pathways on the level of fast chemical changes at amino acid side chains play a key role. However, the genetic framework provides a lead idea about rough organizational principles, and here phylogenetic principles may offer important clues (19).

Drug interactions happen to take place at higher levels of deconvolution or the unfolding of the genetic information: subtle structural alterations by small molecules attenuating downstream effects of gene mutations as rapid and reversible modulators of related protein activity and structure. The use of such chemical probes on a genome-wide scale is called "chemical genomics" and has been originally used in yeast but also more recently in metazoan cell lineages. The underlying idea, that deletion of a gene encoding the target of an inhibitory compound should cause cellular effects that are similar to the inhibition of the target by drug treatment is of course simplistic in the light of the arguments developed above, but in a few cases helps shed light on major components of pathways. In one study (20), the chemical–genetic profiles of five different compounds were found to be reasonably similar to the genetic-interaction profiles of the target gene or genes in the target pathway. In another example, synthetic lethal genetic interactions were hopeful for the identification of compounds that target specific pathways and selectively kill cells with defined mutant genotypes in cancer models (21). Using a metazoan retroviral expression system, Burkard et al. (22) identified pleiotropic functions of polo-like kinases 1 throughout mitosis and cytokinesis and demonstrated some usefulness of chemical genetics in dissecting these complex but short-lived events within human cells. A similar strategy was used to replace cyclin-dependent kinase 7, the cyclin-dependent kinase-activating kinase, with an analogue sensitive version in human cells (23). However, given the large number of kinases and the highly redundant biochemical pathways involved in mitosis and cytokinesis, the further application of chemical genetic analysis of the metazoan cell division cycle has to be further validated.

In general, genetic approaches have provided a tremendous amount of largely descriptive information and most often lack to adequately represent the relevant proteomic level, where we observe a considerable inflation of numbers of molecular species due to posttranslational modifications and moreover intricate kinetic features which require tight analytical synchronisation due to huge dynamic ranges of concentration differences and time scales of changes, all this in frameworks of a huge degree of biological stochasticity (24).

2. Multi-target Drugs

There is a broad understanding that the success of target and lead structure identification, optimization, pre-clinical validation and clinical development has always depended on a thorough understanding of underlying biological mechanisms in disease and corresponding treatment. The reality of current pharmaceuticals out in the market, however, includes a significant segment of successful drugs which have low selectivities and affinities and multiple targets (24) where the exact modes of action are far from clear. Given, for example, the complex and polyetiological molecular biological background of central nervous system disorders, functional genomic and in particular proteomic profiling of efficacy models (25, 26) increasingly contribute to drug development (27–30). The enormous functional flexibility of many protein targets results in crosstalk and pleiotropy, generating combinatorial effects and cascades of non-linear cellular functions. A decade ago and based on the hypothesis that 100,000 open reading frames for proteins exist in the human genome, Drews and Ryser proposed roughly 500 molecular targets interacting with marketed drugs (31). Today, after realising that there might be only about 20,000 genes in *Homo sapiens* on the one hand, but that due to posttranslational modifications of proteins on the other hand, there might exist millions of distinct molecular species, the situation is still unsettled. If a target is considered to be "a molecular structure (chemically definable by at least a molecular mass)" that will undergo a specific interaction with given chemicals, we might end up with a bigger number of potential targets, may be 1,000 (32). However, this type of definition as so-called mini-targets based on one major and predominant biochemical interaction may prove to be too reductionist. In fact, most drugs appear to interact with several targets over a range of affinities. Therefore, every substance has a unique "effect-space" in biological homeostasis, which usually is far out of equilibrium and energetically highly "charged." The subtleties imposed on drug/target interactions by chemical modifications of amino acid side chains of proteins like phosphorylation, methylation, and ADP-ribosylation, as well as proteolytic processing, offer new chances of defining drugable substructures beyond the mere amino acid backbone of translated genetic information (24).

Living dynamic systems react to external input on time scales starting in the nanosecond range with consequences potentially lasting years, but research tools used in drug development are generally limited to study targets as static objects (for example, 3D protein structures). A relevant "target" often exhibits varying degrees of freedom depending upon physical and chemical conditions of cellular micro-environments. The sum of these interactions will give as an output an effect profile. So far we are unable to adequately address these so-called macro-targets. Current high through-put in vitro screening techniques appear to identify such targets only by serendipity. Targeting net effects rather than direct high affinity chemical interactions will require novel concepts, integrating genetic and proteomic information about redundant, pleiotropic and recurrent biological signalling (33, 34). In this regard, concepts like gene expression state spaces and attractors have been introduced which try to provide a mathematical and molecular basis for "epigenetic landscapes" (35). Dynamic transient behaviours of cellular reactions have been discussed in terms of chaotic itinerancy (36), and in particular the aspect of multi-degree-of freedom reactions has led to the analysis of high-dimensional variance data. There are approaches to reveal and interpret the structure in such data spaces or "landscapes" or signatures, essentially by geometrical and correlational procedures (37).

The application of these concepts to human diseases and drug development requires general approaches for the quantification of synergies of combinatorial effects (multiple modes of action of a single drug or a combination of drugs). The relevance of non-linear dose response surfaces in complex biological systems has been shown in HIV and cancer chemotherapy by drug combination experiments. In this context, the definition of exact sites of drug interactions with protein target domains in vitro is the first important step. However, related screening usually employs recombinant proteins devoid of posttranslational modifications, ignoring the considerable flexibility in protein structure which can shape signalling decisively. The attractiveness of this reductionist approach, which has set the standards in pharmaceutical research for decades, lies in the fact that high through-put can be achieved in screening, albeit as we have been suspecting for a while, at the cost of relevance. Today, we face new challenges moving forward to develop rationales for the analysis of drugs with multiple modes of action, often with moderate affinities and poor selectivities on the background of a daunting biological complexity. Indeed, new scientific paradigms can be expected to emerge in particular with regard to mathematical treatment and modelling (38–41).

Healthy biological signalling is always far out of equilibrium, consuming constantly high amounts of energy. Control of energy homeostasis is maintained by a first level of key metabolites or second messengers including intracellular calcium concentrations, ATP, and NADH/H^+. The "energy" status of a cellular system is partly integrated by two posttranslational modifications consuming energy-rich substrates: phosphorylation and ADP-ribosylation, which are commonly found to be regulated on very fast time scales in almost every major signalling pathway. In particular, the activation of most types of receptors at some point involves phosphorylation cascades. Cellular stress on the other hand converges on the intrinsic or mitochondrial apoptotic pathway of cell death. Many calcium-driven pathologies, apoptosis, ATP, energy, reactive oxygen species (ROS), and cell death converge at the level of mitochondria (42–45).

These disequilibria result in simultaneous effects of multi-target drugs across several biochemical pathways or in different organs. Because of the nature of living systems, the net result will not be linearly deducible from single effects, but will be nonlinear, with particular "effect spaces or surfaces" with a continuum between beneficial drug effects and toxic side reactions. For drug combinations, this is even more complicated. A mechanism-based simulation of pharmacodynamic drug–drug interactions was described recently (46). For this reason, the principle of blocking a single pharmacological target with high potency has been seducing because it appears to minimize side effects coming with less specific, multi-target drugs.

However, in complex diseases like cancer or neurodegenerative disorders, we observe a manifold of genetic and epigenetic contributions to gradually and accumulating cellular damage. Age is one of the most common risk factors (47) and mitochondrial dysfunction plays a central role. The mode of action of current medications in, for example, the treatment of Alzheimer's disease, like acetylcholine esterase inhibition or NMDA receptor channel blockade only remotely (if at all) correspond to firmly established genetic risk factors such as amyloid precursor protein, presenilins, neuregulin-1 or ApoE4 (48–52).

There is an increasing awareness that mitochondrial and epi- or postgenetic mechanisms like the assembly and processing of receptors and other membrane proteins in cholesterol-rich "rafts" might have a key role in the manifestation of genetic risks (33, 53, 54).

Even in the case of successful drugs like Genentech's breast cancer treatment Herceptin and Novartis' leukaemia drug Gleevec, each of which targets a specific genetic mutation, therapeutic effects are not stable. Eventually, these drugs become inactive in many patients because of gradually increasing resistance of cancer cells. Some of these escape mechanisms are systemic responses far beyond single-target mechanisms, as the emerging example of cancer stem cells shows, where a minority cell population evolves into something like an alien organism inside the mother organism under the pressure of the immune defence (55–58).

The example shows that cancer is not just the result of strayed genes, but merely starts there, proceeding into an ongoing evolutionary battle for survival against host defence. Mutations in at least 189 genes have recently been described in human breast and colorectal cancers (59). These findings support concerns that monofunctional drugs will not cure most of common carcinomas. The number of genes relevant to carcinogenesis in general could be much higher (60), because mutations in non-coding regulatory regions of coding genes may contribute as well. Given the relatively small number of genes, we have to realize that advanced cancer represents a complete systemic reorganisation, with cancer cells often being more alien to the organism, in which they reside, than a closely related species. Curing cancer will require understanding (and interrupting) these complex escape mechanisms. Eventually, cocktails of drugs are more likely than single magic bullets (61, 62). For example, it was shown that the inhibitors of the arachidonic acid pathway and peroxisome proliferator-activated receptor ligands have superadditive effects on lung cancer growth inhibition (63). When given together they are much more efficient in killing cancer cells than each of them given alone. Synthetic oleanane triterpenoids and retinoids, two new classes of multifunctional drugs (64) are neither conventional cytotoxic agents nor are they monofunctional drugs that uniquely target single steps in signal transduction pathways. They have unique molecular and cellular mechanisms of action and might prove to be synergistic with standard anticancer treatments. An opposite example is Vioxx: recent problems with this highly selective cyclooxygenase-2-inhibitor are considered to be due to its extreme selectivity which seems to tip the balance of pro- and antithrombotic mediators in an unfavourable way (65). Also the history of antipsychotics shows that "dirty" or "promiscuous" drugs might be increasingly important in the future (66, 67). The first generation of antipsychotic drugs; discovered about 50 years ago (e.g., chlorpromazine), were described as selective D2 dopamine antagonists, and today are known as "typical antipsychotics" (68, 69). These drugs produce a host of adverse side effects, without having any effect on the negative symptoms of schizophrenia.

Later "atypical antipsychotics" were introduced (70) which cause significantly less extrapyramidal side effects. On the molecular level, the difference between the two classes can be attributed to a different binding profile to D2 dopamine and 5-HT2A serotonin receptors. The second generation of atypical antipsychotics (also known as "dopamine-serotonin system stabilizers") such as aripiprazole, offered further advantage, due to an improved efficacy in treating the negative symptoms of schizophrenia and a decreased incidence and severity of central and peripheral side effects (70). Receptor binding studies revealed that the second-generation of atypical antipsychotics are partial agonists of D2 and 5-HT1A receptors, and antagonists of 5-HT2A receptors but also a number of other targets (24).

Another example is (–)-epigallocatechin gallate (EGCG), the major constituent of green tea, causing induction of apoptosis and cell cycle arrest in many types of cancer cells without affecting normal cells (71, 72). Apoptosis is a highly ordered protective mechanism with clear mitochondrial connotations through which unwanted or damaged cells are eliminated from the system and EGCG has been shown to affect a number of targets (24). A number of subsequent studies have shown similar effects of other dietary constituents (73).

Three strategies are available to the designers of multitarget therapies. The first strategy is to prescribe multiple individual medications. The drawback is patient compliance and the danger of drug–drug interactions. To overcome this problem, a second strategy is the development of multi component medications that contain two or more active ingredients formulated in the same delivery vehicle, such as a single pill. However, due to significant differences in pharmacokinetics, metabolism and bioavailability, the galenics and formulation of drug combinations are no trivial problems. Further, two drugs that are safe when dosed individually cannot be assumed to be safe in combination. Another drawback using drug combinations is that the target space of the current pharmacopeia is limited to approximately 1,200 FDA-approved drugs. The third strategy is to design a single compound acting on multiple targets (74, 75). Dosing with a single compound may have advantages over a drug combination in terms of equitable pharmacokinetics and biodistribution.

3. Analysis of Modes of Action of Multi-target Drugs Using a Network Approach

The rational design of drugs with multiple targets necessitates the molecular analysis of their modes of action in appropriate cellular models. Cellular networks form four major classes: metabolic pathways, gene regulation networks, signal transduction networks and protein interaction networks (76). In a living cell, at least three, possibly all four, of these networks interact: requiring more holistic strategies of investigation. In the network concept, the cell is perceived as a set of interacting elements, which are connected by links. Links have a weight, which characterizes their strength defined for example by affinity or propensity. Links may also be directed links, when one of the elements has a larger influence to the other. Interacting molecules are considered as elements in these networks, and their interactions form the weighted, but not necessarily directed links. Alternatively, we may also envision directed links as representations of signalling or metabolic processes of the functional networks in the cell. Cellular networks contain hubs, that is, elements, which have a large number of neighbours. These networks can be dissected to overlapping modules, which are supposed to form hierarchical communities (77, 78). Computational cellular network models offer a lot of possibilities to identify nodal elements as potential drug targets. In recent years, several experimental and modelling approaches have been attempted to identify targets in a network and systems biology context (79–83). The major analytical challenges result from the complexity of protein expression in terms of linear dynamic range, the number of biochemical species and the dynamics of posttranslational modifications. The time scales range from fractions of seconds (e.g., phosphorylation, regulation by proteolytic cleavage) to many days, months or even longer (e.g., permanent modification of synapses during learning, initiated by an activity-dependent phosphorylation cascade). If we accept the notion, that multi-target drugs have their effects on the level of effect spaces defined mainly by protein isoforms, then the first prerequisite is an apparently simple one: statistically significant differential control of protein expression in appropriately treated biological samples. Recently, it became clear that this is not a trivial task due to huge ranges of abundances, kinetics and chemical diversity of proteins (84).

Dealing with convoluted biological networks of extreme flexibility and redundancy has even let to consider "game theory approaches" as useful for the simplification of complex sets of non-equilibrium conditions by the introduction of "multi-target drug design games" (85). Simple topological network models will only suffice to provide preliminary insight (Fig. 1). In such networks, the modelling of pharmacological modes of action is attempted by representing general elements of the network in a sense of functionally defined target classes (proteins, RNA- or DNA-sequences). Connecting links are meant to represent known interactions of single entities within the cell. However, since thus far only amino acid backbone information of proteins (essentially amounting to the genetic information available) is included, these concepts have not delivered validated predictions yet. Drug-induced effects upon single targets are in an initial step modelled by the elimination of all interactions at the representing element (complete knockout), which essentially ignores any degree of redundancy and pleiotropy. The much more frequent and relevant partial inactivation of drug targets has been modelled in different ways: either by knocking

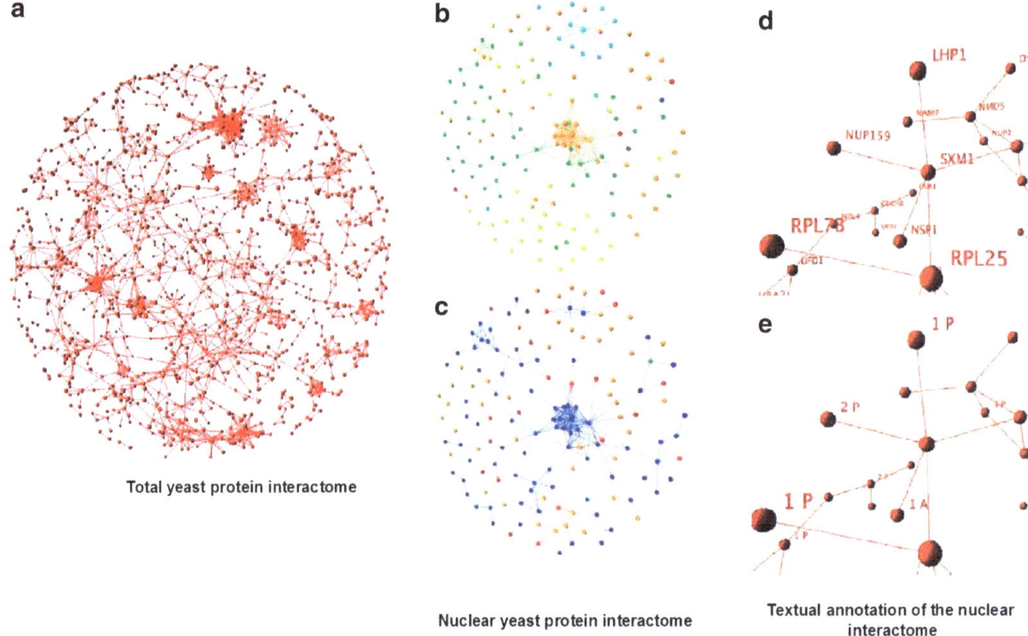

Fig. 1. Topological molecular networks of the yeast cell interactome, adapted from Ho et al. (98). (**a**) Presentation of the full Filtered Yeast Interactome data set. Differences in node size are due to the 3-D layout rendering nodes projected towards the viewer that are larger than those that are projected away from the viewer and are not indicative of any differences in proteomic attributes. (**b**) Protein interactions and associated data in the biggest connected component of the yeast nuclear interactome. *Yellow* RNA metabolism; *orange* organelle organization and biogenesis; *light blue* protein biosynthesis; *dark blue* cell cycle; *green* transcription; *red* process unknown. (**c**) Proteins colored by localization. *Blue* nucleus; *light blue* nucleolus; *green* mitochondrion; *gold* cytoplasm; *red* unknown. (**d**) Textual annotation of a portion of the nuclear interactome, showing nodes labelled by gene name. (**e**) The same nodes annotated for post-translational modification: P is phosphorylation and A is acetylation. The number associated with the modification type indicates the number of known sites per protein.

out complete parts of the interactive options of a given protein or by attenuating more moderately selected interactions of a protein. In simplified models, the "attack" on a network, such as the genetic regulatory networks of *Escherichia coli* (86, 87) or *Saccharomyces cerevisiae* (88, 89), it was reported that multiple but partial attacks on carefully selected targets are more efficient than the complete knockout of a single, though equally well selected, target (90). The largest effect to the *E. coli* regulatory network is reached by removing one element with 72 connections. However, the same effect is achieved by partial inactivation of three to five network elements, despite the observation, that the number of affected interactions does not increase (90). This example was based on network topology and applied to antimicrobial drugs, where network damage corresponds well to desired drug effects. Bacterial models are especially well suited for modelling attempts because of the nearly complete absence of posttranslational modifications in prokaryotes.

In principle, this can be useful for novel concepts about the development of "multidimensional" drugs in much more complex mammalian models.

We have to consider why this is so; on the one hand, microbes are far less complex than mammalian cells due to lack of posttranslational modifications. On the other hand, it seems that during the course of evolution, the complexity observed in higher biological systems is just representing manifold feedback variations of relatively clear-cut "original" and basic enzymatic and structural biological themes; many of them have in principle been invented by prokaryotes (19). The major evolutionary reason for this inflation of molecular species by posttranslational modifications of proteins appears to be the resulting redundancy, cross-talk and compensation and thus an enormous functional flexibility. By small changes of activity-dependent parameters, for example, intracellular calcium, qualitatively completely different, even opposing functional reactions can be elicited. Moreover, redundancy is one of the major cellular insurance mechanisms against jeopardy of the whole system by single detrimental effectors. We simply observe cellular stress-management and it may be not surprising that modes of action of drugs obey similar principles. In many cases the receptors, enzymes and other biomolecules involved in the networks of stress management are the very same responsible for physiological signal transduction (24). For the rational design of multi-target drugs, much more detailed information about kinetics, localisational and biochemical dynamics of specific signalling, metabolic and transcriptional pathways would be required. And although observing a shift of paradigm towards highly efficient low affinity, multi-target drugs, the challenges resulting from the complexity of systems require novel approaches and tools in analytical, computational and modelling efforts (91–93). So far, some progress has been made by opportunistic approaches (25, 26, 76), but a rationale for systematic screening of multiple-target compounds, with moderate affinities and selectivities or even appropriate mixtures of drugs is still in the making. The main difficulty is adequate representation and synchronisation of fast chemical changes and resulting effects on activity and structure of biomolecules.

4. Analysis of Action of Multi-target Drugs Using Systems Biology Tools

The screening of compounds and compound libraries for this type of mechanism requires an intricate combination of advanced technologies: genomic and proteomic analyses correlated directly to biochemical and/or cell biological in vitro assays might be most appropriate and fastest. Before multimodal drug responses can be validated in vivo, functional in vitro models for human or animal tissues, which simultaneously can be submitted to molecular analysis or genetic manipulation, are mandatory. Models like embryonic stem cells and their organ-specific derivatives can be submitted to functional kinetic read-outs by fluorescent techniques, and at the same time being correlated directly to molecular snapshots of biomarker signatures obtained by microarray or differential proteomic techniques (94, 95).

The complexity of biological systems thus also requires convoluted analytical procedures, in particular, adequate differential procedures for quantitative kinetic measurements and the subsequent independent validation of protein biomarkers, and there is no high-throughput method or screening readily at hand (19, 84, 96).

4.1. Differential Quantification of Protein Biomarker Signatures and the Study of Interactomes

Proteomics provides the parallel analysis of large numbers of proteins (19, 96). However, most proteomic technologies are relatively poor in exactly measuring protein abundances over the large linear dynamic range of typical biological samples (24). There are essentially only three methods for a reliable quantification of differential proteins in complex mixtures: labelling with stable or radioactive isotopes or with fluorescent dyes, usually Cy-dyes. The statistical treatment of differential quantities measured for different sample conditions is crucial and depends on the dynamic range and resolution of the separation method. The underlying statistical concepts have been discussed in depth recently (24). In essence, raw differential quantities of labelled and separated proteins have to be normalized, the reproducibility of the detection method and potential labelling bias have to be controlled by statistical procedures, like Bland-Altman and MA-plots. Subsequently, repeat measurements of inversely labelled samples need to be statistically analysed in terms of significance of amplitude differences (Volcano plots, showing the relation between *p*-values and intensity differences). These procedures serve to identify robust protein biomarker candidates from complex mixtures of proteins and surrogate patterns of multidimensional LC-MS or 2D-PAGE analysis (84). These considerations clearly point to the dynamic range of the detection method and resolution of separation technique as the most crucial parameters for finding valid biomarkers in complex biological or clinical samples (84, 97).

The increasingly recognized role of posttranslational modifications justifies laborious "wet" proteomic approaches, because only in this way relevant molecular content information can be generated from native biological material. Given the functional significance of minute structural modifications of proteins in multidimensional interactomes, it is frankly questionable which artifactual effects might result from the interpretation of non-linear data by the introduction of recombinant affinity or fluorescent tags into interacting proteins. The essential failure of high-throughput array data or in silico approaches to contribute to novel modes of action and innovative drug design shows the necessity of quantitative proteomic data from appropriate native models before approaching a realistic modelling of drug effects. The degree of stochastic and non-linear behaviour in living systems is amazing and the complexity of biological systems as well. The codes of kinetically resolved and statistically significant protein signatures or molecular snap shots necessary for the screening of novel multifunctional drugs so far are beyond our understanding. However, we can use protein signature information for predictions. The mathematical concepts for appropriate modelling tools and the analytical procedures needed for protein signature-based prediction are currently in the making and under hot debate.

The main caveat of current methodology which is impeding computational efforts towards an integrative treatment of protein interactions, protein abundances and localizations, is based in the intrinsic structural variability by posttranslational modifications and the fact that recombinant methods only distantly relate to and reflect biological reality (98). Protein turnover is largely contributing to protein signature snapshots and cannot be assessed by recombinant methods. Turnover of proteins after traditional and non-traditional ubiquitinylation, SUMO-ylation and subsequent degradation and regulation are increasingly recognized as major epi- or postgenetic principles (99–101).

We see an increasing focus on the integration of quantitative "native" proteomic data, with novel concepts regarding the acquisition of quantitative biological data, their functional synchronisation, statistics, validation and modelling (102).

4.2. Data Storage and Integration, Modelling

As an example for current problems and approaches to data storage and modelling, the authors provide an outline for strategies in creating joint data repositories for complex data sets generated in the frame of a number of projects which are part of the European Community Sixth Framework Programme. These are ageing- or toxicology-related projects assembling genomic, transcriptomic, proteomic and functional data from a variety of models (MIMAGE; www.mimage.org; Reprotect; www.reprotect.eu; ESNATS; www.esnats.eu).

Within MIMAGE, age-dependent protein maps and signalling pathways emerging from common traits in mitochondrial proteins isolated from five species in a variety of ageing models have provided a first glimpse on initial key events in fundamental ageing mechanisms. One example is the age-dependent oxidation of tryptophan-residues to N-formyl-kynurenine by ROS in SAM-dependent O-methyltransferase (PaMTH1, in *Podospora anserina*, a fungal ageing model). The human homologue of PaMTH1, COMT plays a key role in the etiology of Parkinson's disease.

Parkinson's disease is an age-dependent disease characterized by protein aggregation (α-synuclein-plaques) in dopaminergic neurons. Mitochondrial dysfunction has been identified as a major factor contributing to the onset of Parkinson's disease. Also, COMT and dopamine metabolism are major therapeutic targets in schizophrenia (they are important in cognition).

An integration of such a result into models of underlying signalling requires that data sets of age-dependent molecular and functional events, such as differential quantitative proteomic data (images, numbers, mass spectra), functional correlates (enzymatic activities, native multi protein complexes, calcium, electrochemical potentials, ATP), genomic information (longevity mutants, site-directed mutagenesis, knock-in-knock-out's) and other biological background (medline) are first organized in an appropriate standardized fashion.

Traditional proteomic differential analysis compares the proteomes of two experimental conditions to each other. With more advanced techniques, (19) a multitude of different conditions or experimental interventions can be compared at once. As a result, matched vectors of several hundred protein abundances ("proteomic inventories") under a number of experimental conditions become available.

This new quality of experimental data promotes a demand for joint analyses of several individual comparisons:

- How do experimental conditions relate to each other?
- Is the proteomic inventory of one condition similar to or largely different from the inventory of another one?
- Which conditions exhibit the most diverse proteomic inventory?
- Which combinations of conditions may be expected to represent extreme points of protein abundances?
- Can subsets of all protein spots be defined that show similar (or most different) abundances in certain (combinations of) experimental conditions?

With these perspectives in mind, we suggest the following structure for data repository for advanced systems biology projects.

4.2.1. Data Repository Structure

When planning a systems biology data repository, several requirements are to be met:

- The intrinsic structure of all data layers needs to be represented by the data storage format in a natural fashion.
- Data exchange with outside researchers and existing data storage solutions need to be easily possible.
- The effort necessary for setting up and maintaining the structure of the database solution must not be unreasonably high.

- Data need to be made available in a format that is suitable to serve as input for system biological studies.

The last requirement calls for employing standardized data formats at least for data types of higher complexity. During the past years, a substantial effort has been spent on defining and agreeing on universally accepted minimum information requirements for the description of typical experimental procedures in various biological sub-fields. Examples of the results of such efforts are MIAME (Minimum Information about a Microarray Experiment (103)), MIBBI (Minimum Information for Biological and Biomedical Investigations (104)), MIFlowCyt (Minimum Information about a Flow Cytometry Experiment), MINI (Minimum Information about a Neuroscience Investigation), or MIAPE (Minimum Information about a Proteomics Experiment (105)). From verbose descriptions of minimum information required to describe certain types of experiments, abstract models of corresponding data structures have been developed. For example, the Functional Genomics Experiment Object Model (FuGE-OM (106)) arises from MIAME, or the mzData and analysisXML data models for capturing peak list information and representing mass spectrometry informatics data are (partial) materializations of MIAPE. Both models are advanced by the HUPO Proteomics Standards Initiative (107, 108) (http://www.psidev.info/index.php?q=node/80). Especially in projects, where a large body of proteomic data is generated, the MIAPE standards have a particular value.

To our experience, data repositories of the type sketched above require an integration of a number of existing database solutions to keep investments of man power and other costs in acceptable limits.

We suggest combining two separate database frameworks specialized in storing different aspects of biological experimental result data. For the general description of the experimental conditions and for simply structured input and output data, the FuGE compliant "SYstems and Molecular Biology data and metadata Archive" (SyMBA) is used (http://symba.sourceforge.net/SyMBA-Mged2008.pdf; developed by CISBAN, Centre for Integrated Systems Biology of Ageing and Nutrition, Newcastle University, http://www.cisban.ac.uk/). The data resulting from proteomic investigations, on the other hand, are stored in a modified "PRoteomics IDentifications database" (PRIDE) (109) (EBI, European Bioinformatics Institute; http://www.ebi.ac.uk/pride/) which is based on the mzData and analysisXML data models.

Technically, the SyMBA installations are implemented as a Hibernate Software ToolKit (STK) encompassing a FuGE-structured relational backend database (based on PostgreSQL), a Java Object-Relational Persistence layer (Hibernate and Spring), a set of Java Object entity classes representing FuGE UML entities, and a set of Data Access Object (DAO) classes that facilitate and encapsulate access to entity classes. SyMBA employs FuGE (http://fuge.sf.net) as the core of its database structure, and provides means for using both MIBBI checklists (http://www.mibbi.org/), OBI controlled vocabulary (http://purl.obofoundry.org/obo/obi). As a particular strength, SyMBA features highly elaborate built-in appliances for standardizing, structuring and storing experimental metadata for a range of omics experiments which include the following packages (106):

- Audit (storing contacts, auditing and security settings for all objects),
- Description (to allow for additional annotations and free-text descriptions for all objects),
- Measurement (defines slots for providing atomic, Boolean, range and complex values with appropriate units, sourced from an ontology),
- Ontology (provides a mechanism for referencing external ontologies or terms from a controlled vocabulary),
- Protocol (represents a model of procedures, software, hardware and parameters and can define workflows by relating input and output materials and/or data to the protocols that act on them),
- Reference (enables external bibliographic or database references that can be applied to many objects across the FuGE model),
- ConceptualMolecule (captures database entries of biological molecules such as DNA, RNA or amino acid sequences and provides an extension point for other molecule types, such as metabolites or lipids),
- Data (sets the dimensions of data and storage matrices, or references to external data formats),
- Investigation (defines an overview of the investigation structure by capturing the overall design and the experimental variables and by providing associations to related data),
- Material (models material types such as organisms, samples or solutions. Materials are characterized by ontology terms or by the extension of the Material package).

By means of these packages, SyMBA offers data types for modelling experimental objects such as samples, protocols, instruments, or software modification histories. Finally, data describing specific input and output values of the respective experiments may be represented by defining suitable elements in FuGE's Data Collection object.

PRIDE is a database for protein and peptide identifications that may potentially have been published in the literature, which will typically arise from specific species, tissues and subcellular locations, perhaps under specific disease conditions, and for which post-translational modifications may have been located on individual peptides (109). The PRIDE database definition and Java classes may be downloaded from EBI and built using Maven. Employing the Hibernate persistence framework, the code can be modified to accommodate additional fields and to allow for joining PRIDE and SyMBA database applications. In particular, it is necessary,

- To include further fields in the protein identification table describing the quality of the identification,
- To define/supplement structures for the specification of the material which the identification originates from (electrophoresis gel and protein spot identifier), and
- To store the outcome of quantitative comparisons of protein abundances between two different biological conditions (typically two age groups).

The SyMBA and PRIDE parts can be connected as follows: The experimental details, in particular, the specific conditions entering each proteomic comparison are described in the SyMBA database. Each one of these comparisons is labelled by a unique identifier. Similarly, each result of such a comparison is labelled by the same unique identifier in the PRIDE part of the database. This allows formulating queries over both parts of the combined database in a unified fashion. For example, by matching these unique identifiers, all results of a given comparison in the SyMBA branch can be retrieved from the PRIDE branch, or the comparison details for a given result in the PRIDE branch can be retrieved from the SyMBA branch.

Both parts of such a database may be generated from (more or less) abstract model descriptions via an automated process including MAVEN. All code is controlled using the Subversion version control system. Basic experimental data can be entered via web interfaces. More complex data may be uploaded as files (*.txt, *.doc, *.xls, *.csv, …). For proteomic data, suitable import routines have been programmed.

Data may be accessed via versatile search interfaces which allow for, for example, finding all occurrences of a given protein in any one of the experiments in the data repository. More extensive data collections may be retrieved from the database via file download.

4.2.2. Employing Proteomic Data in Mathematical Modelling

Enzymes are organized in metabolic networks, and overall network function depends on the function of an individual enzyme in a highly intricate fashion. Therefore, it often is not at all obvious how activity- or age- (or generally input-) related changes in enzyme abundance (that have been diagnosed in the course of a differential proteomic analysis) affect network performance. To answer this type of questions, one can use a mathematical model of the respective metabolic pathways to study the effects of changes in abundance or activity of an individual enzyme on the entire network.

For purposes of modelling metabolic pathways, a number of tools are publicly available:

- The KEGG electronic pathway maps (110) (http://www.genome.ad.jp/kegg/pathway/map/map01100.html) give detailed information about the enzymes involved in metabolic networks. KEGG maps are editable and may be modified/annotated for presentation (111). However, KEGG does not provide for direct access to enzyme kinetic data.
- For retrieving numerical values for the enzyme kinetic constants, web-based database services are available which can be searched for the kinetic constants of an enzyme of interest, for example, "BRENDA – The Comprehensive Enzyme Information System" (112) (http://www.brenda-enzymes.org/).
- Setting up the code for computing network behaviour is greatly facilitated by modelling frameworks like the "Systems Biology Workbench" (SBW) (113) (http://www.sys-bio.org/sbwWiki/doku.php?id=sysbio:sbw). The generated models are stored in systems biology markup language (SBML) (114) which may be imported into a variety of other modelling packages. The SBW framework includes the graphic network editor JDesigner for setting up metabolic network geometry and kinetics.

In the process of modelling a number of prerequisites need to be met:

- The elements of the metabolic network need to be sufficiently well known.
- Sub-networks of the overall metabolic network need to be identified, which are largely self-contained but sufficiently comprehensive to generate biologically meaningful models.
- Realistic numerical values for the kinetic constants of all involved enzymes need to be acquired from the literature, or more easily from a web-based database service like BRENDA (112).

  However, great care must be taken in the factual choice of the numerical values because it is not uncommon that enzyme databases report functional parameters for exactly the same enzyme deviating by up to two orders of magnitude. Moreover, kinetics may differ greatly between isoforms of the same enzyme, related species and even tissues. Therefore, it is generally not sufficient to just look up the values of interest in the database and plug them into the model. Rather, the exact conditions and methods of measurement need to be assessed from the original publication and adequately accounted for when fixing model parameters. This may also be achieved by statistical approaches (115). Moreover, it is a good idea to perform and take into account model runs for a number of sensibly selected parameter ranges.

- Similarly as enzyme kinetic constants, enzyme volume-related activities need to be input in the models. For some enzymes, specific activities may be found in BRENDA. In other cases, data may be obtained from original investigations of enzyme kinetics in which material from animal tissues was used. Frequently, series of measured activities during the purification process are published which allow for computing the desired tissue activities.
  Alternatively, enzyme tissue activities may be computed if the specific activity (per gram of enzyme) and enzyme concentration in the tissue of interest are known.
- Frequently, measurements of enzyme kinetic data have been performed under conditions different from the ones in living tissues. Therefore, suitable corrections of measured kinetic data need to be applied, in particular for deviations in pH and temperature.
- After a model has been set up, it needs to be validated. To that end, physiological parameters like local substrate concentrations, partial pressures, or turnover rates must be identified the numerical values of which are known from the literature or can be easily measured on the one hand and which can also be predicted by the model on the other. Validation is performed by comparing the behaviour of measured and predicted values for a number of locations, physiological conditions or the like.
- In order to identify suitable test cases for the model, data and accession tools of such data repositories need to be structured in an appropriate fashion. In particular, the fields describing identified differential proteins need to be rendered searchable and a search mechanism has to be implemented that allows finding all experimental conditions under which a given protein of interest is up- or downregulated, changes its activity, undergoes certain modifications, etc.

In a recent example, in age-related comparisons of a number of model organisms, several enzymes of the dopamine metabolism turned out to be differential (19). Therefore, as a first model case, degradation of dopamine was selected. Dopamine formation and decomposition is part of the tyrosine metabolism. The dopamine-related details of the pertinent pathway map by KEGG (110) (http://www.genome.jp/dbget-bin/get_pathway?org_name=hsa&mapno=00350) is shown in Fig. 2. Dopamine is produced from l-dopa by the enzyme dopa decarboxylase (4.1.1.28). The subsequent dopamine metabolism is fairly complicated and involves the formation of the hormones epinephrine and norepinephrine. One of the enzymes found to be affected by ageing was a homologue of mammalian catecholamine O-methyltransferase (COMT). Also involved is monoamine oxidase (MAO) which plays an important role in the pharmacological modification of dopamine degradation.

As becomes apparent from Fig. 2, dopamine metabolism is only linked to the rest of tyrosine metabolism via the enzyme dopa decarboxylase (4.1.1.28), so all reactions downstream of dopa decarboxylase may be viewed as a (largely) independent sub-network.

Dopamine metabolism was modelled using the "Systems Biology Workbench" and JDesigner (113). The SBW framework allows for graphically setting up networks of chemical species

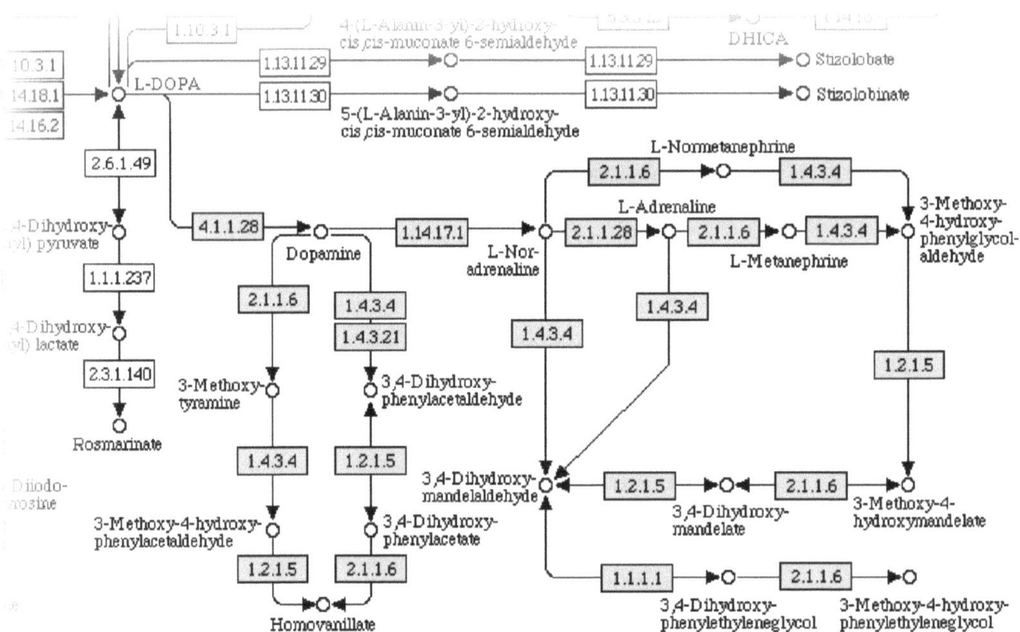

Fig. 2. Detail of KEGG pathway map of tyrosine metabolism (http://www.genome.jp/dbget-bin/get_pathway?org_name=hsa&mapno=00350). Only reactions relating to dopamine formation and metabolism are shown. Dopamine is produced from l-dopa by the enzyme dopa decarboxylase (4.1.1.28). The subsequent dopamine metabolism is fairly complicated and involves the formation of the hormones epinephrine and norepinephrine. A homologue of catecholamine O-methyltransferase (COMT, 2.1.1.6) has been found to be upregulated in ageing. Also involved is monoamine oxidase (MAO, 1.4.3.4) which plays an important role in the pharmacological modification of dopamine degradation. Note that COMT as well as MAO participate in no fewer than six distinct reactions in the degradation of dopamine, epinephrine and norepinephrine.

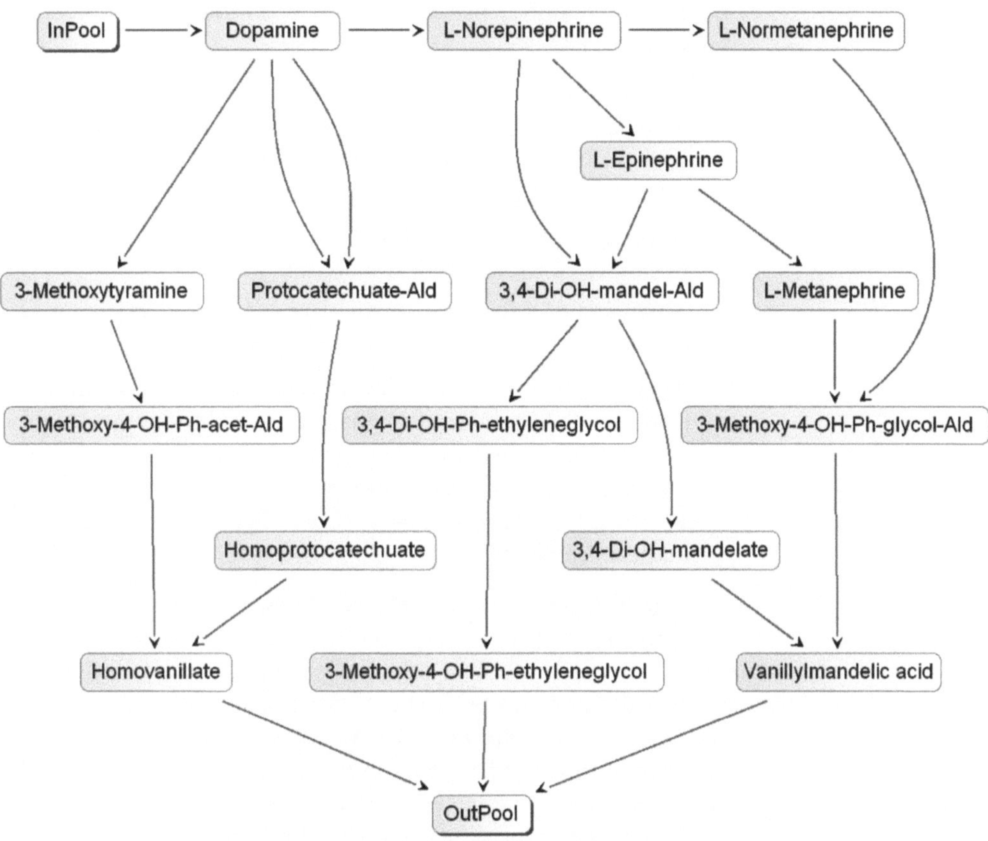

Fig. 3. Graphical representation of the mathematical model of dopamine metabolism as displayed in JDesigner/Systems Biology Workbench. Each node in the network denotes a chemical species which takes part in the reactions. Each *arrow* corresponds to an enzyme, the reaction kinetics of which can be chosen from a variety of different kinetic laws with deliberately definable kinetic constants.

which take part in the reactions (Fig. 3). Each arrow represents an enzyme, the reaction kinetics of which can be chosen from a variety of different kinetic laws (or completely re-defined by the user) with kinetic constants determinable at one's own discretion.

In conclusion, proteins change compartments or active states due to posttranslational modifications like methylation, phosphorylation, glycosylation and non-enzymatic oxidation. Moreover, proteolytic processing occurs, inactivating or mobilizing effector domains. The kinetics of these events include and integrate fast (activity-dependent) and slow (developmental) time frames. Even in the apparent absence of posttranslational modifications dramatic localisational effects in completely different cells or compartments regulate cellular biology on the level of the organism.

In order to reflect these various kinetic changes in protein function more closely, a number of prerequisites need to be met in future approaches to mathematical modelling of metabolic networks:

- Necessity of kinetic control of biological material. We need time-resolved recordings of kinetically evolving synchronous, statistically significant differences in the proteomes of the compared samples.
- Necessity of precise knowledge concerning fractionation strategies of the biological material.
- Necessity of high resolution, large dynamic range, precise quantification, and meaningful statistics (84).
- Necessity of novel strategies in data analysis and interpretation. We need to identify signatures of redundant protein isoforms which are to be viewed as effect "spaces" or "landscapes." On top of traditional bioinformatic analysis, the stochastic and non-linear properties of these effect spaces need mathematical attention.

These new approaches are expected to have most important consequences in drug discovery where we may face a paradigm shift towards low affinity/low selectivity drugs or even multidimensional (= dirty) drugs or drug combinations. Over and beyond the role of the proteome in pharmacological research, many novel treatments promoted under the heading "Biologics" or the like are anticipated to be proteins themselves.

As a consequence, future system biologic strategies will need to cope with the following problems:

- How to screen for multi-faceted kinetic proteome data?
- How to extract higher level explicative information by applying advanced mathematical models to the measured data?
- How to interpret the greatly complex, multidimensional modelling results?

5. Chemical Proteomics

An important recent development in addressing problems associated with limited experimental windows due to poor resolution or dynamic range of separation and detection methodologies is the application of affinity-based approaches. Specific antibodies have been used to pull down targets and associated proteins. However, antibodies usually are selected against non-modified peptides and the complexity resulting from posttranslational modifications cannot be addressed. Moreover, cross-reactivity of antibodies and the fact that they are proteins themselves is causing problems in mass spectrometry-based identification procedures.

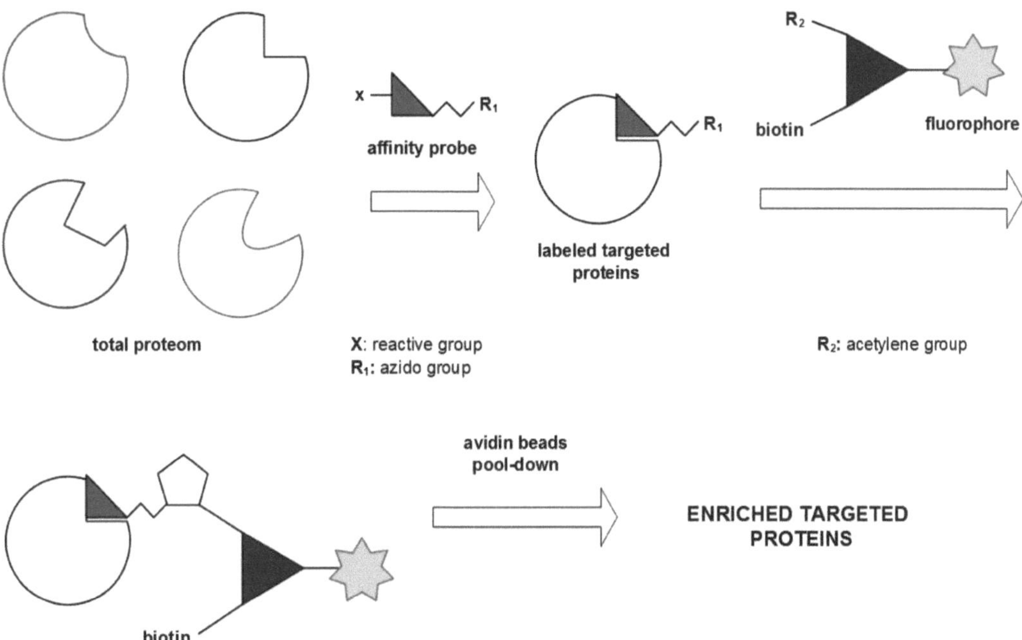

Fig. 4. General principle of affinity-based chemical proteomics. Protein labelling is three-step process. In the first step, affinity probe that consists of: (1) a reactive group, typically an electrophilic or photoreactive group, for covalent labelling of protein targets; (2) a binding group, which directs the reactive element towards specific classes of proteins; and (3) azide group that facilitates application bioorthogonal "click chemistry," react with protein. In the second step, analytical probe that consists of: (1) fluorophore or (and) biotin, for detection and/or enrichment of probe-labelled proteins; and (2) acetylene group as counterpart to azide group for "click chemistry," react with tagged protein. And, finally, in the third step labelled proteome can be selectively detected and enriched.

"Chemical proteomics," is employing active site-directed chemical affinity probes to explore "interacting proteomes" for drugs/ligands as a special embodiment of affinity fractionation (30, 116) (Fig. 4). These affinity probes can either be generated by synthesizing ligands which are covalently bound to solid phases to create affinity resins for protein-binding partners or as soluble compounds consisting of at least two general elements: (1) a reactive group for binding and covalently modifying the binding site of ligand, and (2) a reporter tag for the detection, enrichment, and identification of probe-labelled proteins. Typical reporter tags include fluorophores and/or biotin for in-gel detection and avidin-based enrichment of probe-labelled enzymes or receptors, respectively. To date, this type of technology has been used successfully to target all major classes of proteases, kinases, phosphatases, glycosidases, GSTs, oxidoreductases, phosphatidyl-inositol 3-kinase signalling and histone deacetylases as well as for post-translationally modified proteins (24, 117–119). Bulky reporter tags, such as fluorophores or biotin, can be replaced with much smaller azide or alkyne groups which enable profiling of enzyme activities in living cells and animals (30, 120). In these experiments, post-labelling ex vivo conjugation to reporter tags is accomplished by a bioorthogonal reaction like the azide–alkyne cycloaddition or the Staudinger ligation (121). A recent example is the exploration of an additional mode of action of 4-azasteroids, like Dutasteride, which have been shown to interfere with the mitochondrial permeability transition pore (30).

6. Conclusion

Drug development is currently moving forward from the nucleic acid-based high-throughput screening strategies of the past decades and examining novel concepts like systems biology beyond the "one disease-one-target and drug" thinking. On the contrary, latest genomic screening of ever increasing human populations rather indicate that many if not most pathological phenotypes may be converging from a multitude of subtle genomic changes. In a context of the resulting discussion about "personalized medicine" (122), it is increasingly acknowledged that successful compounds do not exert their effects through a single target, but instead have multiple targets and also rather have their effects on systemic phenotypes or key nodal points of integrative molecular/biochemical networks, downstream of genes and/or SNP's. There is hope that these additional modes of action of compounds with lower selectivities and affinities might have a better chance of positive effects in complex equilibriums of whole cellular networks (118). However, the question of rational design and screening of such multi-target drugs remains open (119). On a first level, the challenges concerning robust tools for screening and characterization of corresponding protein biomarkers are a hurdle. Once statistically and biologically validated biomarker signatures for compounds across a relevant set of biological conditions have been firmly established, it is not yet decided how to employ this type of complex information for screening. The integration of experimental and modelling approaches is the major challenge of the emerging field of systems biology. The conceptual frameworks for generating quantitative data relevant on a systems level, the appropriate statistical and mathematical treatment of such data sets and subsequent structural and pathways modelling still do not exist. The mere quantitative reconstruction of the dynamics of cellular molecular changes is still lacking generally accepted experimental approaches and strategies. The degrees of freedom of metastable biological systems and the information codes underlying these disequilibrium states are maintained by energetically extremely costly regulation (ATP, NADH).

These tools could help to identify a suitable set of parallel targets and multi-target substances for specific conditions. The integrative character of systems biology with a strong focus on the proteomic level is already having some impact on the drug discovery process. Future screening procedures will probably be not so much high throughput but will include opportunistic wet laboratory optimization by combining in vitro models with chemical proteomics and affinity fractionation (120, 123). Drugs will be characterized by selected endpoints of more complex (76) (Fig. 3) dynamic signatures of "interacting proteomes" (84). Many of them will be biologics, due to the modular structural flexibility of biomolecules.

References

1. Boone C, Bussey H, Andrews BJ (2007) Exploring genetic interactions and networks with yeast. Nat Rev Genet 8:437–449
2. Davierwala AP, Haynes J, Li Z, Brost RL, Robinson MD, Yu L, Mnaimneh S, Ding H, Zhu H, Chen Y, Cheng X, Brown GW, Boone C, Andrews BJ, Hughes TR (2005) The synthetic genetic interaction spectrum of essential genes. Nat Genet 37:1147–1152
3. Tong AH, Lesage G, Bader GD, Ding H, Xu H, Xin X, Young J, Berriz GF, Brost RL, Chang M, Chen Y, Cheng X, Chua G, Friesen H, Goldberg DS, Haynes J, Humphries C, He G, Hussein S, Ke L, Krogan N, Li Z, Levinson JN, Lu H, Menard P, Munyana C, Parsons AB, Ryan O, Tonikian R, Roberts T, Sdicu AM, Shapiro J, Sheikh B, Suter B, Wong SL, Zhang LV, Zhu H, Burd CG, Munro S, Sander C, Rine J, Greenblatt J, Peter M, Bretscher A, Bell G, Roth FP, Brown GW, Andrews B, Bussey H, Boone C (2004) Global mapping of the yeast genetic interaction network. Science 303:808–813
4. Hughes TR, Robinson MD, Mitsakakis N, Johnston M (2004) The promise of functional genomics: completing the encyclopedia of a cell. Curr Opin Microbiol 7:546–554
5. Dolinski K, Botstein D (2005) Changing perspectives in yeast research nearly a decade after the genome sequence. Genome Res 15:1611–1619
6. Lehner B, Crombie C, Tischler J, Fortunato A, Fraser AG (2006) Systematic mapping of genetic interactions in *Caenorhabditis elegans* identifies common modifiers of diverse signaling pathways. Nat Genet 38:896–903
7. Schlabach MR, Luo J, Solimini NL, Hu G, Xu Q, Li MZ, Zhao Z, Smogorzewska A, Sowa ME, Ang XL, Westbrook TF, Liang AC, Chang K, Hackett JA, Harper JW, Hannon GJ, Elledge SJ (2008) Cancer proliferation gene discovery through functional genomics. Science 319:620–624
8. Silva JM, Marran K, Parker JS, Silva J, Golding M, Schlabach MR, Elledge SJ, Hannon GJ, Chang K (2008) Profiling essential genes in human mammary cells by multiplex RNAi screening. Science 319:617–620
9. Berns K, Horlings HM, Hennessy BT, Madiredjo M, Hijmans EM, Beelen K, Linn SC, Gonzalez-Angulo AM, Stemke-Hale K, Hauptmann M, Beijersbergen RL, Mills GB, van de V, Bernards R (2007) A functional genetic approach identifies the PI3K pathway as a major determinant of trastuzumab resistance in breast cancer. Cancer Cell 12:395–402
10. Cordell HJ (2002) Epistasis: what it means, what it doesn't mean, and statistical methods to detect it in humans. Hum Mol Genet 11:2463–2468
11. Luo B, Cheung HW, Subramanian A, Sharifnia T, Okamoto M, Yang X, Hinkle G, Boehm JS, Beroukhim R, Weir BA, Mermel C, Barbie DA, Awad T, Zhou X, Nguyen T, Piqani B, Li C, Golub TR, Meyerson M, Hacohen N, Hahn WC, Lander ES, Sabatini DM, Root DE (2008) Highly parallel identification of essential genes in cancer cells. Proc Natl Acad Sci USA 105:20380–20385
12. Ito T, Chiba T, Ozawa R, Yoshida M, Hattori M, Sakaki Y (2001) A comprehensive two-hybrid analysis to explore the yeast protein interactome. Proc Natl Acad Sci USA 98:4569–4574
13. Uetz P, Giot L, Cagney G, Mansfield TA, Judson RS, Knight JR, Lockshon D, Narayan V, Srinivasan M, Pochart P, Qureshi-Emili A, Li Y, Godwin B, Conover D, Kalbfleisch T, Vijayadamodar G, Yang M, Johnston M, Fields S, Rothberg JM (2000) A comprehensive

analysis of protein-protein interactions in *Saccharomyces cerevisiae*. Nature 403:623–627

14. Ho Y, Gruhler A, Heilbut A, Bader GD, Moore L, Adams SL, Millar A, Taylor P, Bennett K, Boutilier K, Yang L, Wolting C, Donaldson I, Schandorff S, Shewnarane J, Vo M, Taggart J, Goudreault M, Muskat B, Alfarano C, Dewar D, Lin Z, Michalickova K, Willems AR, Sassi H, Nielsen PA, Rasmussen KJ, Andersen JR, Johansen LE, Hansen LH, Jespersen H, Podtelejnikov A, Nielsen E, Crawford J, Poulsen V, Sorensen BD, Matthiesen J, Hendrickson RC, Gleeson F, Pawson T, Moran MF, Durocher D, Mann M, Hogue CW, Figeys D, Tyers M (2002) Systematic identification of protein complexes in *Saccharomyces cerevisiae* by mass spectrometry. Nature 415:180–183
15. Krogan NJ, Cagney G, Yu H, Zhong G, Guo X, Ignatchenko A, Li J, Pu S, Datta N, Tikuisis AP, Punna T, Peregrin-Alvarez JM, Shales M, Zhang X, Davey M, Robinson MD, Paccanaro A, Bray JE, Sheung A, Beattie B, Richards DP, Canadien V, Lalev A, Mena F, Wong P, Starostine A, Canete MM, Vlasblom J, Wu S, Orsi C, Collins SR, Chandran S, Haw R, Rilstone JJ, Gandi K, Thompson NJ, Musso G, St OP, Ghanny S, Lam MH, Butland G, Altaf-Ul AM, Kanaya S, Shilatifard A, O'Shea E, Weissman JS, Ingles CJ, Hughes TR, Parkinson J, Gerstein M, Wodak SJ, Emili A, Greenblatt JF (2006) Global landscape of protein complexes in the yeast *Saccharomyces cerevisiae*. Nature 440:637–643
16. Gavin AC, Bosche M, Krause R, Grandi P, Marzioch M, Bauer A, Schultz J, Rick JM, Michon AM, Cruciat CM, Remor M, Hofert C, Schelder M, Brajenovic M, Ruffner H, Merino A, Klein K, Hudak M, Dickson D, Rudi T, Gnau V, Bauch A, Bastuck S, Huhse B, Leutwein C, Heurtier MA, Copley RR, Edelmann A, Querfurth E, Rybin V, Drewes G, Raida M, Bouwmeester T, Bork P, Seraphin B, Kuster B, Neubauer G, Superti-Furga G (2002) Functional organization of the yeast proteome by systematic analysis of protein complexes. Nature 415:141–147
17. Winzeler EA, Shoemaker DD, Astromoff A, Liang H, Anderson K, Andre B, Bangham R, Benito R, Boeke JD, Bussey H, Chu AM, Connelly C, Davis K, Dietrich F, Dow SW, El BM, Foury F, Friend SH, Gentalen E, Giaever G, Hegemann JH, Jones T, Laub M, Liao H, Liebundguth N, Lockhart DJ, Lucau-Danila A, Lussier M, M'Rabet N, Menard P, Mittmann M, Pai C, Rebischung C, Revuelta JL, Riles L, Roberts CJ, Ross-MacDonald P, Scherens B, Snyder M, Sookhai-Mahadeo S, Storms RK, Veronneau S, Voet M, Volckaert G, Ward TR, Wysocki R, Yen GS, Yu K, Zimmermann K, Philippsen P, Johnston M, Davis RW (1999) Functional characterization of the *S. cerevisiae* genome by gene deletion and parallel analysis. Science 285:901–906
18. Maxwell CA, Moreno V, Sole X, Gomez L, Hernandez P, Urruticoechea A, Pujana MA (2008) Genetic interactions: the missing links for a better understanding of cancer susceptibility, progression and treatment. Mol Cancer 7:4
19. Groebe K, Krause F, Kunstmann B, Unterluggauer H, Reifschneider NH, Scheckhuber CQ, Sastri C, Stegmann W, Wozny W, Schwall GP, Poznanovic S, Dencher NA, Jansen-Durr P, Osiewacz HD, Schrattenholz A (2007) Differential proteomic profiling of mitochondria from *Podospora anserina*, rat and human reveals distinct patterns of age-related oxidative changes. Exp Gerontol 42:887–898
20. Parsons AB, Brost RL, Ding H, Li Z, Zhang C, Sheikh B, Brown GW, Kane PM, Hughes TR, Boone C (2004) Integration of chemical-genetic and genetic interaction data links bioactive compounds to cellular target pathways. Nat Biotechnol 22:62–69
21. Sharom JR, Bellows DS, Tyers M (2004) From large networks to small molecules. Curr Opin Chem Biol 8:81–90
22. Burkard ME, Randall CL, Larochelle S, Zhang C, Shokat KM, Fisher RP, Jallepalli PV (2007) Chemical genetics reveals the requirement for Polo-like kinase 1 activity in positioning RhoA and triggering cytokinesis in human cells. Proc Natl Acad Sci USA 104:4383–4388
23. Larochelle S, Merrick KA, Terret ME, Wohlbold L, Barboza NM, Zhang C, Shokat KM, Jallepalli PV, Fisher RP (2007) Requirements for Cdk7 in the assembly of Cdk1/cyclin B and activation of Cdk2 revealed by chemical genetics in human cells. Mol Cell 25:839–850
24. Schrattenholz A, Soskic V (2008) What does systems biology mean for drug development? Curr Top Med Chem 15:1520–1528
25. Ashburn TT, Thor KB (2004) Drug repositioning: identifying and developing new uses for existing drugs. Nat Rev Drug Discov 3:673–683
26. Youdim MB (2007) Magic bullets or novel multimodal drugs with various CNS targets for Parkinson's disease? Nat Rev Drug Discov 6:499–500
27. Natt F (2007) siRNAs in drug discovery: target validation and beyond. Curr Opin Mol Ther 9:242–247

28. Burbaum J, Tobal GM (2002) Proteomics in drug discovery. Curr Opin Chem Biol 6:427–433
29. Keskin O, Gursoy A, Ma B, Nussinov R (2007) Towards drugs targeting multiple proteins in a systems biology approach. Curr Top Med Chem 7:943–951
30. Soskic V, Klemm M, Proikas-Cezanne T, Schwall GP, Poznanovic S, Stegmann W, Groebe K, Zengerling H, Schoepf R, Burnet M, Schrattenholz A (2007) A connection between the mitochondrial permeability transition pore, autophagy and cerebral amyloidogenesis. J Proteome Res 7:2262–2269
31. Drews J, Ryser S (1997) The role of innovation in drug development. Nat Biotechnol 15:1318–1319
32. Imming P, Sinning C, Meyer A (2006) Drugs, their targets and the nature and number of drug targets. Nat Rev Drug Discov 5:821–834
33. Schrattenholz A, Soskic V (2006) NMDA receptors are not alone: dynamic regulation of NMDA receptor structure and function by neuregulins and transient cholesterol-rich membrane domains leads to disease-specific nuances of glutamate-signalling. Curr Top Med Chem 6:663–686
34. Apic G, Ignjatovic T, Boyer S, Russell RB (2005) Illuminating drug discovery with biological pathways. FEBS Lett 579:1872–1877
35. Huang S, Ingber DE (2006) A non-genetic basis for cancer progression and metastasis: self-organizing attractors in cell regulatory networks. Breast Dis 26:27–54
36. Tsuda I, Fujii H (2007) Chaos reality in the brain. J Integr Neurosci 6:309–326
37. Schoner G, Scholz JP (2007) Analyzing variance in multi-degree-of-freedom movements: uncovering structure versus extracting correlations. Mot Control 11:259–275
38. Southern J, Pitt-Francis J, Whiteley J, Stokeley D, Kobashi H, Nobes R, Kadooka Y, Gavaghan D (2008) Multi-scale computational modelling in biology and physiology. Prog Biophys Mol Biol 96:60–89
39. Price ND, Shmulevich I (2007) Biochemical and statistical network models for systems biology. Curr Opin Biotechnol 18:365–370
40. Konopka AK (2007) Surrogacy theory and models of convoluted organic systems. Proteomics 7:846–856
41. Huang S, Wikswo J (2006) Dimensions of systems biology. Rev Physiol Biochem Pharmacol 157:81–104
42. Boelsterli UA, Lim PL (2007) Mitochondrial abnormalities – a link to idiosyncratic drug hepatotoxicity? Toxicol Appl Pharmacol 220:92–107
43. Dykens JA, Marroquin LD, Will Y (2007) Strategies to reduce late-stage drug attrition due to mitochondrial toxicity. Expert Rev Mol Diagn 7:161–175
44. Lecellier G, Brenner C (2007) Genomic and proteomic screening of apoptosis mitochondrial regulators for drug target discovery. Curr Med Chem 14:875–881
45. Roses AD, Saunders AM, Huang Y, Strum J, Weisgraber KH, Mahley RW (2007) Complex disease-associated pharmacogenetics: drug efficacy, drug safety, and confirmation of a pathogenetic hypothesis (Alzheimer's disease). Pharmacogenomics J 7:10–28
46. Jonker DM, Visser SA, van der Graaf PH, Voskuyl RA, Danhof M (2005) Towards a mechanism-based analysis of pharmacodynamic drug-drug interactions in vivo. Pharmacol Ther 106:1–18
47. Harman D (1956) Aging: a theory based on free radical and radiation chemistry. J Gerontol 11:298–300
48. Ward M (2007) Biomarkers for Alzheimer's disease. Expert Rev Mol Diagn 7:635–646
49. Filley CM, Rollins YD, Anderson CA, Arciniegas DB, Howard KL, Murrell JR, Boyer PJ, Kleinschmidt-DeMasters BK, Ghetti B (2007) The genetics of very early onset Alzheimer disease. Cogn Behav Neurol 20:149–156
50. Sonnen JA, Keene CD, Montine KS, Li G, Peskind ER, Zhang J, Montine TJ (2007) Biomarkers for Alzheimer's disease. Expert Rev Neurother 7:1021–1028
51. Ertekin-Taner N (2007) Genetics of Alzheimer's disease: a centennial review. Neurol Clin 25:611–667, v
52. Chai CK (2007) The genetics of Alzheimer's disease. Am J Alzheimers Dis Other Demen 22:37–41
53. Mancuso M, Coppede F, Murri L, Siciliano G (2007) Mitochondrial cascade hypothesis of Alzheimer's disease: myth or reality? Antioxid Redox Signal 9:1631–1646
54. Reid PC, Urano Y, Kodama T, Hamakubo T (2007) Alzheimer's disease: cholesterol, membrane rafts, isoprenoids and statins. J Cell Mol Med 11:383–392
55. Argyle DJ, Blacking T (2008) From viruses to cancer stem cells: dissecting the pathways to malignancy. Vet J 177:311–323
56. Sales KM, Winslet MC, Seifalian AM (2007) Stem cells and cancer: an overview. Stem Cell Rev 3:249–255

57. Alison MR, Murphy G, Leedham S (2008) Stem cells and cancer: a deadly mix. Cell Tissue Res 331:109–124
58. Erenpreisa J, Cragg MS (2007) Cancer: a matter of life cycle? Cell Biol Int 31:1507–1510
59. Sjoblom T, Jones S, Wood LD, Parsons DW, Lin J, Barber TD, Mandelker D, Leary RJ, Ptak J, Silliman N, Szabo S, Buckhaults P, Farrell C, Meeh P, Markowitz SD, Willis J, Dawson D, Willson JK, Gazdar AF, Hartigan J, Wu L, Liu C, Parmigiani G, Park BH, Bachman KE, Papadopoulos N, Vogelstein B, Kinzler KW, Velculescu VE (2006) The consensus coding sequences of human breast and colorectal cancers. Science 314:268–274
60. Venkatesan RN, Bielas JH, Loeb LA (2006) Generation of mutator mutants during carcinogenesis. DNA Repair (Amst) 5:294–302
61. Araujo RP, Liotta LA, Petricoin EF (2007) Proteins, drug targets and the mechanisms they control: the simple truth about complex networks. Nat Rev Drug Discov 6:871–880
62. Lehar J, Zimmermann GR, Krueger AS, Molnar RA, Ledell JT, Heilbut AM, Short GF III, Giusti LC, Nolan GP, Magid OA, Lee MS, Borisy AA, Stockwell BR, Keith CT (2007) Chemical combination effects predict connectivity in biological systems. Mol Syst Biol 3:80
63. Avis I, Martinez A, Tauler J, Zudaire E, Mayburd A, Abu-Ghazaleh R, Ondrey F, Mulshine JL (2005) Inhibitors of the arachidonic acid pathway and peroxisome proliferator-activated receptor ligands have superadditive effects on lung cancer growth inhibition. Cancer Res 65:4181–4190
64. Liby KT, Yore MM, Sporn MB (2007) Triterpenoids and rexinoids as multifunctional agents for the prevention and treatment of cancer. Nat Rev Cancer 7:357–369
65. Grosser T, Fries S, FitzGerald GA (2006) Biological basis for the cardiovascular consequences of COX-2 inhibition: therapeutic challenges and opportunities. J Clin Invest 116:4–15
66. Frantz S (2005) Drug discovery: playing dirty. Nature 437:942–943
67. Millan MJ (2006) Multi-target strategies for the improved treatment of depressive states: conceptual foundations and neuronal substrates, drug discovery and therapeutic application. Pharmacol Ther 110:135–370
68. Farah A (2005) Atypicality of atypical antipsychotics. Prim Care Companion J Clin Psychiatry 7:268–274
69. Roth BL, Sheffler DJ, Kroeze WK (2004) Magic shotguns versus magic bullets: selectively non-selective drugs for mood disorders and schizophrenia. Nat Rev Drug Discov 3:353–359
70. Stahl SM (1998) What makes an antipsychotic atypical? J Clin Psychiatry 59:403–404
71. Ahmad N, Feyes DK, Nieminen AL, Agarwal R, Mukhtar H (1997) Green tea constituent epigallocatechin-3-gallate and induction of apoptosis and cell cycle arrest in human carcinoma cells. J Natl Cancer Inst 89: 1881–1886
72. Khan N, Afaq F, Saleem M, Ahmad N, Mukhtar H (2006) Targeting multiple signaling pathways by green tea polyphenol (–)-epigallocatechin-3-gallate. Cancer Res 66:2500–2505
73. Khan N, Afaq F, Mukhtar H (2007) Apoptosis by dietary factors: the suicide solution for delaying cancer growth. Carcinogenesis 28:233–239
74. Hopkins AL (2008) Network pharmacology: the next paradigm in drug discovery. Nat Chem Biol 4:682–690
75. Dessalew N, Workalemahu M (2008) On the paradigm shift towards multitarget selective drug design. Curr Comput Aided Drug Des 4:76–90
76. Korcsmaros T, Szalay MS, Bode C, Kovacs IA, Csermely P (2007) How to design multi-target drugs: target search options in cellular networks. Expert Opin Drug Discov 2:1–10
77. Barabasi AL, Albert R (1999) Emergence of scaling in random networks. Science 286:509–512
78. Palla G, Derenyi I, Farkas I, Vicsek T (2005) Uncovering the overlapping community structure of complex networks in nature and society. Nature 435:814–818
79. Nacher JC, Schwartz JM (2008) A global view of drug-therapy interactions. BMC Pharmacol 8:5
80. Csermely P, Agoston V, Pongor S (2005) The efficiency of multi-target drugs: the network approach might help drug design. Trends Pharmacol Sci 26:178–182
81. Strong M, Eisenberg D (2007) The protein network as a tool for finding novel drug targets. Prog Drug Res 64:191, 193–215
82. Morphy R, Rankovic Z (2007) Fragments, network biology and designing multiple ligands. Drug Discov Today 12:156–160
83. Hyduke DR, Jarboe LR, Tran LM, Chou KJ, Liao JC (2007) Integrated network analysis identifies nitric oxide response networks and dihydroxyacid dehydratase as a crucial target in *Escherichia coli*. Proc Natl Acad Sci USA 104:8484–8489

84. Schrattenholz A, Groebe K (2007) What does it need to be a biomarker? Relationships between resolution, differential quantification and statistical validation of protein surrogate biomarkers. Electrophoresis 28:1970–1979
85. Kovacs IA, Szalay MS, Csermely P (2005) Water and molecular chaperones act as weak links of protein folding networks: energy landscape and punctuated equilibrium changes point towards a game theory of proteins. FEBS Lett 579:2254–2260
86. Shen-Orr SS, Milo R, Mangan S, Alon U (2002) Network motifs in the transcriptional regulation network of *Escherichia coli*. Nat Genet 31:64–68
87. Jiang R, Tu Z, Chen T, Sun F (2006) Network motif identification in stochastic networks. Proc Natl Acad Sci USA 103:9404–9409
88. Milo R, Shen-Orr S, Itzkovitz S, Kashtan N, Chklovskii D, Alon U (2002) Network motifs: simple building blocks of complex networks. Science 298:824–827
89. Milo R, Itzkovitz S, Kashtan N, Levitt R, Shen-Orr S, Ayzenshtat I, Sheffer M, Alon U (2004) Superfamilies of evolved and designed networks. Science 303:1538–1542
90. Agoston V, Csermely P, Pongor S (2005) Multiple weak hits confuse complex systems: a transcriptional regulatory network as an example. Phys Rev E Stat Nonlin Soft Matter Phys 71:051909
91. Sivachenko AY, Yuryev A (2007) Pathway analysis software as a tool for drug target selection, prioritization and validation of drug mechanism. Expert Opin Ther Targets 11:411–421
92. Herwig R, Lehrach H (2006) Expression profiling of drug response – from genes to pathways. Dialogues Clin Neurosci 8:283–293
93. Elrick MM, Walgren JL, Mitchell MD, Thompson DC (2006) Proteomics: recent applications and new technologies. Basic Clin Pharmacol Toxicol 98:432–441
94. Schrattenholz A, Klemm M (2006) How human embryonic stem cell research can impact in vitro drug screening technologies of the future. In: Marx U, Sandig V (eds) Drug testing in vitro: breakthroughs and trends in cell culture technology. Wiley/VCH, New York, pp 205–228
95. Schrattenholz A, Klemm M (2007) Neuronal cell culture from human embryonic stem cells as in vitro model for neuroprotection. ALTEX 24:9–15
96. Malmstrom J, Lee H, Aebersold R (2007) Advances in proteomic workflows for systems biology. Curr Opin Biotechnol 18:378–384
97. Hunzinger C, Schrattenholz A, Poznanovic S, Schwall GP, Stegmann W (2006) Comparison of different separation technologies for proteome analyses: isoform resolution as a prerequisite for the definition of protein biomarkers on the level of posttranslational modifications. J Chromatogr A 1123:170–181
98. Ho E, Webber R, Wilkins MR (2007) Interactive three-dimensional visualization and contextual analysis of protein interaction networks. J Proteome Res 7:104–112
99. Grillari J, Katinger H, Voglauer R (2006) Aging and the ubiquitinome: traditional and non-traditional functions of ubiquitin in aging cells and tissues. Exp Gerontol 41:1067–1079
100. Tanaka K (2009) The proteasome: overview of structure and functions. Proc Jpn Acad B Phys Biol Sci 85:12–36
101. Ulrich HD (2009) The SUMO system: an overview. Methods Mol Biol 497:3–16
102. Schrattenholz A, Šoškić V, Groebe K (2010) Synchronisation of posttranslational modifications during ageing: time is a crucial biological dimension. Ann NY Acad Sci 1197:118–128
103. Brazma A, Hingamp P, Quackenbush J, Sherlock G, Spellman P, Stoeckert C, Aach J, Ansorge W, Ball CA, Causton HC, Gaasterland T, Glenisson P, Holstege FC, Kim IF, Markowitz V, Matese JC, Parkinson H, Robinson A, Sarkans U, Schulze-Kremer S, Stewart J, Taylor R, Vilo J, Vingron M (2001) Minimum information about a microarray experiment (MIAME)-toward standards for microarray data. Nat Genet 29:365–371
104. Taylor CF, Field D, Sansone SA, Aerts J, Apweiler R, Ashburner M, Ball CA, Binz PA, Bogue M, Booth T, Brazma A, Brinkman RR, Michael CA, Deutsch EW, Fiehn O, Fostel J, Ghazal P, Gibson F, Gray T, Grimes G, Hancock JM, Hardy NW, Hermjakob H, Julian RK Jr, Kane M, Kettner C, Kinsinger C, Kolker E, Kuiper M, Le NN, Leebens-Mack J, Lewis SE, Lord P, Mallon AM, Marthandan N, Masuya H, McNally R, Mehrle A, Morrison N, Orchard S, Quackenbush J, Reecy JM, Robertson DG, Rocca-Serra P, Rodriguez H, Rosenfelder H, Santoyo-Lopez J, Scheuermann RH, Schober D, Smith B, Snape J, Stoeckert CJ Jr, Tipton K, Sterk P, Untergasser A, Vandesompele J, Wiemann S (2008) Promoting coherent minimum reporting guidelines for biological and biomedical investigations: the MIBBI project. Nat Biotechnol 26:889–896
105. Taylor CF, Paton NW, Lilley KS, Binz PA, Julian RK Jr, Jones AR, Zhu W, Apweiler R, Aebersold R, Deutsch EW, Dunn MJ, Heck AJ,

Leitner A, Macht M, Mann M, Martens L, Neubert TA, Patterson SD, Ping P, Seymour SL, Souda P, Tsugita A, Vandekerckhove J, Vondriska TM, Whitelegge JP, Wilkins MR, Xenarios I, Yates JR III, Hermjakob H (2007) The minimum information about a proteomics experiment (MIAPE). Nat Biotechnol 25:887–893

106. Jones AR, Miller M, Aebersold R, Apweiler R, Ball CA, Brazma A, Degreef J, Hardy N, Hermjakob H, Hubbard SJ, Hussey P, Igra M, Jenkins H, Julian RK Jr, Laursen K, Oliver SG, Paton NW, Sansone SA, Sarkans U, Stoeckert CJ Jr, Taylor CF, Whetzel PL, White JA, Spellman P, Pizarro A (2007) The Functional Genomics Experiment model (FuGE): an extensible framework for standards in functional genomics. Nat Biotechnol 25:1127–1133
107. Kaiser J (2002) Proteomics. Public-private group maps out initiatives. Science 296:827
108. Orchard S, Hermjakob H, Apweiler R (2003) The proteomics standards initiative. Proteomics 3:1374–1376
109. Jones P, Cote RG, Martens L, Quinn AF, Taylor CF, Derache W, Hermjakob H, Apweiler R (2006) PRIDE: a public repository of protein and peptide identifications for the proteomics community. Nucleic Acids Res 34:D659–D663
110. Kanehisa M, Goto S, Kawashima S, Okuno Y, Hattori M (2004) The KEGG resource for deciphering the genome. Nucleic Acids Res 32:D277–D280
111. Junker BH, Klukas C, Schreiber F (2006) VANTED: a system for advanced data analysis and visualization in the context of biological networks. BMC Bioinform 7:109
112. Schomburg I, Chang A, Schomburg D (2002) BRENDA, enzyme data and metabolic information. Nucleic Acids Res 30:47–49
113. Sauro HM, Hucka M, Finney A, Wellock C, Bolouri H, Doyle J, Kitano H (2003) Next generation simulation tools: the Systems Biology Workbench and BioSPICE integration. OMICS 7:355–372
114. Hucka M, Finney A, Sauro HM, Bolouri H, Doyle JC, Kitano H, Arkin AP, Bornstein BJ, Bray D, Cornish-Bowden A, Cuellar AA, Dronov S, Gilles ED, Ginkel M, Gor V, Goryanin II, Hedley WJ, Hodgman TC, Hofmeyr JH, Hunter PJ, Juty NS, Kasberger JL, Kremling A, Kummer U, Le NN, Loew LM, Lucio D, Mendes P, Minch E, Mjolsness ED, Nakayama Y, Nelson MR, Nielsen PF, Sakurada T, Schaff JC, Shapiro BE, Shimizu TS, Spence HD, Stelling J, Takahashi K, Tomita M, Wagner J, Wang J (2003) The systems biology markup language (SBML): a medium for representation and exchange of biochemical network models. Bioinformatics 19:524–531
115. Borger S, Liebermeister W, Klipp E (2006) Prediction of enzyme kinetic parameters based on statistical learning. Genome Inform 17:80–87
116. Simon GM, Cravatt BF (2008) Challenges for the 'chemical-systems' biologist. Nat Chem Biol 4:639–642
117. Salisbury CM, Cravatt BF (2008) Optimization of activity-based probes for proteomic profiling of histone deacetylase complexes. J Am Chem Soc 130:2184–2194
118. Rexach JE, Clark PM, Hsieh-Wilson LC (2008) Chemical approaches to understanding O-GlcNAc glycosylation in the brain. Nat Chem Biol 4:97–106
119. Martin BR, Cravatt BF (2009) Large-scale profiling of protein palmitoylation in mammalian cells. Nat Methods 6:135–138
120. Codelli JA, Baskin JM, Agard NJ, Bertozzi CR (2008) Second-generation difluorinated cyclooctynes for copper-free click chemistry. J Am Chem Soc 130:11486–11493
121. Ovaa H, van Swieten PF, Kessler BM, Leeuwenburgh MA, Fiebiger E, van den Nieuwendijk AM, Galardy PJ, van der Marel GA, Ploegh HL, Overkleeft HS (2003) Chemistry in living cells: detection of active proteasomes by a two-step labeling strategy. Angew Chem Int Ed Engl 42:3626–3629
122. Jorgensen JT (2009) New era of personalized medicine: a 10-year anniversary. Oncologist 14:557–558
123. Kurpiers T, Mootz HD (2009) Bioorthogonal ligation in the spotlight. Angew Chem Int Ed Engl 48:1729–1731

Chapter 3

Systems Biology "On-the-Fly": SILAC-Based Quantitative Proteomics and RNAi Approach in *Drosophila melanogaster*

Alessandro Cuomo and Tiziana Bonaldi

Abstract

Stable isotope labeling with amino acids in cell culture (SILAC) has become increasingly popular as a quantitative proteomics (qProteomics) method. In combination with high-resolution mass spectrometry (MS) and new efficient algorithms for the analysis of quantitative MS data, SILAC has proven to be a potent tool for the in-depth characterization of functional states. QProteomics extends transcriptomics analysis in providing comprehensive and unbiased protein expression profiles. In this chapter, we describe the use of SILAC procedure in combination with RNA interference (RNAi) to characterize loss-of-function phenotypes, an example to illustrate how qProteomics can address many of the systems-wide approaches previously restricted to the mRNA level.

Furthermore, by explaining the adaptation of SILAC to a novel cellular model, the *Drosophila melanogaster* Schneider cells SL2, we aim to offer an example enabling the readers to apply the same strategy to any other cell culture, specific for their need.

Key words: Quantitative proteomics, RNA interference, SILAC, Schneider cells, Mass spectrometry, Orbitrap, MaxQuant

1. Introduction

Seeking differences between one functional state of a biological system versus another has been the source of many seminal discoveries. In order to increase the wealth of information achievable by such comparative analyses, comprehensive or system-wide screening approaches are needed. Differences between cellular states are reflected in changes in gene expression, at the level of both the message (mRNA) and the final product (protein). The first "genome-wide" method for expression analysis was offered by mRNA-based microarrays, which allow measuring thousands of transcripts in a single experiment. The major drawback of

Qing Yan (ed.), *Systems Biology in Drug Discovery and Development: Methods and Protocols*, Methods in Molecular Biology, vol. 662, DOI 10.1007/978-1-60761-800-3_3, © Springer Science+Business Media, LLC 2010

transcriptomics analysis is the lack of information at the level of proteins, the effectors of most biological functions. Protein levels, in fact, depend not only on the levels of the corresponding messages but also on a panel of translational controls and regulated degradation (1, 2). These factors may be as essential as mRNA synthesis, but they cannot be measured by microarrays (3–5). Thus, the information provided by proteomics offers a much closer description of the phenotype of a cell in a specific state (6–8).

In spite of this enormous potential and of the remarkable progresses in mass spectrometry (MS), MS-based "shotgun" proteomics still faces great challenges: sensitivity and dynamic range are the most striking and have been thoroughly discussed (9–11). In case of comparative studies, not only identification, but also accurate and comprehensive quantitation of protein levels is of utmost importance. Whereas stable isotope labeling with amino acids in cell culture (SILAC) has proved extremely successful in several quantitative proteomics studies, the accurate and robust quantitation of proteins from MS-data still remains a very challenging task, limiting the applicability of quantitative strategies in system-wide screenings.

Recently, the release of MaxQuant, a new efficient algorithm specifically developed for the analysis of high-resolution, quantitative MS data, represented a major breakthrough for the quantitation of proteins in SILAC-based experiments (12). Apart from achieving a considerable increase in peptide identification rates, MaxQuant allows the automatic quantification of several hundreds of thousands peptides per experiment and, consequently, the statistically robust quantification of several thousands proteins in complex proteomes. With such computational tool at hand, proteomics appears as the most informative instrument for "functional genomics" experiments, where microarrays have been so far the method of choice.

One example of functional genomics studies, where system-wide screening produce a comprehensive picture of the cellular response to a perturbation, is represented by profiling gene expression upon ablation of a gene, or a gene product. Loss-of-function is a popular approach to gain functional insight in the large subset of annotated genes, whose physiological role is still elusive. Loss-of-function, traditionally achieved by classical forward genetics methods, has been facilitated by the development of RNA interference (RNAi) approach. Model systems such as *Caenorhabditis elegans* and *Drosophila melanogaster* are excellent organisms to perform RNAi, due to the possibility to deliver long double-stranded RNA (dsRNA). The easiness of RNA transfer, the high penetrance of target depletion, and the lower incidence of "off-targets" (13) rendered interference in those model organisms highly successful. Recently, a genome-wide library of double-stranded RNAs targeting every gene in the *Drosophila* genome was published, paving the way to high-throughput genetic screens (14).

In a systems biology perspective, comprehensive screenings describing the cellular response to RNAi are required: while microarrays have been largely employed in the past, quantitative proteomics is emerging as the elective choice for high-content phenotypization. Thus, a specific investment of qProteomics in model organisms such as worm and fly is desirable: with the possibility to use cell-based assays, SILAC is preferred. Recently, we published the first SILAC-labeling of *D. melanogaster* Schneider SL2 cultured cells, with the acquisition of quantitative transcriptome and proteome upon RNAi of a specific chromatin remodeling activity (ISWI) (15).

We refer to complete reviews for the description of the basic principles of SILAC, for the recipe of standard SILAC media formulation (DMEM, RPMI) and for the conceptual explanation of MS-based quantitation (16, 17). We also strongly advise reading the article introducing the rationale and the algorithms for the quantitative analysis of SILAC data by MaxQuant (12). Instead in this chapter, we choose to focus on possible troubleshooting faced when adapting SILAC to novel model systems. For instance, we will discuss the reduction of cellular growth due to dialysis of complements in the medium, the limited incorporation and/or interconversion of heavy amino acids as potential sources of quantitation errors and the challenges in increasing the proteome coverage to gain statistical confidence in quantitation. Basing on our direct experience, we offer the "hints and tips" that facilitate the successful establishment of this quantitative strategy.

Furthermore, we will describe the rationale for combining SILAC with RNAi in order to gain a systems biology view of loss-of-function phenotypes, which is somewhat analogous to the microarray-based strategies, already extensively described. As such, we intend to present a "reference protocol" that enables the readers to adapt SILAC to their own cell culture models and for their own specific needs. The approach described is, in fact, generic and can be easily extended to other gene-silencing system (full knock-out, microRNA over expression, inhibitors) and/or to other model cell lines.

2. Materials

2.1. Standard Culture of SL2 Drosophila Embryonic Cells

1. Drosophila S2 Cells Schneider's *Drosophila* Medium.
2. Fetal Bovine Serum (Heat Inactivated, insect cell culture tested).

3. Penicillin 50 units/ml (Sigma).
4. Streptomycin 50 mg/ml (Sigma).
5. Bottles for growth of cells in suspension.
6. Cell culture dishes.
7. Incubator at 26°C, without CO_2 supply.

2.2. SILAC Labeling of SL2 Cells

2.2.1. One Liter Schneider Culture Medium

1. *Solution A*: NaCl, $Na_2HPO_4 \cdot 2H_2O$, KH_2PO_4, KCl, $MgSO_4 \cdot 7H_2O$, β-ketoglutaric acid, succinic acid, fumaric acid, malic acid.
2. *Solution B*: glucose, trehalose.
3. *Solution C*: β-alanine, L-asparagine, L-aspartic acid, L-cysteine, L-glutamic acid, glycine, L-histidine, L-isoleucine, L-leucine, L-methionine, L-phenylalanine, L-proline, L-serine, L-threonine, L-tryptophane, L-valine, L-cystine, L-tyrosine.
4. *Solution D*: Yeastolate.
5. *Solution E*: $CaCl_2$.
6. Spectra/Por® Cellulose Ester Membranes.

2.2.2. Normal and Heavy Isotope-Enriched Amino Acids for SILAC Metabolic Labeling

1. L-lysine ($H_2N(CH_2)_4CH(NH_2)CO_2H \cdot HCl$, Sigma) (light, K0).
2. L-arginine hydrochloride ($C_6H_{14}N_4O_2 \cdot HCl$, Sigma) (light, R0).
3. L-Lysine-$^{13}C_6$, $^{15}N_2$ hydrochloride ($H_2{}^{15}N({}^{13}CH_2)_4{}^{13}CH({}^{15}NH_2){}^{13}CO_2H \cdot HCl$, Sigma) (heavy, K8).
4. L -Arginine$^{13}C_6$, ^{15}hydrochloride ($H_2{}^{15}N^{13}C({}^{15}NH){}^{15}NH({}^{13}CH_2)_3{}^{13}CH({}^{15}NH_2){}^{13}COH$ HCl, Sigma) (heavy, R10) (see Note 1).

2.2.3. Assembly of Schneider Media

1. Prepare *Solution A* as follows:

   Dissolve in 250 ml ddH_2O:

NaCl	2.1 g
$Na_2HPO_4 \cdot 2H_2O$	0.43 g
KH_2PO_4	0.68 g
KCl	1.6 g
$MgSO_4 \cdot 7H_2O$	3.7 g
α-ketoglutaric acid	0.2 g
Succinic acid	0.10 g
Fumaric acid	0.10 g
Malic acid	0.10 g

2. Prepare *Solution B* dissolving 2.0 g glucose and 2.0 g Trehalose in 50 ml ddH_2O.
3. Prepare *Solution C* as follows
   (a) Dissolve in 450 ml ddH2O:

β-alanine	0.55 g
L-asparagine	0.04 g
L-aspartic acid	0.44 g
L-cysteine	0.07 g
L-glutamic acid	0.88 g
Glycine	0.27 g
L-histidine	0.44 g
L-isoleucine	0.16 g
L-leucine	0.16 g
L-methionine	0.88 g
L-phenylalanine	0.16 g
L-proline	1.87 g
L-serine	0.27 g
L-threonine	0.38 g
L-tryptophane	0.11 g
L-valine	0.33 g

   (b) Dissolve 0.10 g L-cystine in 50 ml hot acidified water (pH 2, with HCl, e.g., 5 ml 1N HCl/45 ml ddH_2O).
   (c) Dissolve 0.50 g L-tyrosine in 50 ml alkaline water (pH 9 with NaOH, e.g., 5 ml 1N NaHO/45 ml ddH_2O).
   (d) Add the cystine solution (b) and the tyrosine solution (c) slowly (drop-wise) to the general amino acid solution (a) (see Note 2).
4. Prepare *Solution D* dissolving 2.0 g Yeastolate in 50 ml ddH_2O. Dialyze over night against 5 L 0.9% NaCl in MCWO 3,500 Da tubes (see Note 3).
5. Prepare *Solution E* dissolving 0.60 g $CaCl_2$ in 50 ml ddH_2O.
6. Assemble the "minimal Schneider medium" as follows:
   (a) Combine sequentially solutions *A–E.*
   (b) Adjust slowly pH to 6.7 with 1N KOH (~15 ml).
   (c) Bring the titrated *A–E* mix to a final volume of 1 L.
   (d) Sterilize by filtration and keep at 4°C.

7. Stock solution (100×) of the labeling amino acids (either light or heavy):

L-arginine	40 g/L in PBS
L-lysine HCl	165 g/L inPBS

8. "Complete SILAC-Schneider Medium" (500 ml)

   Combine:

   (a) 440 ml "minimal Schneider medium".

   (b) 5 ml arginine/lysine (either light or heavy) stock (100×).

   (c) 3 ml Glutamine.

   (d) 5 ml Pennicilline/Streptomycine (100×).

   (e) 50 ml Serum (dialyzed against 0.9% NaCl in MWCO 3,500 Da tubes) (see Note 3).

   (f) Filter-sterilize the SILAC medium and store at 4°C for up to 3 months.

2.3. RNAi Interference of SL2 Cells

2.3.1. Double-Stranded dsRNA Preparation

1. Primers for PCR amplification of target gene, containing a 5′ T7 RNA Pol binding site, followed by sequences specific for the target gene.
2. High Pure PCR Purification Kit (Roche, Molecular Biochemicals).
3. MEGASCRIPT T7 transcription kit (Ambion, Austin, TX).
4. PCR Thermocycler.

2.3.2. RNAi Interference Experiment

1. $5–10 \times 10^6$ cells seeded in 10 cm culture dishes.
2. 15–30 μg dsRNA.

2.4. Sample Preparation

2.4.1. Cells Extraction and Fractionation

1. Swelling buffer: 20 mM HEPES-KOH pH 7.5, 250 mM Sucrose, 0.5 mM EDTA 1 mM DTT. (Protease Inhibitor Cocktail Tablets from Roche Applied Science to be added just before use).
2. PBS: 2 g/L NaCl, 2 g/L KH_2PO_4, 11.5 g/L Na_2HPO_2.
3. RIPA buffer: 50 mM Tris-HCl pH 7.5 150 mM NaCl, 1% NP-40, 0.5% deoxycholate, 0.1% SDS, 0.4 mM EDTA, 10% glycerol (see Note 4).
4. TCA: trichloroacetic acid.
5. Cell-homogenizer (Isobiotec, Heidelberg, Germany).
6. Vacufuge Concentrator 5301.

2.4.2. SDS-PAGE for Protein Separation

1. Precast 4/12% Tris-HCl polyacrylamide gel: NuPAGE® Novex Bis-Tris gels (Invitrogen).
2. NuPAGE® LDS Sample Buffer (4×) (Invitrogen) (see Note 5).
3. Colloidal Blue Staining Kit (Invitrogen).

2.4.3. In-Gel Digestion

1. Acetonitrile (ACN) (HPLC grade).
2. Ammonium hydrogen carbonate (ABC) (Sigma).
3. Trifluoroacetic Acid (TFA) (Sigma).
4. Milli-Q® Ultrapure Water Purification Systems.
5. Digestion buffer: 50 mM ABC in water (pH 8.0).
6. Distaining buffer: 25 mM ABC/50% ACN.
7. Trypsin solution: 12.5 ng/μL sequencing grade trypsin (Promega Corporation, Madison, WI) in 50 mM ABC (see Note 6).
8. Extraction buffer: 3% TFA/30% ACN.
9. StageTips (18): Empore SPE C18 disks (3 M).

2.5. High Performance Liquid Chromatography–Tandem Mass Spectrometry

1. Acetic Acid (AA).
2. HPLC solvent "A": 0.5% AA in water.
3. HPLC solvent "B": 0.5% AA/100% ACN in water.
4. Reversed-phase material for nano-HPLC column: Reprosil-Pur C18-AQ, 3 μm (Dr. Maisch).
5. Agilent 1100 Series (Agilent Technologies), comprising a solvent degasser, a nanoflow pump, and a thermostated microautosampler.
6. Silica transfer line 20-cm long, 25-μm inner diameter (Composite Metals).
7. Micro Tee-connector (Upchurch).
8. LTQ orbitrap mass spectrometer (Thermo Fisher Scientific).
9. Ion Source Kit for Thermo LTQ-FT (Proxeon Biosystems).

2.6. Data Acquisition and Analysis

1. Thermo Fisher Scientific Xcalibur software 2.0.5.
2. RAWMSM 2.1.
3. MaxQuant 1.0.13.13.
4. MASCOT 2.2.
5. PC with 2 GB of RAM, dual core processor, Windows XP (32-bit version).

3. Methods

3.1. Establishment of SILAC in Drosophila SL2 Cells

SILAC has originally been developed using mammalian cells, where full incorporation can easily be achieved using isotope-coded essential amino acids in the media. Normally, *Drosophila* cells are grown in Schneider medium, differing from mammalian media for the presence and concentration of various components. Schneider medium is not purchased in a formula adapted for

SILAC labeling (e.g., specifically depleted of the amino acids arginine and lysine), thus we assemble it from each component, based on the published receipt (16, 17). Furthermore, serum and total yeast extract, supplemented to the broth during culture, can be both source of free amino acids interfering with labeling efficiency. As such, we introduce a specific step of dialysis (see Notes 3 and 7).

Cell growth and viability in SILAC medium are carefully compared with those of cells growing in standard medium, to detect any alteration from physiology that might be caused by the poorer growth conditions:

1. Inspect cells at the microscope to uncover potential gross morphological alterations.
2. Count cells and plot growth curves of cells in SILAC versus standard medium.
3. Compare cell mortality by Trypan Blue staining that allows distinguishing viable from dead cells at the microscope: at each day, the percentage of viable versus dead/dying cells is calculated (see Note 8).

3.2. Cell Extraction and Subcellular Fractionation

Fractionating the total cellular extracts reduce sample complexity with consequent increase in the dynamic range of protein identification by LC-MSMS (see Notes 9 and 10).

1. Mix equal number of cells for each sample (heavy and light).
2. Wash in PBS.
3. Incubate in 4 ml swelling buffer at 4°C for 10 min.
4. Pass the cell suspension four times through a cell-homogenizer (Isobiotec) containing a ball allowing for 10 μm clearance (see Note 11).
5. Centrifuge the cell lysate at 1,500 × *g* for 10 min.
6. Dissolve the pellet containing the nuclei (N) in 200 μl of RIPA buffer 7. Centrifuge the supernatant at 120,000 × *g* for 15 min.
7. Concentrate the supernatant (cytosolic fraction, C) by TCA (15%) precipitation and directly resuspend in SDS-PAGE loading buffer.
8. Wash the Pellet (membrane fraction, M) with PBS and resuspend in RIPA buffer.

3.3. SDS-PAGE and In-Gel Digestion

1. Resolve the proteins mixture in polyacrylamide gel. In this protocol, protein separation was achieved on a 4–12% gradient NuPAGE Novex Bis-Tris gels (Invitrogen, Carlsbad, CA) (see Note 12).
2. Cut the lanes in 10–15 slices (~5 mm broad), and each slice in small cubes, 1 mm wide.

3. Destain the gel slices with 25 mM ABC/50% ACN and add absolute ACN to shrink the gels. Repeat until the gels are completely destained. Washes are carried out in a Thermomixer, with strong shaking (>1,000 rpm) at RT.
4. Add reduction buffer (10 mM DTT) to the gel pieces; incubate for 1 h at 56°C.
5. Remove reduction buffer and add alkylation buffer (55 mM IAA); incubate for 45 min at RT, in the dark.
6. Remove alkylation buffer and repeat twice the washes (**step 3**).
7. Remove ACN by aspiration and dry the gels in vacuum centrifuge.
8. Rehydrate the gel pieces with ice-cold 12.5 ng/μL trypsin solution in 25 mM ABC and incubate on ice till the gels are fully swollen. Remove trypsin solution in excess.
9. Add 50 mM ABC to completely cover the gel pieces. Incubate overnight at 37°C.
10. After overnight, spin down evaporated buffer and collect liquid part in a new tube.
11. Add the extraction buffer (30% ACN, 2% TFA) to the gel pieces; incubate in a Thermomixer with strong agitation for 20 min at RT. Repeat twice.
12. Pool all supernatants. Lyophilize the peptide mixture in a vacuum centrifuge.
13. Reconstitute dried samples in 1% TFA.
14. Desalt and concentrate peptides on a reversed phase C18 microcolumn (StageTip or equivalent) as previously described (18).
15. Elute peptides from the C18 using 80% ACN/0.5% AA.
16. Remove the organic component by evaporating in a vacuum centrifuge and resuspend the peptides in a suitable injection volume (typically 5–10 μL) of 0.5% AA.
17. Inject approximately 1–5 μg of peptide into a column for nanoLC–MS analysis.

3.4. LC–MS Analysis

3.4.1. Liquid Chromatography Analysis

In this step, the peptides from in-gel digestion are separated by HPLC and introduced into MS via an on-line nanoelectrospray system.

1. Pack an analytical column in a 15-cm fused silica emitter (Proxeon Biosystems, 75-μm inner diameter), with methanol slurry of reverse-phase C18 resin at a constant helium pressure (50 bar) using a bomb-loader device (Proxeon Biosystems), as described previously (19).
2. Connect the packed emitter (C18 RP HPLC column) directly to the outlet of the 6-port valve of the HPLC autosampler

through a 20-cm long (25-μm ID) fused silica without using precolumn or split device.

3. Load the tryptic peptides mixture onto C18 column at a flow of 500 nL/min.
4. After sample loading, apply a gradient of 3–60% mobile phase B at 250 nL/min over 120 min, for peptide mixture elution.

3.4.2. Mass Spectrometry Analysis

Mass spectrometry analysis is carried out on an LTQ Orbitrap mass spectrometer (Thermo Fisher Scientific) (see Note 13).

1. Operate the mass spectrometer in the data-dependent mode to automatically switch between MS and MSMS using the Xcalibur 2.4 software package.
2. Use the following settings in the "Tune" acquisition software:

   (a) FT full scan: accumulation target value 1×10^6; maximum filling time 1 s.

   (b) IT MSn: accumulation target value 10×10^3; maximum filling time 250 ms.
3. In the "Xcalibur Instrument Setup," create a data-dependent acquisition method in which full scan MS spectra, typically in the m/z range from 300 to 1,800 amu, are acquired by the orbitrap detector with resolution $R = 60{,}000$ (see Note 14).
4. For accurate mass measurements, enable the "lock mass" option in both MS and MSMS mode (20) (see Note 15).
5. Standard acquisition method settings:

   (a) Electrospray voltage, 2.4 kV.

   (b) No sheath and auxiliary gas flow.

   (c) Ion transfer (heated) capillary temperature, 190°C.

   (d) Collision gas pressure, 1.3 mtorr.

   (e) Dynamic exclusion of up to 300 precursor ions for 90 s upon MSMS; exclusion mass width of 10 ppm.

   (f) Normalized collision energy using wide-band activation mode 35%.

   (g) Ion selection thresholds: 1,000 counts.

   (h) Activation $q = 0.25$; Activation time = 30 ms.

3.5. Data Analysis

All raw data files acquired are analyzed with in-house developed quantitative proteomics software MaxQuant, version 1.0.13.13 (12).

MaxQuant is designed to analyze large and high-resolution MS data sets. While for the detailed description of rationale and algorithms, we advise reading the dedicated publication (12), we refer to Note 16 for a synthetic description of the modules composing this analytical platform.

The data are processed according to the following parameters:

1. Mascot search:
   (a) Fixed modification: Carbamidomethylation (see Note 17).
   (b) Variable modification: *N*-acetyl (Protein), Oxidation (M).
   (c) Missed cleavages up to 3.
   (d) Mass-accuracy of the parent ions in the initial Mascot search: 7 ppm.
   (e) Mass-accuracy for MSMS peaks in the initial Mascot search: 0.5 Da.
2. Parameters for MaxQuant:
   (a) Peptide false discovery rates (FDR) (see Note 18): 0.01.
   (b) Protein false discovery rates (FDR) (see Note 18): 0.01.
   (c) Maximum posterior error probability (12) (see Note 19): 1.
   (d) Minimum peptide length: 5.
   (e) Minimum number of peptides: 2.
   (f) Minimum number of unique peptides: 1.

3.6. Incorporation Test (see Note 20)

Usually cells are cultured in heavy medium for at least six cell doublings to allow the incorporation of the heavy amino acids. The incorporation level is measured by the proportion of remnant light peptides found in the heavy-labeled cell sample.

1. Grow cells in "heavy" medium at 26°C for at least six replications.
2. Resolve whole cell extracts from heavy-labeled cells by SDS/PAGE.
3. Follow the protocol for in-gel digestion and subsequent MS analysis, as described (Subheadings 3.3 and 3.4).
4. Ideally, peptides identified from this pool should contain only heavy amino acids, without detectable signal at m/z values corresponding to the light peptide; however, in reality, light peptides are remaining and nonnormalized ratio H/L is therefore measured (see Notes 21–23).

3.7. One-to-One Mixture of Differentially Labeled Cells (see Note 24)

For the acquisition of a "one-to-one Schneider SL2 proteome":

1. Grow SL2 cells in both light (L) and heavy (H) media, for the optimal number of replications indicated by the incorporation test (in this case about six replications).
2. Harvest the cells and mix in equal amount.
3. Resolve the protein extract in SDS-PAGE.
4. Digest proteins with Trypsin, as described (Subheading 3.3).

5. Analyze the peptide mixture by mass spectrometry.

Plot the histogram of normalized log2(H/L) ratios for the one-to-one SILAC proteome (Fig. 2a) (see Note 25).

3.8. RNAi Interference in SILAC-Labeled SL2 Cells

3.8.1. dsRNA Production

1. Amplify by PCR the individual DNA fragments, approximately 700 bp long and containing coding sequences for the proteins to be silenced. Each primer used in the PCR contained a 5′ T7 RNA polymerase binding site, followed by sequences specific for the targeted genes.
2. Purify the PCR products with the High Pure PCR Purification Kit (Roche).
3. Produce dsRNA from the purified PCR products using a MEGASCRIPT T7 transcription kit (Ambion).
4. Ethanol-precipitate the dsRNA products and redissolve in water.
5. Reannealing can be improved by incubation at 65°C for 30 min, followed by slow cooling to room temperature (see Note 26).

3.8.2. RNA Interference of SILAC-Labeled SL2 Cells (see Note 27)

1. Seed $\pm 8 \times 10^6$ SL2 cells in 10 cm dishes in custom serum-free medium w/o serum but w/ amino acid isotopes.
2. Add ±25 μg dsRNA to the cells and incubate for 1 h at 26°C.
3. Add serum-containing heavy or light media (equal volume) and let cells grow till harvest (see Note 28).

3.9. Experimental Set-Up for the SILAC/RNAi Combined Approach

1. The experimental set-up for a hypothetical experiment where a specific gene product (in this case the chromatin remodeling factor ISWI) is depleted by RNAi in SL2 cells differentially SILAC-labeled is schematized in Fig. 1a.
2. In a typical “forward” labeling setting, the protein of interest is knocked-down in heavy medium, while a “mock” RNAi against GST (or any other nonendogenously expressed protein) in light medium serves as a control sample.
3. Cells are switched to SILAC medium at the time of dsRNA incubation.
4. Cells are harvested after 7 days, mixed in equal amounts.
5. Samples are fractionated in nuclear, cytosolic, membrane factions (N, C, and M respectively) (Fig. 1b).
6. Proteins are separated, digested, and analyzed by MS (Fig. 1c–e).
7. The mass spectrometry-based readout of the RNAi-SILAC experiment is schematically depicted in Fig. 1f as follows: the target protein itself is depleted from the heavy population

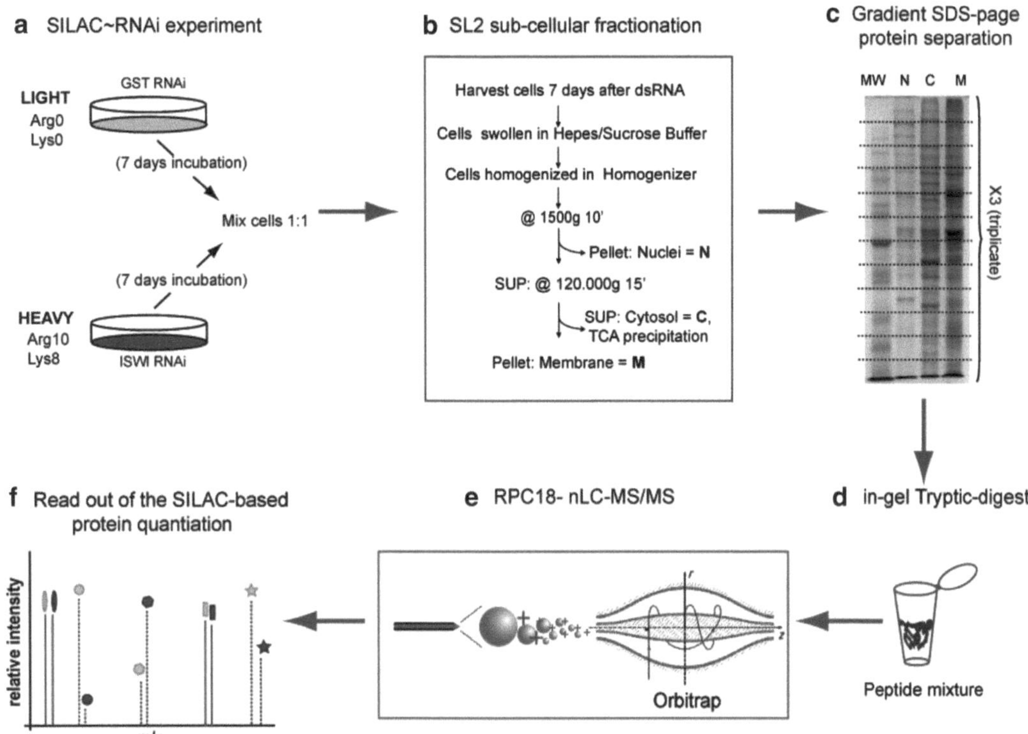

Fig. 1. Overview of the analytical strategy combining SILAC with RNAi (**a**) heavy (Arg10, Lys8) and light (Arg0, Lys0) SILAC-labeled cells are subjected to ISWI and GST RNAi, respectively. Seven days after dsRNA incubation, cells are harvested and mixed in equal amount (1:1). (**b**) Cells are extracted and subcellular fractionated. (**c**) The three fractions are loaded, in triplicate, on SDS-PAGE, and each lane is cut into ten slices. (**d**) Proteins are subjected to tryptic in-gel digestion, (**e**) peptides mixtures analyzed by nLC-MSMS on a LTQ-Orbitrap. (**f**) The schematized mass spectrum shows peak-pairs which corresponds to light- and heavy-peptides: significant responders were identified by peak ratios differing from one (*dotted lines*, *circles*, *stars* and *hexagons*), while nonresponders had a ratio of one (*continued lines*, *ovals* and *rectangles*).

relative to the light one and this should result in a peptide fold change corresponding to the knock-down efficiency (dotted lines + circles). Other significant responders should also show peak ratios differing from value 1, either increased or decreased (dotted lines, stars and hexagons). By contrast, non responding proteins should be present in similar amounts in both, heavy and light form, with a ratio of 1 (continued lines, ovals and rectangles) (see Note 29).

The statistical analysis of the ratio distribution for the whole SILAC ~ RNAi proteome will lead the identification of significant responders, both up- and down-regulated upon depletion of a specific gene product (Fig. 2b, gray-dotted lines represent statistically significant outliers, 5% of the total).

However, the direct comparison between the ratio distributions of the "RNAi experiment" and of the "one-to-one experiment" appears to be more informative of the global response of

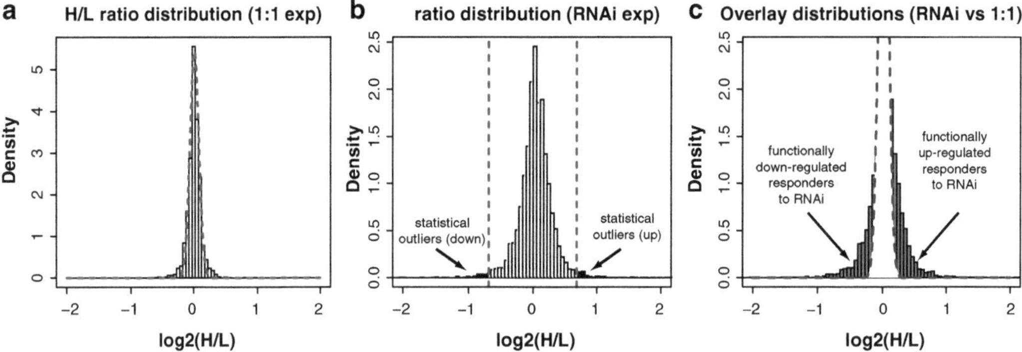

Fig. 2. Ratio distributions of "one-to-one" and "RNAi" experiments, and their comparison. (**a**) Distribution of H/L ratios in the one-to-one mixture. The histogram of log-transformed normalized H/L ratios (n = 2,449) fits a normal distribution, with standard deviation 0.12. (**b**) Distribution of H/L ratios in the ISWI/GST-RNAi SL2 proteome. The histogram of log-transformed normalized ratios (n = 3,976) fits a normal distribution, with standard deviation 0.19. The *gray-dotted lines* represent STDEV = 2 and define statistical up- and downregulated outliers (*dark-gray bins*, about 5% of the total) (**c**) Overlay of the fitted and centered H/L ratio distributions of the RNAi and one-to-one control (*dotted*) experiments. The overlay allows establishing the threshold of the fold change upon RNA interference, compared to the background noise measured by the 1:1. Responders that changes slightly but significantly upon the perturbation are detected and marked with indicated by *light-gray bins*.

the system to the gene product ablation, especially when a proteome-wide perspective is adopted. As exemplified in Fig. 2c, the broader distribution of the ISWI-RNAi experiment compared to that of the one-to-one control suggests that the cells respond with a slight but significant abundance modulation of a large number of proteins, in both directions (up- and down-regulation, light-gray bins) (see Note 30). A plain statistical analysis would miss these responders, because it identifies only the "extreme outliers", frequently not representative of the global response of a system.

4. Notes

1. Heavy isotope amino acids used in SILAC are not radioactive and thus do not require special handling precautions.
2. The final volume of solution C should be 550 ml; however, it is advisable to limit the volume as much as possible, to avoid the risk of exceeding the maximum volume during the pH titration step.
3. Different dialysis protocols need to be tested to find the optimal compromise between the removal of free amino acids and the efficient growth of cells. In this case, about 20 h of dialysis using a MW-cut off (MCWO) of 3,500 Da for serum and yeastolate guarantee both cell growth and efficient heavy amino acids incorporation.

4. Prepare as a stock solution and store at room temperature (25°C). Fresh protease inhibitors should be added just before use. Here we use the Protease Inhibitor Cocktail Tablets Complete® (Roche Applied Science).
5. Add DTT to a final concentration of 100 mM, just before boiling.
6. Trypsin stock solution can be stored at –80°C for 6 months.
7. In some cases, more sensitive cell types might suffer from the use of dialyzed serum or other components, due to the loss of small peptides working as growth factors. If so, supplementing the SILAC medium with single purified growth factors or with a small percentage of normal serum might compensate this.
8. Generating a detailed growth curve is a useful tool to evaluate the growth characteristics of a cell line. From a growth curve, the lag time, population doubling time, and saturation density can be determined. Plot the cell number on a log-linear scale: the population-doubling time can be determined by identifying a cell number along the exponential phase of the curve, tracing the curve until that number has doubled, and calculating the time between the two. In our case, cell growth is slightly reduced in SILAC, when compared to normal medium. However, when cells are cultured in SILAC medium for prolonged periods, up to 2 months, a detailed investigation revealed neither reduction in viability nor morphological alterations.
9. The full characterization of a given proteome by LC/MSMS analysis can be limited by three instrumental factors: sensitivity, sequencing speed, and dynamic range (21). Sensitivity can be a limiting factor, depending on the starting amount of sample: when the amount is low, sensitivity decreases accordingly, in any case. Another major limitation is due to the "scan time" of the mass analyzer, which is relatively long (~1 s) and thus might not be competitive with the elution time scale of each peptides during chromatographic separation, with a consequent decrease in the sequencing capacity of the instrument and in the identification rate of peptides. Finally, proteome coverage can be limited by the "dynamic range" of the instrument, in relation to the dynamic range of the proteome to be analyzed: protein copy number is distributed from 7 to 8 orders of magnitude in cells, a range that cannot be covered by a mass spectrometer. New generation hybrid mass spectrometers started to overcome these limitations, as in the case of LTQ-Orbitrap (see Note 12); however, an efficient sample prefractionation remains an excellent strategy to improve the dynamic range of LC/MSMS analysis (see Notes 10 and 11).

10. Several techniques alternative to classical SDS-PAGE can be used to obtain better sample prefractionation. In our experience, a recently introduced strategy for the separation of digested peptide from total extracts by isoelectric focusing along immobilized IPGs, named OFFGEL (Agilent), represents an attractive method for high throughput analysis; in fact, it demonstrated optimal performance, combining a higher number of protein identified by MS with diminished work-up requirements (15, 22).
11. Subcellular fractionation can be efficiently achieved also by means of alternative fractionation protocols, including kits from different brands. However, fractionation should be optimized to the type of cells employed, before setting up the SILAC experiment.
12. Reduce the risk of keratin contaminations that could interfere with LC/MS by following the procedures for sample preparation previously described (23, 24).
13. The availability of new generation of high-performance hybrid mass spectrometers such as LTQ Orbitrap allows an average absolute measurement mass accuracy (MMA) less than one parts-per-million (ppm). This is of utmost importance for two main reasons: first, high MMA of precursor ions enable using very low ppm tolerance in database search, which increases identification confidence; second, high MMA allows high-resolution MS, leading to a better peak definition with a consequent positive effect on quantitation in full MS.
14. In our case, for better sampling of the higher m/z ions, the LTQ-Orbitrap mass spectrometer operate in a "2-range" regime: an acquisition cycle consisted of two survey scans in the Orbitrap analyzer (mass ranges of 300–1,000, and 950–1,650, at resolution $R=60{,}000$), each followed by MSMS of the five most intense ions in the LTQ.
15. Use the background polydimethylcyclosiloxane (PCM) ions generated from ambient air (e.g., $m/z=445.120025$) for internal recalibration in real time. If the fragment ion measurements are performed in the Orbitrap, use the PCM ion at $m/z=429.088735$ (PCM with neutral methane loss).
16. MaxQuant works in combination with Mascot engine software (25) and is composed of two modules: "Quant.exe" assembles isotope patterns into SILAC pairs before MSMS data submission to Mascot. This model includes a sophisticated three-dimensional peak and isotope pattern detection. The output files consist of a "peak list" (.msm) of MSMS spectra, also including information relative to the Liquid Chromatography/full MS runs, such as retention times, peak intensity, etc. The ".msm files" produced by "Quant.exe" are subjected to Mascot search engine; "Identify.exe" module

takes the search results, the raw files as well as all results from the "Quant.exe" and performs robust statistical validation, to obtain an accurate assembling of peptides into proteins and, finally, to quantify proteins. Ratios for proteins are determined as the median over all measured peptide ratios for a given protein. The significance of protein ratios is measured as:

(a) Significance-A, calculated by first estimating the variance of the distribution of all protein ratios in a robust way, and then reporting the error function for the *z*-score corresponding to the given ratio.

(b) Significance-B, which uses the same strategy, but also takes into account the dependence of the distribution on the summed protein intensity.

17. The modifications corresponding to arginine and lysine labeled with heavy stable isotopes (arginine $^{13}C_6{}^{15}N_4$ and $^{13}C_6{}^{15}N_2$) can be treated as fixed modifications in the Mascot search, if applicable, after identification of SILAC pairs in the "Quant.exe" module.

18. False positive rates for peptides are calculated as described in (26). In this case, we fixed 1% false discovery rate, both at the level of peptides and proteins.

19. PEP is the probability of a false hit, given the peptide identification score and length of peptides (12).

20. Quantification errors will arise if proteins and peptides are not fully labeled with heavy SILAC amino acids at the end of the adaptation phase. Therefore, an analysis to check the degree of incorporation of the heavy SILAC amino acid is critical when working with a new cell line or medium formulation.

21. In practice, after 7 days of SL2 growth, the average of heavy to light (H/L) nonnormalized peptide ratios is about 30, corresponding to more than 95% incorporation efficiency (1-1/AV Ratio).

22. It is also possible to use an in-solution digestion of whole cell extract from heavy sample as previously described (27): in fact, a complete proteome is not needed to check the incorporation level and a statistically significant subset is enough.

23. In some cell types grown in standard media, the metabolic interconversion between arginine and proline can occur when arginine is provided to cells in excess. The reverse metabolic conversion of Pro-to-Arg can also occur when cells are not provided with enough arginine. As such, when assembling a new SILAC medium, the optimum concentration of arginine must be determined experimentally for the cell line under investigation. Titration of arginine in SILAC Schneider medium is followed by measuring the frequency of heavy proline in heavy SILAC-growing cells. An interesting alternative

is also increasing the concentration of light Proline in the medium, to counteract Arg conversion.

24. SILAC-based quantitative proteomics relies on the assumption that the abundance of specific proteins is not affected by the isotopic composition of the SILAC medium employed. The distribution of protein ratios in the quantitative proteome from the one-to-one "light" and "heavy" mixture allows validating this hypothesis. The normal distribution of the histogram informs about the homogeneity of labeled amino acids incorporation and about any potential effect of the isotopic composition of the medium on protein levels. Furthermore, the width of the distribution correlates with the precision of the SILAC-ratio measurement for identified and quantified proteins, over the background noise. The background noise results from several factors: the biological variability among cell cultures, the random variation of protein turn-over, the precision of the instrument and others. As such, the normalized H/L ratio distribution from the one-to-one experiment represents an indicator of the overall sensitivity of the approach in detecting protein level changes upon perturbation, over the background noise.
25. In our case, the histogram of log-transformed normalized H/L ratios ($n = 2,449$) in the one-to-one mixture fits a normal distribution with a standard deviation of 0.12 (dotted line, Fig. 2a).
26. To ensure that the majority of the dsRNA exists as a single band of about 700 bp, a small aliquot of dsRNA is analyzed by agarose gel electrophoresis.
27. *Drosophila* cells are uniquely suited to RNAi-mediated gene knockdown (28, 29). Target genes can be depleted by simply incubating the cells with long dsRNA molecules, in serum-free medium. Long dsRNA are directly up-taken by pinocytosis by cells and subsequently they enter the interference splicing and processing pathway. Albeit to a minor extent, potential off-targets effects can be observed in fruit fly system, due to the unspecific silencing of targets by a certain dsRNA (30). Thus, biological replicates using alternative dsRNAs sequences are highly recommended.
28. Validation of the silencing of the protein of interest (ISWI) by RNAi is carried out by standard Western Blot.
29. To increase the biological significance of the analysis, different experimental replicates should be planned; a common strategy is to perform one experiment in a "forward" set-up, followed by at least one of the replicates in a "reverse" set-up, where the heavy and light media are "switched" between the real and the control experiment. The MS-based readout of the changes in protein ratio will consequently be inverted as a result.

30. Importantly, the overlay between fitted and centered H/L ratio distributions of the RNAi and the one-to-one control experiment allows establishing the threshold of the fold change to be considered for a comprehensive quantitation analysis. As such, a more comprehensive picture of the global response of a perturbed system can be achieved. Noteworthy, the strategy described allows detecting fine but significant changes at the protein level. In this scenario, we claim that this approach is useful for robust and reliable comparative analysis, as in the case of proteome versus transcriptome studies (31).

Acknowledgments

The work in T.B. laboratory is supported by an Armenise-Harvard foundation career development program grant, a grant from the Associazione Italiana Ricerca sul Cancro (AIRC) (REF. # 6011), a grant from the Association of International Cancer Research (AICR) (REF. # 09-0281) and a grant from Cariplo Foundation (REF. # 2009-2721).

References

1. Gygi SP, Rochon Y, Franza BR, Aebersold R (1999) Correlation between protein and mRNA abundance in yeast. Mol Cell Biol 19:1720–1730
2. Lu P, Vogel C, Wang R, Yao X, Marcotte EM (2007) Absolute protein expression profiling estimates the relative contributions of transcriptional and translational regulation. Nat Biotechnol 25:117–124
3. de Groot MJ, Daran-Lapujade P, van Breukelen B, Knijnenburg TA, de Hulster EA, Reinders MJ, Pronk JT, Heck AJ, Slijper M (2007) Quantitative proteomics and transcriptomics of anaerobic and aerobic yeast cultures reveals post-transcriptional regulation of key cellular processes. Microbiology 153:3864–3878
4. Tian Q, Stepaniants SB, Mao M, Weng L, Feetham MC, Doyle MJ, Yi EC, Dai H, Thorsson V, Eng J, Goodlett D, Berger JP, Gunter B, Linseley PS, Stoughton RB, Aebersold R, Collins SJ, Hanlon WA, Hood LE (2004) Integrated genomic and proteomic analyses of gene expression in Mammalian cells. Mol Cell Proteomics 3:960–969
5. Nie L, Wu G, Culley DE, Scholten JC, Zhang W (2007) Integrative analysis of transcriptomic and proteomic data: challenges, solutions and applications. Crit Rev Biotechnol 27:63–75
6. Tyers M, Mann M (2003) From genomics to proteomics. Nature 422:193–197
7. Dorus S, Busby SA, Gerike U, Shabanowitz J, Hunt DF, Karr TL (2006) Genomic and functional evolution of the *Drosophila melanogaster* sperm proteome. Nat Genet 38:1440–1445
8. Newman JR, Ghaemmaghami S, Ihmels J, Breslow DK, Noble M, DeRisi JL, Weissman JS (2006) Single-cell proteomic analysis of *S. cerevisiae* reveals the architecture of biological noise. Nature 441:840–846
9. Domon B, Aebersold R (2006) Mass spectrometry and protein analysis. Science 312(5771):212–217
10. Kuster B, Mann M (1998) Identifying proteins and post-translational modifications by mass spectrometry. Curr Opin Struct Biol 8:393–400
11. Nielsen ML, Savitski MM, Zubarev RA (2005) Improving protein identification using complementary fragmentation techniques in fourier transform mass spectrometry. Mol Cell Proteomics 4(6):835–845
12. Cox J, Mann M (2008) MaxQuant enables high peptide identification rates, individualized p.p.b.-range mass accuracies and proteome-wide protein quantification. Nat Biotechnol 26:1367–1372

13. Boutros M, Bras LP, Huber W (2006) Analysis of cell-based RNAi screens. Genome Biol 7:R66
14. Boutros M, Kiger AA, Armknecht S, Kerr K, Hild M, Koch B, Haas SA, Paro R, Perrimon N (2004) Genome-wide RNAi analysis of growth and viability in *Drosophila* cells. Science 303:832–835
15. Bonaldi T, Straub T, Cox J, Kumar C, Becker PB, Mann M (2008) Combined use of RNAi and quantitative proteomics to study gene function in *Drosophila*. Mol Cell 31:762–772
16. Ong SE, Mann M (2007) Stable isotope labeling by amino acids in cell culture for quantitative proteomics. Methods Mol Biol 359:37–52
17. Ong SE, Mann M (2006) A practical recipe for stable isotope labeling by amino acids in cell culture (SILAC). Nat Protoc 1:2650–2660
18. Rappsilber J, Ishihama Y, Mann M (2003) Stop and go extraction tips for matrix-assisted laser desorption/ionization, nanoelectrospray, and LC/MS sample pretreatment in proteomics. Anal Chem 75:663–670
19. Olsen JV, Ong SE, Mann M (2004) Trypsin cleaves exclusively C-terminal to arginine and lysine residues. Mol Cell Proteomics 3:608–614
20. Olsen JV, de Godoy LM, Li G, Macek B, Mortensen P, Pesch R, Makarov A, Lange O, Horning S, Mann M (2005) Parts per million mass accuracy on an Orbitrap mass spectrometer via lock mass injection into a C-trap. Mol Cell Proteomics 4:2010–2021
21. de Godoy LM, Olsen JV, de Souza GA, Li G, Mortensen P, Mann M (2006) Status of complete proteome analysis by mass spectrometry: SILAC labeled yeast as a model system. Genome Biol 7:R50
22. Hubner NC, Ren S, Mann M (2008) Peptide separation with immobilized pI strips is an attractive alternative to in-gel protein digestion for proteome analysis. Proteomics 8:4862–4872
23. Shevchenko A, Wilm M, Vorm O, Mann M (1996) Mass spectrometric sequencing of proteins silver-stained polyacrylamide gels. Anal Chem 68:850–858
24. Shevchenko A, Tomas H, Havlis J, Olsen JV, Mann M (2006) In-gel digestion for mass spectrometric characterization of proteins and proteomes. Nat Protoc 1:2856–2860
25. Perkins DN, Pappin DJ, Creasy DM, Cottrell JS (1999) Probability-based protein identification by searching sequence databases using mass spectrometry data. Electrophoresis 20:3551–3567
26. Nesvizhskii AI, Vitek O, Aebersold R (2007) Analysis and validation of proteomic data generated by tandem mass spectrometry. Nat Methods 4:787–797
27. Ong SE, Mann M (2006) Identifying and quantifying sites of protein methylation by heavy methyl SILAC. Curr Protoc Protein Sci Chapter 14, Unit 14.19
28. Armknecht S, Boutros M, Kiger A, Nybakken K, Mathey-Prevot B, Perrimon N (2005) High-throughput RNA interference screens in *Drosophila* tissue culture cells. Methods Enzymol 392:55–73
29. Sachse C, Krausz E, Kronke A, Hannus M, Walsh A, Grabner A, Ovcharenko D, Dorris D, Trudel C, Sonnichsen B, Echeverri CJ (2005) High-throughput RNA interference strategies for target discovery and validation by using synthetic short interfering RNAs: functional genomics investigations of biological pathways. Methods Enzymol 392:242–277
30. Ma Y, Creanga A, Lum L, Beachy PA (2006) Prevalence of off-target effects in *Drosophila* RNA interference screens. Nature 443: 359–363
31. Schrimpf SP, Weiss M, Reiter L, Ahrens CH, Jovanovic M, Malmstrom J, Brunner E, Mohanty S, Lercher MJ, Hunziker PE, Aebersold R, von Mering C, Hengartner MO (2009) Comparative functional analysis of the *Caenorhabditis elegans* and *Drosophila melanogaster* proteomes. PLoS Biol 7:e48

Chapter 4

Systems Biology of Cell Behavior

Najl V. Valeyev, Declan G. Bates, Yoshinori Umezawa, Antonina N. Gizatullina, and Nikolay V. Kotov

Abstract

Systems Biology approaches to drug discovery largely focus on the increasing understanding of intracellular and cellular circuits, by computational representation of a molecular system followed by parameter validation against experimental data. This chapter outlines a universal approach to systems biology that allows the linking of intracellular molecular machinery and cellular activity. This procedure is achieved by applying mathematical modeling to molecular modules of a cell in the light of systems biology techniques.

Key words: Cell activity, Cell behavior, *Paramecium*, Mathematical, Modeling, Systems biology

1. Introduction

The term "Systems Biology" has a very deep meaning. Essentially, it is an attempt to represent Biology as a System and to apply to it the full power of various system-specific techniques developed in physical, engineering and philosophical sciences. In drug discovery, systems biology models can be used to rank compounds or optimize treatment protocols. A number of complex diseases are due to the altered "tuning" of the system, and they can be caused by a combination of environmental and internal factors, rather than by the up- or down-regulation of individual genes. In such cases, cells would still have normal phenotypes. However, impaired signaling between cells can be sufficient to cause a disease, even if it leaves insufficient signs to easily identify diseased cells or pathways. The cause of such diseases lies at the systems level. In the cases of systems biology diseases, the treatment

Qing Yan (ed.), *Systems Biology in Drug Discovery and Development: Methods and Protocols*, Methods in Molecular Biology, vol. 662, DOI 10.1007/978-1-60761-800-3_4, strategies require understanding, classifying and modeling the corresponding mechanisms of cell activity.

There are at least three levels where systems approaches can be utilized in pharmaceutical development: signal reception, intracellular responses, and intercellular communication between various cell type populations (1). A meaningful way of representing a biological system is to divide it into functional modules (2). A classical analysis that highlights the differences between traditional sets of tools employed in biology and engineering was published by Yuri Lazebnik (3).

2. Philosophical Foundations of Systems Biology

It has recently been recognized that robust mathematical representation of biological systems can be achieved by utilizing a number of philosophical principles (4). Any attempt to model a biological system always faces a significant degree of ambiguity, since there are always multiple possibilities for the mathematical representation of a biological system. This ambiguity can be significantly reduced by focusing on the most essential core biological processes which are identified when a cell is analyzed in the light of critical tasks it requires for its functionality and survival. Here, we propose a universal approach for developing a systems biology framework which links intracellular molecular machinery with cellular activity. Such a procedure can be achieved by applying mathematical modeling to molecular modules of a cell in the light of philosophical systems biology approaches (4, 5). Behavior in the most general sense is described in the introduction to the Tinbergen book: "If life is the most complex state of a substance, the behavior, undoubtedly, is the most complex representation of vital activity. Everything that takes place in an organism, chemical, physical and physiological processes, ultimately results in the external activity, in cell behavior" page 7 in (6). This citation is applicable to the cell behavior as well as to multicellular organisms. It has been demonstrated that the behavior of single cell organisms as well as that of individual cells of multicellular organisms can be very complex (7).

Currently, a large number of methodologies have been developed to analyze complex systems in different areas of human activity. For example, engineering deals with systems of arguably comparable levels of complexity as a single cell or at least a pathway. Many engineering principles have been reported to be transferable to biology, with a number of successful examples now appearing in the literature (8–10). The cell behavior mechanisms have continuously been optimizing during evolution. One way to understand the underlying molecular mechanisms behind cell behavior is to reconstruct the development of molecular circuits in relation to cell activity.

A philosophical diagram for the systems biology representation of cell behavior is shown in Fig. 1. It is adopted from a general methodological approach of object representation as a system developed by (5). The diagram shows a number of possible representations of a cell, by "illuminating" it from a number of angles. One representation, which is the most well practiced in biology, is the morphological representation of the cell, which shows the internal protein composition as well as other "parts" connected with each other. Cell behavior can also be seen from a "substrate"-oriented point of view. Another efficient representation of a cell is as a combination of functional modules. The next level of analysis implies the existence of signaling processes between the connected parts. The multitude of representations shown in Fig. 1 is in many cases already well developed in various areas of engineering, but still insufficiently recognized in biology. It is one of the reasons why most faulty objects ranging from a simple radio to a modern space shuttle can always be fixed by a qualified engineer. The same cannot be said about people with disease. Each projection of a cell representation needs to be analyzed under idealized conditions. Such idealized representations have a long history in science. For example, Galileo introduced

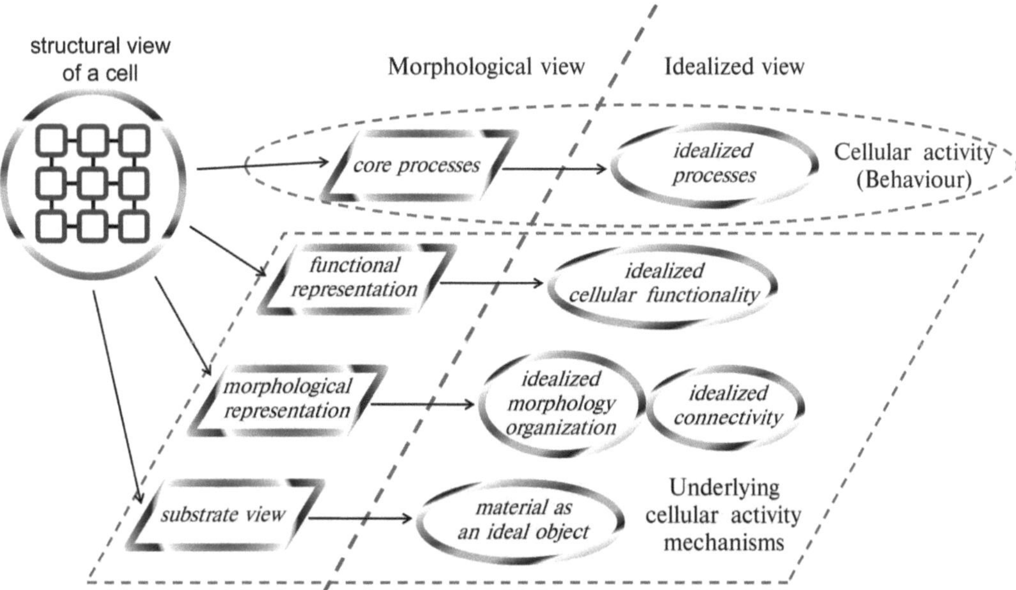

Fig. 1. Functional decomposition of a living cell. A cell is highlighted from a number of angles that can reveal core fundamental processes defining cell behavior. Such decomposition is performed in the morphological and idealized representations. The outlined range of cellular "views" provides guidance for the direction of mathematical model development that would link the molecular mechanism with physiological phenotype.

basic principles in physics by stating that any object would move continuously and indefinitely with the same speed, if it undergoes such motion under idealized conditions with no friction and no influences from other objects.

According to Plato, each engineered system has a backbone idea, which can be converted and represented as functional and procedural diagrams (11) relating the "parts" of the system to each other. The connectivity of those parts would create a system satisfying a predefined set of properties that include the requirements of reliability, immunity, consistency, and criteria of quality. Biological evolution from an engineering point of view can be seen as a development of a system that would allow the organism's survival in a given ecological niche. The Systems Biology approach is based on this backbone idea. Understanding biology from a philosophical system's point of view implies reconstructing the backbone idea of evolutionary engineered systems on a given biological material, that is, proteins, genes and other cell "components". For example, the backbone idea of television is the representation of pictures and sounds as linear sequence of signals, translation of the signals over a distance followed by the picture and sound reconstruction. The backbone idea in the *Paramecium* system example proposed in this chapter is the survival of species mediated by cilium-dependent movements in a complex heterogeneous environment, by feeding on small organisms, organizing reproduction, and active defense from predators while at the same time remaining a single cell organism. This idea is realized in the cell morphology with unique reception and effector systems creating the molecular machine organizing the overall cell behavior. The backbone processes found in any living organism include reproduction, defense, evolutionary development, and functional maintenance.

First consider cell behavior in the light of the idealized backbone processes projection as shown in Fig. 1. As mentioned, reproduction can be considered as one of the basic "core" processes for any living organism. The necessity of reproduction is one of the key features that make living systems different from any other physical objects or human engineered machines. Most of the biological parts, proteins, genes, etc., are unreliable. Proteins can easily misfold, due to a mutation or a physical damage, whereas many stretches of DNA sequence can be highly polymorphic. Reproduction or replacement allows the creation of reliable biological systems composed of unreliable parts. Reproduction is required on a number of levels: reproduction of populations of species, individual organisms, tissues, cells, intracellular substructures, forming cells. On the cellular level, this general reproduction process can be further divided into a number of subprocesses: cell division, cell growth and development (protein synthesis and assembly, cell type differentiation), substrate and energy supply, preprogrammed cell death, the disassembly of proteins as well as other cell parts. The mechanism of reproduction is governed by a number of standard cellular activity "programs", determined by the specifics of the ecological niche and the cell itself.

An excellent example for such analysis can be drawn from the behavior of the unicellular organism *Paramecium caudatum*, since both the intracellular morphological composition and cell behavior under various physiological conditions have been extensively elucidated. The relatively "simple" intracellular machinery is a more "convenient" subject to study in comparison with more complex cells from multicellular organisms. Single cell organisms are also an excellent model for drug screening studies. For example, *Paramecium* has been used as an eukaryotic model system to study cellular effects of neuroactive drugs (12) and investigate cytotoxicity of organic solvents and organophosphorus insecticides (13).

3. Phenomenological Analysis of *Paramecium* Behavioral Reactions

The fact that *Paramecium* behavior is a combination of a limited number of behavioral reactions was observed in the beginning of the nineteenth century (14). The reported reactions include: avoidance reaction, the escape reaction, thigmotaxis (full halt of movement), trichocyst release, conjugation, chemotaxis, galvanotaxis, gravitaxis, and cell division (15). Eight more *Paramecium* behavioral reactions were identified recently: unlimited backward movement, the reverse motion by undergoing the cell shape folding (under conditions of limited space), the "search" reaction, which is based on numerous avoidance reaction initiations, randomized movements, the non-spiraling movement with the oral grove oriented toward the bacteria concentration, flash-dependent movement initiation from thigmotaxis reaction, local trichocyst release, and the directed movement.

Consider the *Paramecium* behavior main underlying process related to the substrate and energy supply in the context of a functional decomposition, as shown in Fig. 1. Paramecia feed on any microorganisms with size smaller than its grove. In real conditions, the feeding substrate for *Paramecium* can be distributed very unevenly, for example localized either on the surface or on the bottom, in the vicinity of decaying flora remains. Therefore, the mechanism of feeding associated behavior needs to contain the following functions shown in Fig. 2: (1) search and identification of substrate (Fig. 2a), (2) dynamic and static maintenance in the area of substrate concentration (Fig. 2b), (3) optimization of grove orientation relative to the substrate presence (Fig. 2c),

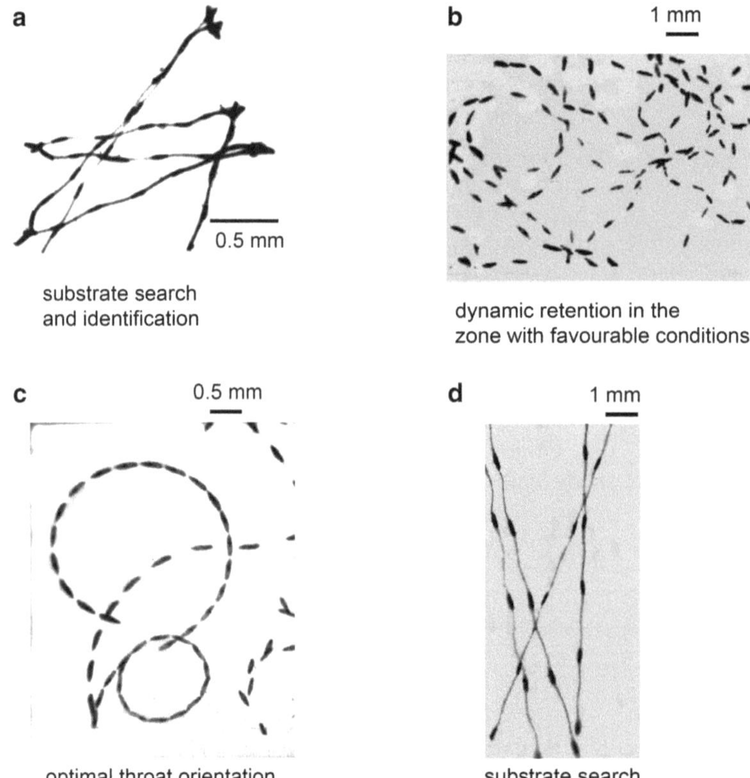

Fig. 2. (**a**) Paramecium "feeding"-associated behavioral movement phenotypes. (**b**) When paramecia identify favorable conditions, Paramecia switch on dynamic behavioral models to remain in the favorable zone. (**c**) Paramecia can also perform special type of movement to optimize the grove orientation to intake maximum amount of bacteria. (**d**) When conditions become unfavorable, cells can move for very significant periods of time without any substrate intake.

(4) food intake, and (5) initiation of the search process upon substrate completion (Fig. 2d). Under favorable conditions, *Paramecium* can also attach itself to an available surface.

The second core process is the evolution of species, organisms, tissues, cells, and cell structure maintaining parts. The mechanisms of this process include mutations, sexual behavior, and genetic exchange between different species, selection, and survival of the fittest. Cells that have sexual behavior possess a number of activity programs allowing the genetic material exchange. The sexual activity of *Paramecium* can be schematically represented by the following scheme (Fig. 3).

Cells which are prepared to conjugate release special substances to the cell surface that make the surface adhesive to a complementary clone and form a furrow without cilia. The furrows allow cellular membranes to come into close proximity. Consider the conjugation process in the context of the functional diagram shown in Fig. 1. The required procedure is the search for a sexual partner and the contact with complementary zones.

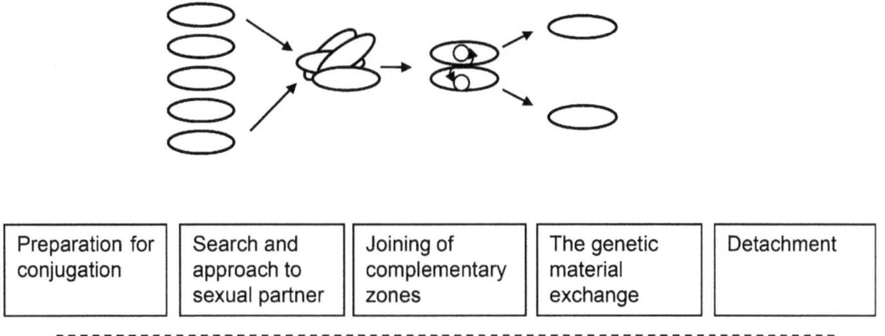

Fig. 3. Schematic representation of the conjugation process in Paramecia.

The analysis of the *Paramecium* motion at the initial stages of conjugation does not reveal any specific motion patterns. The first contact appears to be a random event. Next, sexually prepared cells with adhesive surfaces start random motion with respect to each other. Figure 4 shows the experimental recordings of the Paramecia movements during the conjugation. Initial attachment of multiple cells (Fig. 4a) is followed by a separation of individual cell pairs attached to each other in a randomly oriented fashion. Once separated from the overall group of glued cells, the pair of Paramecia starts performing multiple avoidance reaction by employing abrupt cilia movements causing random mutual shift (Fig. 4b). The random continuous shift of the respective cellular positions occurs until the complementary zones come into contact to exchange genetic material.

Another essential process is defense. Any organism remains alive as long as it is protected. Protection includes the behavioral defense, immunity, and the protection of DNA. Figure 5 shows the *Paramecium* response to danger. Figure 5a shows the *Paramecium* speed in response to contact with a needle. In cases when this reaction is randomly initiated (Fig. 5b) the cellular speed alteration (Fig. 5c) appears to be similar to the contact by a needle. It was also found that paramecia have a light-sensitive reaction. A flash of light causes stationary cells (undergoing the thigmotaxis reaction) to accelerate (16).

3.1. Avoidance Reaction

Avoidance reaction was one of the first observed *Paramecium* movement reactions and was first described in (14). As a result of the avoidance reaction, the cell changes a direction of movement, and continues movement at an angle to the previous motion path trajectory anywhere in the range from 0 to 180°. This reaction is initiated in situations when cells need to overcome mechanical barriers, in the initial stages of the conjugation process and also in order to remain in the "optimum" zone with favorable conditions.

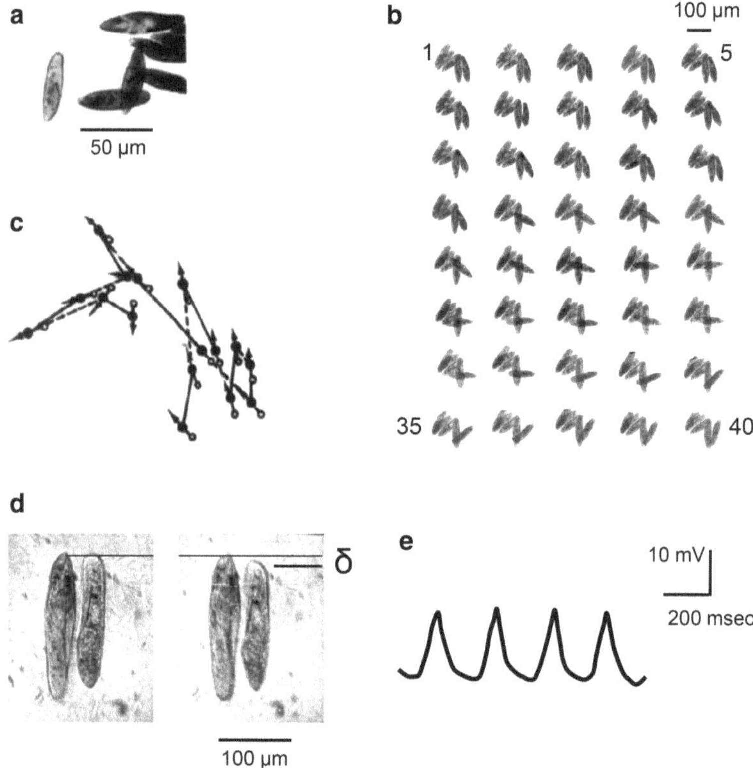

Fig. 4. (**a**) Randomly "glued" cells observed at initial stages of conjugation before individual pairs of cells single out from the overall "tangle". (**b**) Frame by frame images of the mutual cellular movements during the conjugation process. Two cells "glued" together perform multiple avoidance reactions that lead to the change of the body orientation. The process continues until complementary zones come into contact and the exchange of genetic material takes place. (**c**) The cellular trajectory formed as a result of multiple execution of the avoidance reaction. (**d**) The relative movement (δ) of two cells as a result of single avoidance reaction. (**e**) Oscillations of the transmembrane potential on the Paramecium membrane during the initial stages of conjugation.

3.2. Escape Reaction

In response to hyperpolarizing irritation, which can be very short, cells respond by a dramatic speed increase followed by gradual reduction to the original level. Figure 5 shows the cell speed dependence on time during the escape reaction. Electrophysiological studies show that after doubling, speed cells return to the original moving mode in about 0.4 s (17, 18).

3.3. Thigmotaxis

In situations where bacterial substrate is localized on the bottom or any other local volume in the medium, Paramecia have been shown to come to a complete halt of movement. For the first time, this reaction was described in (14, 19). In thigmotaxis, most of the cell surface cilia move with a low frequency, whereas the

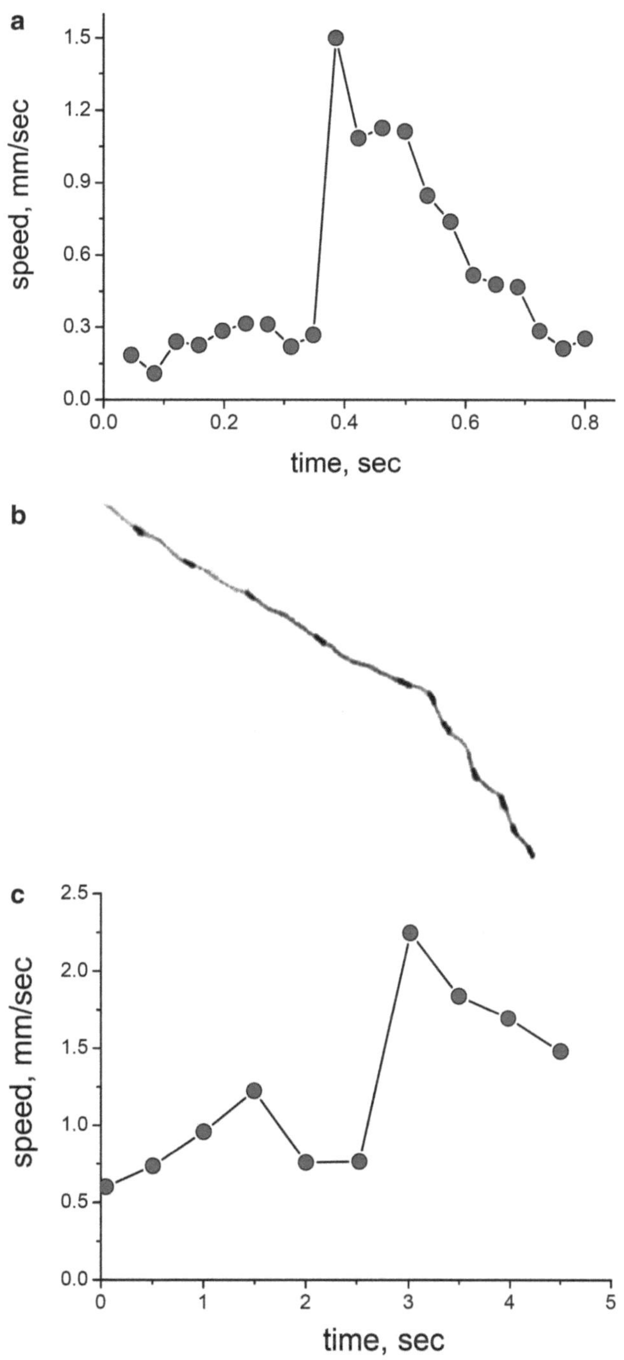

Fig. 5. (**a**) Escape reaction. (**b**) The cellular speed as a function of time in response to an irritation caused by a needle. (**c**) The trajectory is shown during randomly initiated escape reaction with the corresponding cellular speed.

cilia distributed around the oral grove continue intensive strokes. According to Dogel (20), the only possibility for a cell to remain completely motionless is by being attached to a surface by locally released trichocysts.

3.4. Conjugation

Under certain conditions, Paramecia cells create cell couples with a bridge connecting between cytoplasmic membranes, through which the genetic material exchange occurs (21).

3.5. Chemotaxis

Paramecium cells cultivated in an experimental tube gather in the upper layer of the medium, where the oxygen and substrate levels are optimal. Cellular presence in the upper layer has an active dynamic mechanism. Cells can stay attached to the tube surface for some time, but once they start moving, the initiated avoidance reaction returns them into the area with favorable conditions. Paramecia are also sensitive to a number of chemical compounds that alter the parameters of their motion (22).

3.6. Galvanotaxis

Paramecia have been shown to move along spiral trajectories in applied electromagnetic field, with the spiral trajectory axes directed in parallel with the electromagnetic field (23–25). In the described cases, Paramecia move in the cathode direction. Paramecia have also been reported to move at an angle to the electric field or in circles.

3.7. Trichocyst Release

Paramecium body is covered by special organelles that can release thread-like fibers, in situations where cells are being attacked by a predator or in order to create an attachment to a surface.

3.8. Cell Division

Paramecium caudatum reproduces by division. It is clear that the cell is highly asymmetrical, in front it has an oral grove, and an organelle that releases substances at the rear. Shortly after division (in about 10–15 min) the two asymmetrical halves reconstruct the remaining part of the cell.

3.9. Behavior in a Dendrite

Observations of *Paramecium* movement trajectories reveal eight additional behavioral programs. In situations when cells find themselves in a channel type dendrite cavity with a diameter smaller than the cell length, it becomes nearly impossible to turn around by alternating the direction and frequency of cilia beat only. There are two potential possibilities in those situations; one is to start backward movement, another to turn by folding the body. Both types of movement appear to be present in Paramecia.

3.10. Stochastic Ball Movement

A special type of *Paramecium* movement constitutes a slow motion with constantly changing parameters of trajectory, such as spiral radius, spiral step, speed and direction of rotation with periodic spontaneous initiation of the escape reaction. In this case, trajectories are almost "entangled" in a ball. The cell does not change its location over a significant period of time. This behavior is observed in very well adapted cells in cultural medium under favorable conditions.

3.11. Movement in the Vicinity of a Limiting Surface

Paramecium has a specific type of motion in the vicinity of a surface. While contacting a surface, the cell is moving relatively slowly without usual body rotation, constantly contacting the surface with the front part of the body. This type of movement is observed when a thin layer of bacteria is evenly distributed on the bottom or on the top area of an experimental tube.

3.12. The Search Reaction

At initial stages of conjugation, Paramecia form an aggregate from randomly glued up cells. After a period of time, pairs of cells single out from others with a strict body alignment. Analysis of the cell movements during the conjugation process shows that cells make abrupt movements as a result of abrupt cilia strokes as observed during the avoidance reaction. Cells thereby keep adjusting the mutual orientation until creating a contact by complementary zones and forming a bridge between the cellular cytoplasms. If the cell accidentally becomes unattached from the cellular aggregate, it moves in a trajectory that can be observed during the constant initiation of avoidance reaction.

3.13. The "Going Away" Reaction

Under unfavorable positions, with absent bacterial substrate or low KCL concentration, Paramecia follow a spiral trajectory with a very small radius and high speed (up to 2 mm per second). Under such speeding mode, the cellular trajectory represents almost a straight line. A cell moving with an average speed 1 mm per second can swim more than 80 m in 24 h. The substrate, accumulated by cell, can last for several days of movement. By executing this type of movement, *Paramecium* attempts to cover large distances in order to leave the zone with unfavorable conditions.

4. Systems Modeling of Cell Movement

One of the main advantages for systems analysis of *Paramecium* behavior is the relatively well-established intracellular pathways. The two key players for cilia beat regulation are the intracilia Ca^{2+} concentration and the transmembrane potential. While there are a number of channels, notably Na^{+}, K^{+}, and Ca^{2+} channels discovered in the cilia membrane of *Paramecium*, Ca^{2+} regulated Ca^{2+} channels introduce the major contribution into the generation of Ca^{2+} spikes that regulate the frequency and direction of cilia beat in *Paramecium* (26). These Ca^{2+}-dependent Ca^{2+} channels are also found on the endoplasmic reticulum membrane (27, 28) and are considered as important regulators of intracellular signaling in mammalian cells (29, 30).

The Ca^{2+} regulatory system found in *Paramecium* appears to be a highly conserved regulatory system in many multicellular organisms including humans. Multiple Ca^{2+}-dependent pathways are targets for compounds being investigated in multicellular organisms (31–35). A number of mathematical models have been developed to describe the core mechanisms regulating *Paramecium* behavioral phenotypes by employing the philosophical systems biology principles outlined in this chapter.

4.1. Transmembrane Potential and Ca^{2+} Regulation Model

For the first time, a mathematical model for Ca^{2+} and membrane potential in *Paramecium* regulation was proposed in (36). In the most general case, the dynamics of Ca^{2+} alterations in an individual cilium can be described as:

$$V_{\mathrm{R}} \cdot \frac{d\left[\mathrm{Ca}^{2+}\right]}{dt} = \frac{S_{\mathrm{R}}}{z \cdot F} \cdot \left(I^{P}_{\mathrm{Ca}^{2+}} + I^{A}_{\mathrm{Ca}^{2+}} + I^{T}_{\mathrm{Ca}^{2+}} + I^{u}_{\mathrm{Ca}^{2+}}\right) + J([\mathrm{Ca}^{2+}], \cdot [\mathrm{CaM}]) \tag{4.1}$$

where V_{R} is cilium volume, S_{R} is the surface area of a cilium, $I^{P}_{\mathrm{Ca}^{2+}}$, $I^{A}_{\mathrm{Ca}^{2+}}$ and $I^{T}_{\mathrm{Ca}^{2+}}$ are the Ca^{2+} currents due to passive Ca^{2+} transport, active Ca^{2+} transport, and Ca^{2+} leak from cilium into the cell body, respectively. $I^{u}_{\mathrm{Ca}^{2+}}$ is the current generated by Ca^{2+} leakage, $\frac{A}{m^2}$. $J\left(\left[\mathrm{Ca}^{2+}\right], [\mathrm{CaM0}]\right)$ describes Ca^{2+} binding and release by calmodulin, major Ca^{2+} binding regulatory protein in *Paramecium* cilia. $z = 2$ is the charge of Ca^{2+} ions. F is Faraday's constant.

4.2. Intracilia Ca^{2+} Concentration-Dependent Paramecium Movement

The simplified equations for the Ca^{2+} concentration (u) and transmembrane potential (ψ) alterations are given by:

$$\frac{d\psi}{d\tau} = g_{\mathrm{Na}^+}(\mathrm{CaM}_{0,3}) \cdot \left(\psi - \ln\left(\frac{C^{00}_{\mathrm{Na}^+}}{C^{in}_{\mathrm{Na}^+}}\right)\right) + g_{\mathrm{Ca}^{2+}}(\mathrm{CaM}_{0,3}) \cdot \left(\psi - \frac{1}{2} \cdot \ln\left(\frac{u_0}{u}\right)\right) - g_{K^+}(\mathrm{CaM}_4) \cdot \left(\psi - \psi_{K^+}\right) - g_l \cdot (\psi - \psi_l) + I_0,$$

$$b \cdot \frac{du}{d\tau} = g_{\mathrm{Ca}^{2+}}(\mathrm{CaM}_{0,3}) \cdot \left(\psi - \frac{1}{2} \cdot \ln\left(\frac{u_0}{u}\right)\right) + \gamma(\mathrm{CaM}_4) \cdot \frac{u}{K_A + u} - \alpha \cdot \exp(-\beta \cdot \psi) \cdot \left(\zeta - \frac{1}{2} \cdot \ln\left(\frac{u}{u_T}\right)\right) + \lambda \cdot \sum_{i=0}^{4} f_i(\mathrm{CaM}_i) \tag{4.2}$$

where $\tau = t \cdot k_{-1}$, $u = \frac{\mathrm{Ca}^{2+}}{K_{\mathrm{Ca}^{2+}}}$, $\psi = \frac{\varphi_M \cdot F}{R \cdot T}$, $b = \frac{z \cdot F^2 \cdot V \cdot K_{\mathrm{Ca}^{2+}}}{C \cdot S_M \cdot R \cdot T}$. is the capacity of the membrane, S_M is the membrane surface, V is the cilium volume, $K_{\mathrm{Ca}^{2+}}$ is the maximum dissociation constant for the Ca^{2+}–CaM binding reaction. k_{-1} is the Ca^{2+} ion dissociation constant from CaM, φ_M is membrane potential. ψ_{K^+} and ψ_l are the nondimensional equilibrium potentials of K^+ and leakage currents, respectively. $g_{Na^+}(CaM_i)$, $g_{Ca^{2+}}(CaM_i)$, $g_{K^+}(CaM_i)$, g_l are nondimensional conductances. ζ is the nondimensional potential between cell body and cilium. u_T is the nondimensional Ca^{2+} concentration in the cell body, $f_i(CaM)$ is function that describes Ca^{2+} uptake by CaM. α, β, λ, γ, are constants and I_0 is an external nondimensional current.

$$g_{Na^+}(CaM_{0,3}) = \bar{g}_{Na^+} \cdot \frac{CaM_0 + CaM_3}{K_{Na^+} + CaM_0 + CaM_3},$$
$$g_{Ca^{2+}}(CaM_{0,3}) = \bar{g}_{Ca^{2+}} \cdot \frac{CaM_0 + CaM_3}{K_{Ca^{2+}} + CaM_0 + CaM_3},$$
$$g_{K^+}(CaM_4) = \bar{g}_{K^+} \cdot \frac{CaM_4}{K_{K^+} + CaM_4}, \gamma(CaM_4) = \bar{\gamma} \cdot \frac{CaM_4}{K_A + CaM_4},$$

where $\bar{g}_{Na^+}$, $\bar{g}_{K^+}$, $\bar{g}_{Ca^{2+}}$ are the nondimensional conductances, K_{Na^+}, $K_{Ca^{2+}}$, K_{K^+}, K_A - are the nondimensional equilibrium dissociation constants for the corresponding CaM-channel complex formation reactions.

The simplified description of CaM conformational dynamics is given by:

$$\frac{dCaM_i}{d\tau} = \Im_i(CaM_0, CaM_1, \ldots CaM_4, u), i = 0,1 \quad (4.3)$$

The analysis of the dynamic properties of the described equations reveals that the system possesses excitable properties even in the absence of any potential dependent channels. CaM conformations have distinct and selective activation properties when bound to different numbers of Ca^{2+} ions (37, 38). The phase diagram of coupled Eq. (2) shows that when Ca^{2+} and N^+ channels are activated by CaM conformations with one and three Ca^{2+} ions, simultaneously with K^+ channels and active Ca^{2+} transport regulation by fully Ca^{2+} occupied CaM, the system Eq. (2) has three dynamic states: (1) generation of a single Ca^{2+} spike, (2) stationary oscillations, (3) trigger. Figure 6 shows the phase diagram for the trigger type mode of the system.

The alterations of intracilia Ca^{2+} concentrations regulates the cilia movement via changing the levels of cyclic monophosphates, such as cyclic adenosine monophosphate (cAMP) and cyclic guanosine monophosphate (cGMP) (39), followed by cyclic monophosphate-dependent phosphorylation of effector proteins in the base of *Paramecium* cilia. In *Paramecium*, AC is regulated by CaM bound to one Ca^{2+} ion, PDE for cAMP is activated by fully bound CaM, GC is governed by CaM with three Ca^{2+} ions,

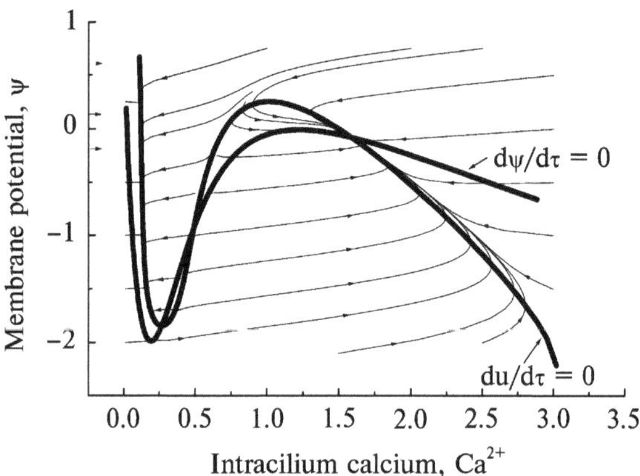

Fig. 6. Phase plot of the relationship between the Ca^{2+} levels and transmembrane potential. The null-clines cross at three points two of which are stable. The solution predicts the trigger-type mode of operation.

whereas the activity of PDE for cGMP is inhibited by apo CaM. The effector proteins (dynein arms) regulating cilia beat frequency are phosphorylated by both cAMP and cGMP-dependent protein kinases, while the proteins regulating the direction of effective cilia beat are phosphorylated by Ca^{2+}-CaM-dependent protein kinase (39, 40). A mechanistic model linking the cilia beat frequency and the direction of effective strike allows calculation of cell movement trajectories as a function of intracilia Ca^{2+} (41).

Mathematical modeling of the Ca^{2+} regulatory system in combination with a physical model for *Paramecium* cilia motion allows the mechanistic understanding of *Paramecium* motion under various external conditions. The combination of the philosophical approaches to systems biology, physiological observations of cell movements, the mathematical modeling of the intracilium and intracellular molecular system, and mechanical consideration of a cell allow the prediction of *Paramecium* reactions according to the external conditions. Figure 7 shows the modeled trajectories of *Paramecium* movement during the performance of individual reactions. The predicted cell movement trajectories are compared with the experimentally observed trajectories at different levels of intracilia Ca^{2+} concentration. Figure 7a–c show the extreme cases for *Paramecium* spiral-type movements, with a very small spiral radius and large spiral step (Fig. 7a): movement along a circle when the spiral step equals to zero (Fig. 7b) and movement when both spiral step and radius equal to zero (Fig. 7c). The experimentally observed *Paramecium* trajectory during the avoidance reaction is compared with the calculated movement in Fig. 7d.

The phase diagram of the coupled Ca^{2+} (u) and transmembrane potential (ψ) variables (Eq. 2) reveals that the system can

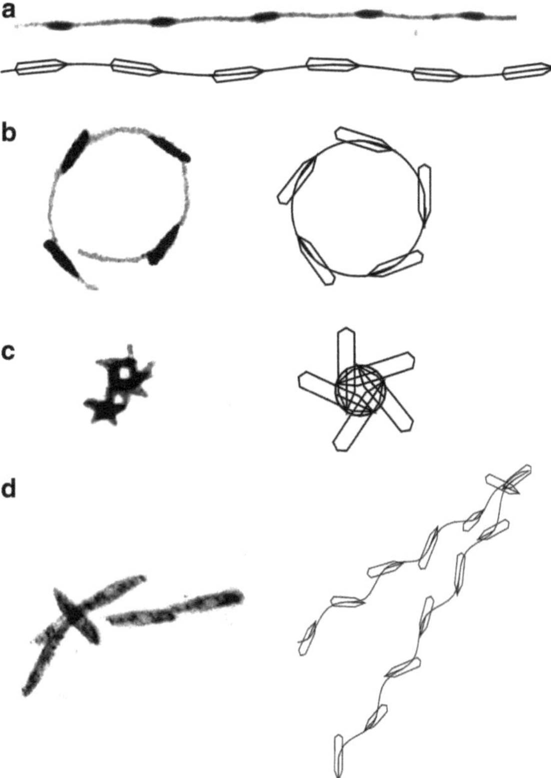

Fig. 7. The calculated *Paramecium* movements in comparison with the experimentally observed trajectories. (**a**) Movement under normal conditions. (**b**) Movement around the circle. (**c**) Stochastic "tangle" type of movement. (**d**) Avoidance reaction.

also generate Ca^{2+} spikes by lowering Ca^{2+} level and reaching the low freshold. The Ca^{2+} spike generated by lowering Ca^{2+} concentration induces the escape reaction described in Fig. 5. Sustained Ca^{2+} oscillations occur during the conjugation reaction. In the mathematical model (Eq. 2), oscillations emerge as a result of non Ca^{2+} increase of depolarizing current (Na^{+} or Ba^{2+}). The trigger mode is observed when the cell is located in the narrow channels, formed, for example, in agar-agar gel. Under these conditions, cells can move forward or backward for a significant period of time. In the model, the trigger mode (Fig. 6) is initiated by the decrease of non Ca^{2+} current.

5. Conclusions

This chapter describes a general methodology for the analysis of cellular activity from a systems biology point of view. We show that any cellular behavior can be understood by applying functional decomposition techniques, analyzing the cell behavioral programs and mathematically modeling the core molecular pathways. The selection of the "core" processes is achieved in the light of functional decomposition of the cellular environment and key requirements for cellular survival. In the given example of *Paramecium caudatum*, the biomolecular system composed of CaM, CaM-dependent channels, cAMP, cGMP and AC, GC, PDEs enzymes reveals sufficient properties to mimic experimentally observed activity modes of the cell. One can suggest that the described system, as one example among many other molecular modules, incorporates in itself biophysical properties, which are sufficient to execute a limited number of essential tasks.

One of the key features of the presented methodology is that it shows that the cell has a *limited* number of behavioral reactions, which are regulated at the molecular level, and executed automatically depending on a combination of external and internal conditions. The development of a systems biology model and the prediction of cellular trajectories that match experimentally observed behavioral reactions offer a range of powerful analysis tools. For example, by utilization of a computational cell, one can test the cellular activity effects of applied compounds. The model further allows compound dosage optimization. The described methodology provides powerful techniques for analyzing altered cellular phenotypes as a result of up- or downregulated protein expression levels, specific mutations, or any other biological "defects".

References

1. Rajasethupathy P, Vayttaden SJ, Bhalla US (2005) Systems modeling: a pathway to drug discovery. Curr Opin Chem Biol 9:400–406
2. Hartwell LH, Hopfield JJ, Leibler S, Murray AW (1999) From molecular to modular cell biology. Nature 402:C47–C52
3. Lazebnik Y (2004) Can a biologist fix a radio? – Or, what I learned while studying apoptosis, (Cancer Cell. 2002 Sep;2(3):179–82). Biochemistry (Mosc) 69:1403–1406
4. Boogerd F, Bruggeman FJ, Hofmeyr J-HS, Westerhoff HV (2007) Systems biology: philosophical foundations. Elsevier Science
5. Schedrovitsky GP (1997) [Phylosophy. Science. Methodology.]
6. Tinbergen N (1969) [Animal behaviour], World
7. Aleksandrov VY (1975) [The behavior of cells and intracellular structures], Knowledge
8. Aldridge BB, Burke JM, Lauffenburger DA, Sorger PK (2006) Physicochemical modelling of cell signalling pathways. Nat Cell Biol 8:1195–1203
9. Ideker T, Winslow LR, Lauffenburger DA (2006) Bioengineering and systems biology. Ann Biomed Eng 34:1226–1233
10. Stelling J, Sauer U, Szallasi Z, Doyle FJ III, Doyle J (2004) Robustness of cellular functions. Cell 118:675–685
11. Plato (1997) Complete works, Hackett Publishing Co, Inc, Indianapolis
12. Wood CR, Hennessey TM (2003) PPNDS is an agonist, not an antagonist, for the ATP receptor of *Paramecium*. J Exp Biol 206:627–636
13. Rajini PS, Krishnakumari MK, Majumder SK (1989) Cytotoxicity of certain organic solvents and organophosphorus insecticides to the ciliated protozoan *Paramecium caudatum*. Microbios 59:157–163
14. Jennings HS (1906) Behavior of the lower organisms. Indiana University, Bloomington
15. Seravin LN (1967) Motile systems of primary organisms. Science, Leningrad

16. Kotov NV, Bukharaeva EA (1977) [*Paramecium* reaction to electromagnetic oscillations], 2953–2977
17. Litvin VG, Samigullin DV, Kotov NV (1999) A study of the defensive acceleration reaction of *Paramecium caudatum*. Biofizika 44:296–302
18. Machemer H, Peyr Y (1977) Swimming sensory cells: electrical membrane parameters, receptor properties and motor control in ciliated Protozoa, *Verh Dtsch Zool Ges*, 86–110
19. Puter A (1900) Studien uber die Tigmotaxis bei Protisten. Arch Anat Physiol, 243–302
20. Dogel VA, Polyansky YI, Kheisin KM (1962) [General protozoology], Soviet Academy of Sciences
21. Vivier E (1962) Demonstration a laide de la microscopic electronigue de echanges cytoplasmiques lors de la conjugaison sher *Paramecium caudatum*. Comp Hend Soc Biol 156:1115–1116
22. Van Houten J (1992) Chemosensory transduction in eukaryotic microorganisms. Annu Rev Physiol 54:639–663
23. Dryl S (1970) Response of ciliate protozoa to external stimuli. Acta Protozool 7: 325–333
24. Machemer H, Suguno K (1989) Electrophysiological control of reversed ciliary betting: a basis of motile behaviour in ciliated protozoa. Comp Biochem Physiol 94:365–374
25. Statkevich P (1903) [Galvanotaxis and galvanotropism of animals. Galvanotaxis and galvanotropism of infusoria]
26. Eckert R, Brehm P (1979) Ionic mechanisms of excitation in *Paramecium*. Annu Rev Biophys Bioeng 8:353–383
27. Bezprozvanny I, Watras J, Ehrlich BE (1991) Bell-shaped calcium-response curves of Ins(1, 4, 5)P3- and calcium-gated channels from endoplasmic reticulum of cerebellum. Nature 351:751–754
28. Hagar RE, Burgstahler AD, Nathanson MH, Ehrlich BE (1998) Type III InsP3 receptor channel stays open in the presence of increased calcium. Nature 396:81–84
29. Ramos-Franco J, Fill M, Mignery GA (1998) Isoform-specific function of single inositol 1, 4, 5-trisphosphate receptor channels. Biophys J 75:834–839
30. Valeyev NV, Downing AK, Skorinkin AI, Campbell ID, Kotov NV (2006) A calcium dependent de-adhesion mechanism regulates the direction and rate of cell migration: a mathematical model. In Silico Biol 6:545–572
31. Li CY, Mao X, Wei L (2008) Genes and (common) pathways underlying drug addiction. PLoS Comput Biol 4:e2
32. Ochi R, Gupte SA (2007) Ryanodine receptor: a novel therapeutic target in heart disease. Recent Pat Cardiovasc Drug Discov 2:110–118
33. Padmanabhan S, Lambert NA, Prasad BM (2008) Activity-dependent regulation of the dopamine transporter is mediated by Ca(2+)/calmodulin-dependent protein kinase signaling. Eur J Neurosci 28:2017–2027
34. Rushlow WJ, Seah C, Sutton LP, Bjelica A, Rajakumar N (2009) Antipsychotics affect multiple calcium calmodulin dependent proteins. Neuroscience 161(3):877–886
35. Zhang GQ, Zhu Z, Zhang W (2009) Inhibitory effect of antihypertensive drugs on calcineurin in cardiomyocytes. Am J Hypertens 22:132–136
36. Hook C, Hildebrand E (1979) Excitation of *Paramecium*. A model analysis. J Math Biology 8:197–214
37. Valeyev NV, Bates DG, Heslop-Harrison P, Postlethwaite I, Kotov NV (2008) Elucidating the mechanisms of cooperative calcium-calmodulin interactions: a structural systems biology approach. BMC Syst Biol 2:48
38. Valeyev NV, Heslop-Harrison P, Postlethwaite I, Kotov NV, Bates DG (2008) Multiple calcium binding sites make calmodulin multifunctional. Mol Biosyst 4:66–73
39. Valeyev NV, Heslop-Harrison P, Postlethwaite I, Gizatullina AN, Kotov NV, Bates DG (2009) Crosstalk between G-protein and Ca^{2+} pathways switches intracellular cAMP levels. Mol Biosyst 5:43–51
40. Davydov DA, Litvin VG, Platov KV, Sadykov IK, Kotov NV (1997) [The analysis of adenylate cyclase system of the cell]. Struct Dyn Mol Syst 4:30–34
41. Kotov NV, Volchenko AM, Davydov DA, Kostyleva EK, Sadykov I, Platov KV (2000) Motor activity of *Paramecium*. Biofizika 45:514–519

Chapter 5

Computational Modeling in Systems Biology

Ravishankar R. Vallabhajosyula and Alpan Raval

Abstract

Interactions among cellular constituents play a crucial role in overall cellular function and organization. These interactions can be viewed as being complementary to the usual "parts list" of genes and proteins and, in conjunction with the expression states of these parts, are key to a systems level understanding of the cell. Here, we review computational approaches to the understanding of the functional roles of cellular networks, ranging from "static" models of network topology to dynamical and stochastic simulations.

Key words: Systems biology, Networks, Protein–protein interaction, Metabolism, Genetic interactions, Regulation

1. Introduction

Recent advances in experimental techniques in molecular biology have led to a high rate of data generation and highlighted the need for computational tools and algorithms to aid in curation, validation, and the analysis of large datasets. These datasets can be examined in silico for consistency, used to predict new data, and to create models of biological mechanisms for the collective functioning of thousands of genes and proteins which, in concert, orchestrate biochemical reactions across numerous pathways. Such efforts hold the key to achieving the potential of systems biology, which aims to redefine the future of medicine by integrating knowledge from experimental and computational approaches (1). These computational steps are essential in order to understand how biological systems achieve and maintain many characteristic properties such as robustness, and why certain perturbations or sequences of perturbations of the genetic and

Qing Yan (ed.), *Systems Biology in Drug Discovery and Development: Methods and Protocols*, Methods in Molecular Biology, vol. 662, DOI 10.1007/978-1-60761-800-3_5, © Springer Science+Business Media, LLC 2010

molecular machinery have system-wide ramifications resulting in disease. While experimental approaches and breakthroughs have dominated molecular biology over the last half century, it is only over the last decade that availability of vast public databases of experimental data have led to the development of models to explain specific biological phenomena by building hypotheses, making inferences, and suggesting more experiments that in turn help refine the models. These quantitative approaches have served to bring the power of predictive modeling, more commonly associated with the physical sciences and engineering, to biology. The focus of this chapter is to review aspects of these computational advances.

This chapter is organized as follows. We first review briefly the advent of microarray technologies and the computational tools used to analyze gene expression data (Subheading 2). We then describe high-throughput experiments for discovering interactions among gene and proteins, and how errors in these experiments can be estimated and reduced by combining data from different experiments (Subheading 3). Next, we review computational approaches to predict regulatory, protein–protein interaction, and genetic networks (Subheading 4). We briefly discuss methods for analyzing network data to elucidate the functional organization of the cell (Subheading 5). Finally, we discuss computational modeling and the simulation of network dynamics, including deterministic and stochastic methods (Subheading 6).

2. Microarrays and the Analysis of Expression Data

Microarray technology permits thousands of genes to be assayed at a very high density, with their expression patterns studied by imaging and analyzing the intensity of hybridization (2). Microarrays opened up a new way to study gene expression patterns on a genomic scale in a high-throughput fashion (3). One of the earliest applications of this technology was the genome-scale profiling of the metabolic and genetic control of gene expression of all known genes in *Saccharomyces cerevisiae* (4).

On a genetic level, adaptation to environmental perturbations is made possible by modifying the expression levels of various genes. Changes in this regulation can be inferred by applying statistical analysis to microarray data (5), leading to the identification of differential expression among genes (see, for example, the reviews (6, 7)). Microarray studies involve experimental design, preprocessing, inference, classification, and validation (8). It should be noted that each of these five steps has its own complexities that are the focus of continued research. For example, one of the major issues with raw microarray data is that spots are not imaged or become corrupted due to noise. This has led to active research in the area of noise removal from spot intensity data (9) and development of methods to impute missing values (see, for example, (10)). Further, the classification step involves many computational aspects where salient characteristics in the data can be identified by applying various algorithms (11, 12).

There are a number of different types of microarray experiments which include cDNA (13, 14), oligonucleotide-based (15) and protein microarrays (16). Microarrays are also widely used in drug discovery to test the effect of drugs by comparing the expression of genes in normal cells to cells treated with drugs (17). They are a particularly important tool in the study of treatments for cancer, as expression profiles of each tumor tends to be different and thus can be used as a molecular signature to track the efficacy of the treatment (18). This technology has sufficiently evolved in the case of some cancers, allowing microarrays to be used to predict outcome (19). With the development of very high density microarrays from Affymetrix, very detailed experiments to unravel the role of noncoding DNA have become feasible and will play a major role in advancing our understanding of gene regulation in the coming years.

3. Computational Approaches for Making Sense of High Throughput Interaction Data

In conjunction with the ability to measure global gene and protein expression, it is clear that the great promise of systems biology lies in the possibility of modeling the topology and the dynamics of biomolecular networks at the genome scale. This can only be possible if we have access to reliable experimental data on the nature of genome-scale networks. As is well known, however, most experiments that uncover interactions at this scale in reasonable laboratory time (high-throughput (HTP) experiments) are error-prone. How can we understand this error, model it, and attempt to alleviate it?

3.1. HTP Experiments: A Brief Summary of Methods

Let us first summarize the methods of the most common HTP experiments that are used to generate genome-scale network data. The yeast 2-hybrid (Y2H) method (20, 21) to find protein–protein interactions consists of the synthesis of two fused protein constructs: one protein (the *bait*) fused with the DNA-binding domain of a transcription factor, and the other protein (the *prey*) fused with the activation domain of the same transcription factor. Physical interaction between the two proteins results in the close proximity of the DNA-binding and activation domains, causing a reporter gene to be expressed. The presence of the physical interaction is thus inferred by the expression of the reporter gene.

Affinity copurification methods for identifying protein–protein interactions actually identify protein complexes by purifying a particular protein and examining, usually by mass spectrometry analysis, all other proteins that are associated with the protein of interest. The purification process itself takes place either by targeting the protein of interest with a specific antibody (coimmunoprecipitation or co-IP (22)), or by fusing the protein of interest with an IgG binding domain attached to a calmodulin binding domain and then carrying out two affinity purification steps (tandem affinity purification or TAP (23)).

Similarly, DNA–protein interactions are identified in a high throughput fashion by a chromatin immuno-precipitation (ChIP) process (24). An antibody is made against the candidate protein, and cells in which the protein is expressed are treated with formaldehyde, which cross-links most proteins bound to the DNA to one another and to the DNA bases. The protein-DNA complex (called chromatin) is then purified away from the rest of the cellular material. Subsequently, the chemical linkage between the protein and DNA is reversed, the DNA is amplified by polymerase chain reaction (PCR) and its sequence is revealed, usually by labeling with a fluorescent molecule and hybridizing to known DNA sequences on a DNA microarray, or by direct sequencing.

Synthetic lethal interactions among nonessential genes are discovered in HTP mode by the Synthetic Genetic Array (SGA) technique in which arrays of single gene deletion mutant yeast cells are mated with all other single gene mutants and the resulting diploid cells are allowed to form haploid double mutants (25). The presence of a viable double mutant shows the lack of synthetic lethality. In this way, over 15% of the total synthetic lethal interaction space of *S. cerevisiae* has now been scanned (26). A measure of phenotype that is finer than simple viability consists of quantitative measurements of growth under certain conditions or the degree of invasiveness into a solid medium. The larger the value of such quantitative measures, the more viable the double mutant is. These types of quantifications on the extent of synthetic lethality are available in the form of EMAP (Epistatic Mini Array Profile) data (27).

3.2. Errors in HTP Data

HTP data are notoriously prone to both false positive as well as false negative errors. In two-hybrid screens, false positives may arise because of nonspecific and unstable "stickiness" of a protein to other proteins. Further, interactions detected in the two-hybrid screen may not exist in vivo because the proteins in question are over-expressed at artificially high levels in the screen, or because the two proteins are not present in the same cellular compartment in vivo, or because the structure of the proteins after fusion with the DNA-binding and activity domains is altered. False negatives in two-hybrid screens, on the other hand, may arise due to low coverage of the two-hybrid experiment and due to under-sampling of bait–prey clones within the proteins that are covered. There are also systematic false negatives due to the fact that a bait–prey mating could prove unsuccessful in a two-hybrid experiment, even though the interaction occurs in vivo (28).

The contribution of nonspecificity to protein–protein interactions discovered in two-hybrid experiments can be explicitly incorporated in a computational model for protein interaction networks, as was carried out in ref. (29). The basic idea behind this model is that the larger the number of hydrophobic residues that reside on the surface of a protein, the higher its propensity to participate in nonspecific interactions – such interactions would be favored because their existence would reduce the number of exposed hydrophobic residues. Thus one can construct a simple computational model where the probability of interaction between two proteins is an increasing function of the total number of hydrophobic residues on both of their surfaces. Deeds et al. (29) used known and predicted secondary structures of yeast proteins to compute the number of exposed hydrophobic residues for them, and simulated many realizations of the yeast protein interaction network using this model. They found, perhaps not surprisingly, that the model was successful at predicting global topological features (such as the degree distribution) of the true yeast network (this finding was further confirmed in (30)). However, they also found that the stochastic component of the model could capture reasonably well the *disagreement* between different yeast two-hybrid datasets. These findings suggest that individual HTP experiments for uncovering protein–protein interaction are dominated by the discovery of nonspecific interactions that are not biologically relevant in vivo. This is not to say, as we discuss below, that HTP experiments taken together cannot produce reliable data.

Affinity purification experiments, while considered definitive tests of direct physical interactions, actually only detect participation in the same protein complex. They too suffer from false positives due to the inflated expression of the bait and due to contaminants. Physical interaction between two proteins in an affinity purification experiment is usually inferred by an intervening "matrix" or "spoke" model, where the matrix model assumes that all pairs of proteins in a complex have pairwise interactions, while the spoke model only assumes pairwise interactions between the bait and each prey. Thus the matrix model has a high propensity for false positives, while the more conservative spoke model may miss out on real prey–prey interactions, leading to false negatives. Further, large-scale affinity purification experiments are biased toward the most abundant proteins, leading to loss of coverage and, therefore, false negatives.

The primary source of error in the ChIP method for detection of DNA–protein interactions is that weak binding sites may remain undetected, and the resolution of the binding site (number of nucleotides bound) is determined by the average DNA fragment size at the immuno-affinity purification step. The latter factor is important for recognizing the correct gene that is bound by the protein. If the distance between two genes is small compared to the resolution, the identity of the bound gene is confounded. This problem is exacerbated when the two genes in question are transcribed in opposite directions and have potentially overlapping regulatory regions. Also, eukaryotic regulatory regions can often be positioned thousands of nucleotides away from the gene of interest, thus further confounding the resolution of gene identity.

3.3. Combining HTP Datasets to Estimate and Alleviate Error

In spite of high potential error rates in individual HTP experiments, it is possible to both estimate these errors as well as correct for them by computational approaches to combining data from different HTP experiments. Here, we briefly discuss how these methods work in the case of protein–protein interaction data, which is particularly prone to error.

Firstly, in any given organism, the true number of protein–protein interactions represents a global unknown parameter to be estimated. Secondly, every HTP experiment has two additional unknown error parameters: the proportion of true interactions that are left undiscovered by the experiment (false negative rate), and the proportion of true noninteractions that are reported by the experiment as true interactions (false positive rate). This picture assumes a no-bias scenario where false positive rates and false negative rates are uniform across all protein pairs that are tested; a modeling bias of this type would complicate the analysis. The observed counts that may be used to estimate these unknown parameters include the number of interactions discovered by each experiment as well as the number of interactions that are common to multiple experiments. Thus, with 1 HTP experiment, one has 3 unknown parameters (total number of true interactions and 2 error rates) and 1 observation (number of observed interactions); with 2 HTP experiments, one has 5 unknown parameters (total number of true interactions and 4 error rates) and 3 independent observations (number of interactions found by each experiment and number of interactions found by both); and with 3 HTP experiments, one has 7 unknown parameters and 7 independent observations (3 counts of interactions reported by each experiment, 3 counts of interactions reported by exactly 2 experiments, and 1 count of interactions reported by all 3 experiments). Furthermore, we have made the additional simplifying assumption that the error rates for each experiment are independent; otherwise, there are additional unknown parameters reflecting dependence among the error rates. Thus, under the assumptions made, at least 3 independent HTP experiments are required to estimate all relevant unknown parameters.

The above method for estimating error rates and the true size of the protein interaction network was pioneered by D'haeseleer and Church (31), who considered in detail situations where the assumption of perfect independence among HTP datasets could be relaxed. They also simplified the problem and increased the reliability of the final estimates by requiring that one of the three datasets had to be a "gold standard" dataset with negligible false positive rate. Using their methods, they estimated that existing HTP experiments to discover protein–protein interactions in yeast had false positive error rates ranging from about 45% to about 90%, and that yeast had about 50,000 true protein–protein interactions. Furthermore, their method allowed for Bayesian reassessment of the accuracy of each protein–protein interaction reported by experiments by finding the posterior probability of the existence of an interaction given that it was reported or not reported by independent HTP experiments. This type of intersection analysis for the estimation of global parameters was also carried out in (32).

More recent work on estimating error rates and improving the reliability of HTP data use more details in HTP experiments than are reflected by the existence (or lack thereof) of an interaction. Specifically, information from raw counts of bait–prey clones is integrated into a computational procedure based on capture-recapture theory (28, 33) and applied to the protein interaction networks of yeast, worm, and fly. Further error rates are not assumed to be unbiased and the models explicitly allow for bait-specific error. This work interestingly finds significantly higher false discovery rates (proportion of predicted interactions that are actually noninteracting) for membrane proteins than other baits. In terms of biological process participation and function, proteins involved in metabolism and those involved in protein binding and transcriptional regulation also tend to have high false discovery rates.

4. Computational Approaches to Predicting Interaction Network Topology

Computational approaches to predict the topology of large biomolecular interaction networks have progressed in concert with experimental methodologies and are, to a large extent, informed by the increasing availability of reliable experimental interactomes. Not unlike the trend in approaches to protein structure prediction, computational network prediction began in the form of ab initio inference of interactions and evolved to sophisticated models that combine vast compendia of data sources, including known interactions themselves, to predict new interactions in a supervised fashion. Here, we briefly summarize the state of predictive methods for regulatory interactions, protein–protein interactions, and genetic interactions.

4.1. Prediction of Regulatory Networks

Ever since the advent of microarray technology, methods to analyze expression data have developed in parallel. While early methods focused on clustering approaches to find functionally related genes, it was quickly realized that microarray data could be used to infer relationships of *influence* among genes, and that the closest approximation to such influence in vivo represented the regulatory relationship. Even today, the most common approaches to predict gene regulatory networks are based on the analysis of microarray or gene expression data, because the inference of regulation from expression represents the classic reverse engineering problem: expression is a *consequence* of the regulatory machinery, therefore it is plausible that expression data should imply the regulatory network.

However, the reverse engineering problem is not simple, and there are a plethora of methodologies that attempt to solve it. There are a number of classic reviews of the literature in this area, including that of van Someren et al. (34), and a more recent survey that includes a discussion of the limitations implicit in the usage of gene expression as the primary predictor of regulatory interactions (35). Here, we simply note that the most fruitful approaches to the prediction of regulatory interactions from expression data can be traced back to the REVEAL (REVerse Engineering ALgorithm) method (36). The input to this method is a binary expression profile for every gene under a set of conditions (expression = 0 indicates that the gene is not expressed under a particular condition, while expression = 1 indicates that the gene is expressed). Briefly, for every gene of interest, REVEAL finds the set of genes whose mutual information with the gene of interest is maximized, where mutual information is computed from the expression profiles. This set of genes is then identified as the set of putative regulators of the gene of interest, and the process is repeated for every gene, thus resulting in a directed network of regulators and genes. In addition, the method also identifies the set of logical rules that determine how the regulators regulate the gene of interest.

The basic REVEAL methodology was simple, yet flexible and powerful enough to admit many extensions. Perhaps the most widely studied extension to the REVEAL method consisted of the use of dynamic Bayesian networks or DBNs (see, for example, (37)), a probabilistic method that can incorporate multiple expression states (as opposed to binary states), time series expression data, noise in the expression data, as well as scoring functions other than mutual information. Furthermore, the DBN methodology can also be extended to model the putative influence of genes whose expression levels are not present in the data at hand, the so-called *hidden factors* (38–40). Many current methods for predicting regulatory networks employ dynamic Bayesian networks in some form. Novel approaches have also been developed to limit the size of the search space when looking for a set of regulators for the gene of interest. These include preclustering the expression profiles into a single cluster or "meta-gene" and only inferring regulatory relationships among the metagenes, but also more sophisticated methods such as clustering putative regulators by the transcriptional time lag between their expression and that of the gene of interest (41).

4.2. Prediction of Protein–Protein Interaction Networks

A number of factors influence the propensity of two proteins to partake in a physical interaction; therefore, the computational prediction of protein–protein interactions usually involves some type of data integration process that combines information from various factors or features. These features are discussed in detail by Skrabanek et al. (42) and Liu et al. (43). Briefly, they include structural "stickiness" (as discussed above), shape complementarity of putative interacting surfaces, close proximity of the protein-coding genes on the chromosome (common gene neighborhood), similarity of evolutionary histories or phylogenetic profiles, existence of a fused protein in another organism that has local sequence similarity to both putative interacting proteins (gene fusion event), history of correlated compensatory mutations, and knowledge of physical interaction in a related organism (existence of an "interolog"). Note that many of these features are also indicative of functional similarity between the pairs of proteins in question. Indeed, proteins that are present in the same complex (hence likely to physically interact) are also often implicated in the same function, namely, the function of the entire complex.

In addition to the "strong signals" for protein–protein interaction mentioned above, there are also a number of "weak signals" that are individually not predictive of protein–protein interaction at all, but are necessary conditions for a protein–protein interaction to take place and are often strongly predictive of protein–protein interaction when combined. These weak signals include: colocalization of the proteins in the same cellular compartment, coexpression of the proteins, common essentiality signature (interacting proteins tend to be both essential or both nonessential since they are functionally related), small distance on the Gene Ontology (GO) tree, and mutual clustering (interacting proteins tend to share the same other interacting partners).

A simple yet powerful method for integrating these weak features (as well as strong ones) is based on posterior odds ratios (44). First, likelihoods for the presence or absence of an interaction are computed as probability distributions of observed feature values among interacting and noninteracting protein pairs, respectively. The ratio of these likelihoods forms the likelihood ratio. Prior probabilities for the presence and absence of interaction are inferred from estimates of sparseness of the protein interaction network. The posterior odds ratio, which is the ratio of the probability of an interaction to the probability of a noninteraction, given observed feature values, is then computed as the product of the likelihood ratio and the prior odds ratio. A posterior odds ratio larger than 1 is then indicative of protein–protein interaction among the proteins in question.

Other methods for data integration include logistic regression (45, 46), random forest models (45), and support vector machines (47). These methods have been surveyed by Liu et al. (43). Furthermore, other methods exist to predict interactions among specific protein domains: these are useful for predicting interactions among novel proteins that contain those domains (48, 49).

4.3. Prediction of Genetic Interaction Networks

Systematic ab initio prediction of genetic interactions was carried out within the framework of the Flux Balance Analysis (FBA) model by Segre et al. (50). In the model, it is assumed that metabolic fluxes within a cell are at steady state and take values so as to optimize a global objective function, in this case, the growth rate of the cell (see discussion of metabolic networks in Subheading 6). Under this assumption, the steady state fluxes can be computed explicitly using linear programming. Furthermore, the effects of gene knockout – modeled by setting the fluxes corresponding to specific enzymes to zero – can be readily incorporated and their effects on growth rate studied. Thus, Segre et al. found epistatic effects among gene pairs by comparing the effect on growth rate of double knockouts with the effect of single gene knockouts of the corresponding genes. They found that, for pairs of metabolic genes in yeast, the epistasis distribution is sharply tri-model, with most gene pairs displaying no epistasis, about 100 gene pairs displaying nearly perfectly buffering epistasis (where one gene knockout completely buffers the deleterious effect of the other gene knockout), and about 200 gene pairs displaying synthetic lethal interactions. They also found striking "monochromaticity" in genetic interactions: gene pairs across two functional classes were likely to have the same type of genetic interaction (synthetic lethal or buffering or nonepistatic). This work thus systematically uncovered, using computational simulation, the extent, nature, and mechanisms behind genetic interactions among metabolic genes.

How does one predict genetic interactions among nonmetabolic genes? Most eukaryotic genes do not participate in metabolism. Also, unlike physical networks like regulatory and protein–protein interactions, genetic interactions are abstract: epistatic effects among genes may occur as a result of a long series of intermediate steps. For example, Kelley and Ideker (51) studied physical mechanisms behind genetic interactions in yeast by mapping genetic interactions to protein–protein, protein–DNA, and metabolic networks, and found that about 40% of the genetic interactions have physical explanations corresponding to both genes belonging to the same pathway or to parallel pathways, with the parallel pathway explanation being more dominant for genetic interactions identified in genome-wide screens such as SGA. Therefore, although genetic interactions are not physical, the fact that they are grounded in physical mechanisms suggests that physical interaction data should form an important component in the in silico prediction of genetic interactions.

Wong et al. (52) implemented a general, supervised method based on decision trees for predicting synthetic sick/lethal (SSL) interactions that uses a number of input features, including participation in protein–protein interaction and participation in the same protein complex. Remarkably, many of the other input features, such as gene fusion, gene neighborhood, sequence homology, and phylogenetic profiles, are in fact indicative of functional similarity and used to also predict protein–protein interactions. On hindsight, this is probably not surprising given the fact that synthetic lethality also measures a type of functional redundancy. This "functional redundancy" interpretation is further reinforced by the fact that the SSL network contains a disproportionately large number of "triangles", that is, if genes A and C have an SSL interaction and genes B and C have an SSL interaction, then it is very likely that genes A and B have SSL interaction. This property can be exploited to predict new SSL interactions, as Wong et al. did. Furthermore, since protein–protein interactions are also suggestive of functional redundancy, they found that two edges of the "triangle" could contain a mixture of SSL and protein–protein interactions in order to predict an SSL interaction for the third edge.

The basic idea of using physical interaction properties to predict SSL interaction was carried further by Paladugu et al. (53), who used a supervised support vector machine method with input features corresponding to various centrality properties of two proteins in a protein interaction network in order to predict SSL interactions between the genes. This method turns out to have comparable cross-validation accuracy to the method of Wong et al., thus suggesting that protein interaction properties have a strong, albeit complex, role in the determination of synthetic genetic interactions.

5. Analysis of Network Function: The Roles of Hubs and Modules

One of the goals of systems biology is that of extracting biologically relevant information from the wealth of genome-scale network data. For example, how may functionally important genes or proteins be identified, and how may we unravel the functional organization of the cell using this type of data? These questions have prompted the identification of the so-called "hub" proteins in protein interaction networks as being biologically special, and have also prompted the development of methods to identify modular structures in these networks that may have a number of desirable properties, including association with function, evolutionary conservation, and coexpression. These developments are briefly reviewed below.

5.1. "Hub" Proteins in Protein Interaction Networks

One of the earliest observations relevant to the topology of large protein–protein interaction networks was that they have "scale-free" degree distributions, with very few proteins having high degree and a large number of proteins having low degree. The top high-degree proteins, termed "hub" proteins, were found to have special biological properties: they tend to be enriched for essential proteins, they may be conserved to a larger extent than nonhubs, and are found to play an important role in the modular organization of the protein interaction network (54, 55). From a network integrity perspective, hubs are important because scale-free networks are robust to random removal of nodes but quickly fragment into disconnected components upon random removal of hubs (54). From the point of view of organization of the network, Han et al. (55) found that hubs could in turn be classified into the so-called "date" and "party" hubs, where party hubs have high coexpression with their interacting partners (and lie at the centers of protein modules), while date hubs have low coexpression with their interacting partners (and mediate connections between modules). Furthermore, the modules defined by party hubs were found to be *functional* modules, in the sense that proteins within a module are functionally similar, while proteins in different modules are likely to have different functions. This work has led to further extensions (56) as well as some controversy regarding the significance of the separation into date and party hubs (57–59), a controversy generated in part due to the lack of a clear definition of a degree cutoff that demarcates the boundary between hubs and nonhubs. Recent developments in this area include attempts to formulate an objective criterion for identifying hub nodes by examining the extent of agreement between different hub definitions based on network topology, essentiality, and coexpression (60).

5.2. Modularity in Biomolecular Networks

The observation that living systems are organized in modular fashion predates systems biology. Clearly, the identification of modules that may play a clean functional role in cells is important because it leads to a simplified picture of organization. The impact and role of modular organization in molecular biology was discussed in a commentary by Hartwell et al. (61), and the notions of signaling pathways and regulatory motifs as functional modules were suggested by Lauffenberger (62), and Rao and Arkin (63).

Modular organization has been found in gene expression networks (64, 65), gene regulatory networks (65, 66), metabolic pathways and protein interaction networks (67, 68).

From the point of view of topological network analysis, Newman and Girvan (69) introduced an explicit quantitative measure of modularity that is an increasing function of the intramodule density of connections and a decreasing function of the intermodule density of connections. This measure was used in conjunction with a simulated annealing method by Guimera and Amaral (70) to discover modules in metabolic networks and classify nodes by their intra- and intermodule connectivity. They found that metabolites that connect different modules tend to be far more conserved than hubs that lie entirely within modules, and later extended their method to identify certain nodes as drug targets (71). By partitioning the proteins in yeast by their age, Fernandez (72) showed that the modularity measure of the yeast protein interaction network has been increasing with evolutionary time in conjunction with a decrease in the amount of assortative mixing (propensity for hub–hub connections).

There was, in fact, earlier evidence based on computational studies that modularity must have evolved: proteins within the same age category tend to have a much larger density of interactions among themselves than with proteins in different age categories, forming "isotemporal clusters" (73). Also, randomized networks show a significantly decreased level of modularity (74, 75). Interestingly, studies of artificial networks found that the evolution of a modular network from a nonmodular one requires adaptation to "modularly" changing environments rather than static environments (74).

Evolutionary conservation can be a complementary tool to topology when identifying modules. Indeed, von Mering et al. explicitly imposed evolutionary conservation as a requirement for functional modules in *Escherichia coli* (76). These modules were identified as clusters of proteins that are functionally associated, with the associations inferred from common phylogenetic histories, conserved gene neighborhoods and gene fusions. Modules so identified were then validated successfully against known pathways and were used to predict additional pathways. On the other hand, Snel and Huynen found that the average degree of evolutionary conservation for known modules in *S. cerevisiae* and *E. coli* (like protein complexes, operons, metabolic pathways, and transcriptional modules) is low primarily due to functional differentiation of duplicate genes (75). These paralogs also contribute to the "fuzziness" in the delineation of modules. Thus, it is found that the more cohesive (less "fuzzy") modules seem to associate with ancient tasks (such as information processing), and rarely contain paralogs, while newly acquired modules enable adaptation to diverse environments (77).

Finally, we should mention that isolated approaches such as those based on topological connectivity or evolutionary conservation are often less fruitful than approaches that seek to identify modules by data integration. A case in point is the simultaneous analysis of protein interaction and gene expression data (55). Petti and Church analyzed the coregulation of pairs of known functional modules in *S. cerevisiae* by examining genes from a given pair of modules and using expression data to assess the possibility that a given transcription factor coregulates genes in both modules (78). This analysis yields a "super-network" of modules such that two modules are connected if they are coregulated by at least one transcription factor. In this analysis, modules corresponding to storage and transmission of genetic information appear as the most highly connected ones. By integrating protein interaction network data with subcellular localization and expression profiles, Lu et al. were able to distinguish between protein complexes and functional modules (defined as proteins that participate in a common cellular process), because these different module types often have different expression and localization signatures (79). Tanay et al. integrated a large number of data sources, including protein interaction, gene expression profiles, regulatory interactions, and phenotypic sensitivity into a single global bipartite network representation in which one set of nodes represents genes and another set represents properties of genes (80). Modules were then identified as locally dense clusters within this global network. They often yield a fine-grained description of a basic biological process that is not revealed by less integrative analyses. Examples include modules corresponding to vesicle transport and ubiquitin-dependent protein degradation.

6. Dynamics of Biological Networks

Life as we know it owes its existence to the ability of cells to grow and divide. The internal organization of cellular functions has evolved over millions of years to achieve maximal growth. Understanding the interconnectedness of cellular subnetworks is an important part of systems biology, and holds the key to unraveling details of functional organization in cells (81). This reflects a change in how biological systems are perceived today in contrast to initial notions of cellular organization driven by early discoveries in molecular biology (61). While our understanding of the organization of cellular functions is based on detailed studies of static networks of genes and proteins, as discussed in previous sections, real biological systems are both spatial and temporal in nature, an important example being the dynamics of cellular signaling (82). Study of the dynamics of such systems is therefore very helpful in elucidating their functional role. Often, the only recourse here is to modeling and simulation, as analytical approaches rapidly become intractable except for the simplest of models.

It has been standard practice to construct models of components of cellular systems that can be improved and refined by means of experimental observations. Such models are representative of the actual biochemical reactions occurring as a part of more complex reaction networks in various pathways, and are therefore simplifications of the original systems they represent. However, in many cases, such simplified models are sufficient to describe the dynamics of many processes involved – for example, a model of intracellular signaling networks that describe binding of ligands to Epidermal Growth Factor (EGF) receptors (83, 84) and result in the activation of downstream proteins in the signaling cascade. Another example is a recent model of the transcriptional dynamics of embryonic stem cells (85), describing a bistable switch that arises due to positive feedback loops and switches on or off by environmental signals. A number of other such models are stored in a database that can be accessed at the internet resource http://www.biomodels.net (86).

6.1. Kinetics of Biochemical Reactions

Biological network models are representations of complex cellular processes that involve the participation of numerous distinct molecules (reactants) binding to each other to form new species of molecules (products). Strictly speaking, both forward (binding) and backward reactions can occur at the same time, with the dominant direction determined by thermodynamics. In most cases, these reactions are catalyzed by enzymes that are, however, not consumed in the reaction. Modeling the biochemical reaction network therefore requires addressing the issue of enzyme kinetics. There is extensive literature devoted to this field, in particular to the Michaelis–Mentin formulation, which is rooted in basic biochemistry (87, 88).

One of the stumbling blocks in extending the success of modeling and simulation to large models of biological systems deals with the lack of kinetics for many reactions occurring in vivo. In cases when experimental data for the missing kinetics is available, parameter estimation methods based on optimization can be used (89) to fit parameters to experimental data. This is indeed an active area of research as the kinetics of most cellular reactions are unknown (90). Most of the approaches used for parameter estimation are based on global optimization methods (91). An alternative approach involves the use of Metabolic Control Analysis (MCA) (92), a theoretical framework that provides a quantitative description of substrate flux in response to changes in system parameters. Although MCA was developed over two decades ago, it is closely linked to the functional genomics of today and has been shown to have implications for drug discovery and disease as it can help identify candidate enzymes in pathways that can be suitable targets for cancer drugs (93). The advantage of using MCA to approximate kinetics of reactions lies in the fact that MCA provides local (elasticities) and global (control coefficients) descriptors of the parameter sensitivity of reaction rates. These descriptors can be used to construct approximate reaction rates using the LinLog approximation (94). This is a better approximation to the nonlinear Michaelis–Mentin kinetics than a linearized approximation around a system steady state value.

6.2. Dynamics of Biochemical Networks: Simulation

For a given biochemical network, based on the discussion in the previous section, all kinetics can be assigned, either by suitable assumptions, or by estimating the parameters in the rate law, or by taking recourse to approximations such as the LinLog formulation. This completely characterizes the description of the network model that involves many reactants participating in a large number of reactions. If the number of molecules for each of the species is sufficiently large, so as to avoid low copy numbers (which results in stochastic effects – see below), a deterministic formulation of the system suffices to describe the time evolution of the network dynamics. In this formulation, the time rate of change of each species is described by Ordinary Differential Equations (ODEs). This rate of change for each species in the network has to equal the sum total of synthesis and degradation of that species. At steady state, the time rate of change for all species is zero.

The steady state value of concentrations has important biological implications. Each steady state value can be shown to be in stable or unstable equilibrium based on the behavior of the system close to the steady state value when slightly perturbed. In case of a stable equilibrium, all trajectories converge back to steady state, while in the case of an unstable equilibrium, they diverge away from it. A number of tools have been developed to allow researchers to build, model, and analyze networks of such systems. Prominent among these are the Systems Biology Workbench (SBW) (95), PySCeS (96) and COPASI (97). These and other software platforms provide simulators that can load models in the Systems Biology Markup Language (SBML) format (98), perform analyses, and carry out simulations. Given the large number of simulators available for this purpose, a recent study was designed to compare all popular simulators (99).

Results from simulations can provide meaningful information when network sizes are small, or if the models are not complex. Here, complexity is defined by the lack of numerical convergence of the simulated solutions. This often arises as a result of ill-conditioned systems where there are fast variables in the system along with slow variables. Most simulators find it difficult to simulate such systems, and often one has to apply techniques to reduce the dimensionality of the system (100, 101).

In the case of networks with a large number of species, there are two ways in which the speed of the simulation can be increased. The first is by means of removing dependent species and simulating only the independent species, which is part of the Structural Analysis problem and will be discussed in the next section. The second means of model reduction involves elimination of the fast variables by projecting them onto the slow variables, and simulating only those slow variables as mentioned above.

6.3. Structural Analysis of Biochemical Network Models

While some features of the system can be found after performing extensive simulations, often, many properties can be deduced by analyzing the underlying stoichiometry matrix of the network. This matrix relates all participants (species) in the network to the reactions they are part of. In particular, the value of each element s_{ij} of the stoichiometry matrix S is the number of molecules of the *i*th species (row) taking part in the *j*th reaction (column). The analysis of the stoichiometry matrix is known as structural analysis, and reveals the structural properties of the network. These structural properties arise due to dependencies in the stoichiometry matrix.

A number of methods are used to extract the structural properties, and are based on matrix factorization approaches. In particular, the analysis can be carried out on the stoichiometry matrix or its transpose, which are not equivalent and the results have different biological interpretations. The first analysis (on S) leads to extraction of the nullspace of the stoichiometry matrix. The rows of the nullspace are steady state solutions and correspond to unique pathways through the network going from a source to a sink in the model. Some of these can also include closed paths, which indicates circulating fluxes. The nullspace is also useful in computing elementary pathways and forms a basis for all the fluxes in the network. Consequently, any realized flux distribution can be projected onto the elementary pathways.

On the other hand, the results obtained by analyzing the transpose of the stoichiometry matrix are known as conserved moieties, and the approach is known as conservation analysis. This approach uncovers dependencies in the network that cause some species concentrations to depend on others and is a consequence of the connectivity inherent in the network. In other words, conservation analysis identifies independent species, or the *rank*, of the stoichiometry matrix. The independent species can be simulated directly ignoring the rest of the species whose concentrations can be reconstructed later, thus speeding up the simulation. The dependent species are related to the independent species by means of a link matrix whose mathematical details can be found in the literature (102, 103).

A number of methods are available to compute conserved moieties and the nullspace for small networks. Among them are the traditionally used Gaussian Elimination, the Row Echelon and the Singular Value Decomposition methods. However, these methods face numerical accuracy issues for large networks, owing to the accumulation of round-off errors. More recently, a new and robust method based on the Householder QR decomposition of the stoichiometry matrix has been developed (103) which has been shown to be capable of carrying out structural analysis on very large networks, such as the whole-genome metabolic network models which could not be analyzed accurately by earlier methods. The Householder QR-based method for structural analysis of biochemical networks has now been implemented by SBW, PySCeS and COPASI, mainly due to its robustness in handling large networks. The importance of the conservation analysis algorithm is that it enables the computation of a nonsingular Jacobian (using the independent species), which in turn can be used to compute control coefficients and elasticities required by Metabolic Control Analysis. Secondly, a nonsingular Jacobian is also necessary in bifurcation studies that decipher how a large biochemical network will behave in the neighborhood of a steady state (104). The tools developed as part of this algorithm, including SBML translators to other languages such as Matlab, C, C#, and Java, are accessible through the SBW platform and support the extraction of independent species (105).

6.4. Modeling Metabolic Networks

An important aspect of biological systems is their ability to convert nutrients to products usable for biosynthesis and growth. This normal cellular operation results in a particular distribution of metabolic fluxes, and is known as the wild type flux distribution. This distribution undergoes a change when there is a change in the nutrients or other perturbations that cause one or more enzymes to be deactivated. This change results from flux being routed through alternate pathways. As opposed to the discussion of prediction of genetic interactions among metabolic genes, the organism in this case cannot be expected to maximize biomass for growth, as it does not satisfy the wild type conditions.

In the ideal case, when kinetics of all reactions in the network are known, the new fluxes can be obtained by means of direct simulation of the network of ODEs using the changed boundary conditions. However, often, the kinetics of many reactions, especially for whole-cell-based models, are unknown. Hence researchers have resorted to using constraint-based optimization techniques to find a new set of fluxes that satisfy the new conditions. This is known as the Flux Balance Analysis method (FBA) (106, 107), which has a number of applications in bioprocessing, discovery of antibiotics, and other areas. Depending on the optimization algorithm used, it is likely that the flux distribution will appear markedly different from the wild type distribution, especially in the case of very large networks. Further, in the case of second order effects such as additional gene deletion or mutation, there are alternatives such as Minimization of Metabolic Adjustment (MoMA) (108), which solves a quadratic optimization to find a new flux distribution that is close to the original mutant's distribution in the flux space. There are also other alternatives such as Regulatory On/Off Minimization (ROOM) (109) that integrates a Boolean regulatory network to find the new flux distributions.

Determination of fluxes in biological networks continues to be an active area of research that has important implications for the drug discovery process (110). Metabolism in general is a wide area that impacts all living organisms from single-celled bacteria to more complex life such as plants and animals. There have been recent initiatives to build on the current knowledge-base and develop means of modeling metabolism of the entire plant (111) as plants are important sources of food and fuel. Increased understanding of plant metabolism will help grow strains that are more drought and pest resistant and have high yield. However, the technical challenges involved are significant, and there are efforts to develop methodologies and computational tools that can cope with the immense combinatorial complexity of fluxes in many pathways by attempting to solve the problem with parallel computing (112).

6.5. Low Copy Numbers and Stochastic Simulation

Stochasticity has been identified as playing an important role in biological networks (113). Simulation tools are a valuable aid to experimental methods in studying the relationship between phenotypic variations and fluctuations of important chemical species at the molecular scale. Often these species exist in small numbers, yet regulate others through many intricate, hierarchical chains of feedback. A number of algorithms have been devised with the goal of modeling the role of stochasticity in biological networks, primary among them being the Gillespie algorithm (114, 115). The time evolution of the complete system in the case of randomly fluctuating numbers of participating species is described by a master equation (116, 117). Obtaining a solution to the master equation is often a difficult task: while solutions of systems with one or two species can be obtained by hand, the problem rapidly becomes intractable for systems with a larger number of participants. The Gillespie algorithm provides an exact solution to the master equation by means of the Stochastic Simulation Algorithm (SSA), which was subsequently modified to increase the simulation speed. Further advances in stochastic simulation of biochemical networks have been addressed elsewhere (118). There are also a number of software tools available for stochastic simulation (119–123). In addition, SBW (95) and COPASI (97) support stochastic simulation.

Acknowledgments

The authors would like to acknowledge support provided by the US National Science Foundation grant FIBR 0527023.

References

1. Kitano H (2002) Systems biology: a brief overview. Science 295:1662–1664
2. Ekins R, Chu FW (1999) Microarrays: their origins and applications. Trends Biotechnol 17(6):217–218
3. Brown PO, Botstein D (1999) Exploring the new world of the genome with DNA microarrays. Nat Genet 21:33–37
4. DeRisi JL, Iyer VR, Brown PO (1997) Exploring the metabolic and genetic control of gene expression on a genomic scale. Science 278:680–686
5. Nadon R, Shoemaker J (2003) Statistical issues with microarrays: processing and delays. Trends Genet 18:265–271
6. Quackenbush J (2002) Microarray data normalization and transformation. Nat Genet 32:496–501
7. Leung YF, Cavalieri D (2003) Fundamentals of cDNA microarray data analysis. Trends Genet 19:649–659
8. Allison DB, Cui X, Page GP, Sabripour M (2006) Microarray data analysis: from disarray to consolidation and consensus. Nat Rev Genet 7:55–65
9. Marshall E (2004) Getting the noise out of gene arrays. Science 306:630–631
10. Troyanskaya O, Cantor M, Sherlock G, Brown PO, Hastie T et al (2001) Missing value estimation methods of DNA microarrays. Bioinformatics 17:520–525
11. Eisen MB, Spellman PT, Brown PO, Botstein D (1998) Cluster analysis and display of genome-wide expression patterns. Proc Natl Acad Sci USA 95(25):14863–14868
12. Quackenbush J (2001) Computational analysis of microarray data. Nat Rev Genet 2:418–427
13. Schena M, Shalon D, Davis RW, Brown PO (1995) Quantitative monitoring of gene expression patterns with a complimentary DNA microarray. Science 270(5235):467–470
14. Duggan DJ, Bittner M, Chen Y, Meltzer P, Trent JM (1999) Expression profiling using cDNA microarrays. Nat Genet 21:10–14
15. Hacia JG (1999) Resequencing and mutational analysis using oligonucleotide microarrays. Nat Genet 21:42–47
16. Haab BB, Dunham MJ, Brown PO (2001) Protein microarrays for highly parallel detection and quantitation of specific proteins and antibodies in complex solutions. Genome Biol 2:1–13
17. Debouck C, Goodfellow PN (1999) DNA microarrays in drug discovery and development. Nat Genet 21:48–50
18. DeRisi JL, Penland L, Brown PO, Bittner ML, Meltzer PS et al (1996) Use of a cDNA microarray to analyze gene expression patterns in human cancer. Nat Genet 14:457–460
19. Michiels S, Koscielny S, Hill C (2005) Prediction of cancer outcome with microarrays: a multiple random validation strategy. Lancet 365:488–492
20. Ma J, Ptashne M (1988) Converting a eukaryotic transcriptional inhibitor into an activator. Cell 55:443–446
21. Fields S, Song OK (1989) A novel genetic system to detect protein–protein interactions. Nature 40:245–246
22. Phizicky EM, Fields S (1995) Protein–protein interactions: methods for detection and analysis. Microbiol Rev 59:94–123
23. Puig O et al (2001) The tandem affinity purification (TAP) method: a general procedure of protein complex purification. Methods 24:218–229
24. Solomon MJ, Varshavsky A (1985) Formaldehyde-mediated DNA-protein crosslinking: a probe for in vivo chromatin structures. Proc Natl Acad Sci USA 82:6470–74
25. Tong AH et al (2001) Systematic genetic analysis with ordered arrays of yeast deletion mutants. Science 294:2364–2368
26. Tong AH et al (2004) Global mapping of the yeast genetic interaction network. Science 303:808
27. Schuldiner M, Collins SR, Weissman JS, Krogan NJ (2006) Quantitative genetic analysis in *Saccharomyces cerevisiae* using epistatic miniarray profiles (E-MAPs) and its application to chromatin functions. Methods 40:344–352
28. Huang H, Jedynak BM, Bader JS (2007) Where have all the interactions gone?

Estimating the coverage of two-hybrid protein interaction maps. PLoS Comput Biol 3:e214
29. Deeds EJ, Ashenberg O, Shakhnovich EI (2006) A simple physical model for scaling in protein–protein interaction networks. Proc Natl Acad Sci USA 103:311–316
30. Przulj N, Higham DJ (2006) Modeling protein–protein interaction networks via a stickiness index. J R Soc Interface 3:711–716
31. D'haeseleer P, Church GM (2004) Estimating and improving protein interaction error rates. Proc IEEE Comput Syst Bioinform Conf, 216–223
32. Hart GT, Ramani AK, Marcotte EM (2006) How complete are current yeast and human protein-interaction networks? Genome Biol 7:120
33. Huang H, Bader JS (2009) Precision and recall estimates for two-hybrid screens. Bioinformatics 25:372–378
34. van Someren EP, Wessels LF, Backer E, Reinders MJ (2002) Genetic network modeling. Pharmacogenomics 3:507–525
35. Margolin AA, Califano A (2007) Theory and limitations of genetic network inference from microarray data. Ann N Y Acad Sci 1115:51–72
36. Liang S, Fuhrman S, Somogyi R (1998) Reveal, a general reverse engineering algorithm for inference of genetic network architectures. Pac Symp Biocomput 1998:18–29
37. Murphy K, Mian S (1999) Modeling gene expression data using dynamic Bayesian networks. Technical report. University of California, Berkeley
38. Perrin B-E, Ralaivola L, Mazurie A, Bottani S, Mallet J, d'Alche Buc F (2003) Gene networks inference using dynamic Bayesian networks. Bioinformatics 19:S138–S148
39. Rangel C, Angus J, Ghahramani Z, Lioumi M, Sotheran E, Gaiba A, Wild DL, Falciani F (2004) Modelling T-cell activation using gene expression profiling and state space models. Bioinformatics 20:1361–1372
40. Beal MJ, Falciani F, Ghahramani Z, Rangel C, Wild DL (2005) A Bayesian approach to reconstructing genetic regulatory networks with hidden factors. Bioinformatics 21:349–356
41. Zou M, Conzen SD (2005) A new dynamic Bayesian network (DBN) approach for identifying gene regulatory networks from time course microarray data. Bioinformatics 21:71–79
42. Skrabanek L, Saini HK, Bader GD, Enright EJ (2008) Computational prediction of protein–protein interactions. Mol Biotechnol 38:1–17
43. Liu Y, Kim I, Zhao H (2008) Protein interaction predictions from diverse sources. Drug Discov Today 13:409–416
44. Jansen R, Yu H, Greenbaum D, Kluger Y, Krogan NJ et al (2003) A Bayesian networks approach for predicting protein–protein interactions from genomic data. Science 302:449–453
45. Lin N, Wu B, Jansen R, Gerstein M, Zhao H (2004) Information assessment on predicting protein–protein interactions. BMC Bioinform 5:154
46. Sharan R, Suthram S, Kelley RM, Kuhn T, McCuine S et al (2005) Conserved patterns of protein interaction in multiple species. Proc Natl Acad Sci USA 102:1974–1979
47. Qi Y, Bar-Joseph Z, Klein-Seetharaman J (2006) Evaluation of different biological data and computational classification methods for use in protein interaction prediction. Proteins 63:490–500
48. Deng M, Mehta S, Sun F, Chen T (2002) Inferring domain-domain interactions from protein–protein interactions. Genome Res 12:1504–1508
49. Kim I, Liu Y, Zhao H (2007) Bayesian methods for predicting interacting protein pairs using domain information. Biometrics 63:824–833
50. Segrè D, DeLuna A, Church GM, Kishony R (2004) Modular epistasis in yeast metabolism. Nat Genet 37:77–83
51. Kelley R, Ideker T (2005) Systematic interpretation of genetic interactions using protein networks. Nat Biotechnol 23:561–566
52. Wong SL, Zhang LV, Tong AHY, Li Z, Goldberg D et al (2004) Combining biological networks to predict genetic interactions. Proc Natl Acad Sci USA 101:15682–15687
53. Paladugu SR, Zhao S, Ray A, Raval A (2008) Mining protein networks for synthetic genetic interactions. BMC Bioinform 9:426
54. Albert R, Jeong H, Barabasi AL (2000) Error and attack tolerance of complex networks. Nature 406:378–382
55. Han JD, Bertin N, Hao T, Goldberg DS, Berriz GF et al (2004) Evidence for dynamically organized modularity in the yeast protein–protein interaction network. Nature 430:88–93
56. Jin G, Zhang S, Zhang XS, Chen L (2007) Hubs with network motifs organize modularity dynamically in the protein–protein interaction network of yeast. PLoS One 2:e1207

57. Batada NN, Reguly T, Breitkreutz A, Boucher L, Breitkreutz BJ et al (2006) Stratus not altocumulus: a new view of the yeast protein interaction network. PLoS Biol 4:1720–1731
58. Batada NN, Reguly T, Breitkreutz A, Boucher L, Breitkreutz BJ et al (2007) Still stratus not altocumulus: further evidence against the date/party hub distinction. PLoS Biol 5:e154
59. Bertin N, Simonis N, Dupuy D, Cusick ME, Han JDJ et al (2007) Confirmation of organized modularity in the yeast interactome. PLoS Biol 5:e153
60. Vallabhajosyula RR, Chakravarti D, Lutfeali S, Ray A, Raval A (2009) Identifying hubs in protein interaction networks. PLoS One 4:e5344
61. Hartwell LH, Hopfield JJ, Liebler S, Murray AW (1999) From molecular to modular cell biology. Nature 402:C47–C52
62. Lauffenberger DA (2000) Cell signaling pathways as control modules: complexity for simplicity? Proc Natl Acad Sci USA 97:5031–5033
63. Rao CV, Arkin AP (2001) Control motifs for intracellular regulatory networks. Annu Rev Biomed Eng 3:391–419
64. Ihmels J, Friedlander G, Bergmann S, Sarig O, Ziv Y, Barkai N (2002) Revealing modular organization in the yeast transcriptional network. Nat Genet 31:370–377
65. Segal E, Shapira M, Regev A, Pe'er D, Botstein D et al (2003) Module networks: identifying regulatory modules and their condition-specific regulators from gene expression data. Nat Genet 34:166–176
66. Tavazoie S, Hughes JD, Campbell MJ, Cho RJ, Church GM (1999) Systematic determination of genetic network architecture. Nat Genet 22:281–285
67. Rives AW, Galitski T (2003) Modular organization of cellular networks. Proc Natl Acad Sci USA 100:1128
68. Ravasz E, Somera AL, Mongru DA, Oltvai ZN, Barabasi AL (2002) Hierarchical organization of modularity in metabolic networks. Science 297:1551–1555
69. Newman MEJ, Girvan M (2004) Finding and evaluating community structure in networks. Phys Rev E Stat Nonlin Soft Matter Phys 69:026113
70. Guimera R, Amaral LAN (2005) Functional cartography of complex metabolic networks. Nature 433:895–900
71. Guimera R, Sales-Pardo M, Amaral LAN (2007) A network-based method for target selection in metabolic networks. Bioinformatics 23:1616–1622
72. Fernandez A (2007) Molecular basis for evolving modularity in the yeast protein in the yeast protein interaction network. PLoS Comput Biol 3:e226
73. Qin H, Lu HHS, Wu WB, Li W-H (2003) Evolution of the yeast protein interaction network. Proc Natl Acad Sci USA 100:12820
74. Kashtan N, Alon U (2005) Spontaneous evolution of modularity and network motifs. Proc Natl Acad Sci USA 102:13773–13778
75. Snel B, Huynen MA (2004) Quantifying modularity in the evolution of biomolecular systems. Genome Res 14:391–397
76. von Mering C, Zdobnov EM, Tsoka S, Ciccarelli FD, Pereira-Leal JB et al (2003) Genome evolution reveals biochemical networks and functional modules. Proc Natl Acad Sci USA 100:15428–15433
77. von Campillos M, Mering C, Jensen LJ, Bork P (2006) Identification and analysis of evolutionary cohesive functional modules in protein networks. Genome Res 16:374–382
78. Petti AA, Church GM (2005) A network of transcriptionally coordinated functional modules in *Saccharomyces cerevisiae*. Genome Res 15:1298–1306
79. Lu H, Shi B, Wu G, Zhang Y, Zhu X et al (2006) Integrated analysis of multiple data sources reveals modular structure of biological networks. Biochem Biophys Res Commun 345:302–309
80. Tanay A, Sharan R, Kupiec M, Shamir R (2004) Revealing modularity and organization in the yeast molecular network by integrated analysis of highly heterogenous genomewide data. Proc Natl Acad Sci USA 101:2981–2986
81. Barabasi AL (2004) Network biology: understanding the cell's functional organization. Nat Rev Genet 5:101–113
82. Kholodenko BN (2006) Cell-signalling dynamics in time and space. Nat Rev Mol Cell Biol 7:165–176
83. Schoeberl B, Eichler-Jonsson C, Gilles ED, Muller G (2002) Computational modeling of the dynamics of the MAP kinase cascade activated by surface and internalized EGF receptors. Nat Biotechnol 20:370–375
84. Wiley SH, Shvartsman Y, Lauffenburger DA (2003) Computational modeling of the EGF-receptor system: a paradigm for systems biology. Trends Cell Biol 13:43–50
85. Chickarmane V, Troein C, Nuber UA, Sauro HM, Peterson C (2006) Transcriptional

dynamics of the embryonic stem cell switch. PLoS Comput Biol 2(9):1080–1092
86. Le Nov`ere N, Bornstein B, Broicher A, Courtot M, Donizelli M (2006) BioModels Database: a free, centralized database of curated, published, quantitative kinetic models of biochemical and cellular systems. Nucleic Acids Res 34:689–691
87. Segel IH (1975) Enzyme kinetics. Wiley, New York
88. Cornish-Bowden A (1979) Fundamentals of enzyme kinetics. Butterworths, London and Boston
89. Mendes P, Kell D (1998) Non-linear optimization of biochemical pathways: applications to metabolic engineering and parameter estimation. Bioinformatics 14:869–883
90. Ko CL, Voit EO, Wang FS (2009) Estimating parameters for generalized mass action models with connectivity information. BMC Bioinform 10:140
91. Moles CG, Mendes P, Banga JR (2003) Parameter estimation in biochemical pathways: a comparison of global optimization methods. Genome Res 13:2467–2474
92. Fell DA (1992) Metabolic control analysis: a survey of its theoretical and experimental development. Biochem J 286(2):313–330
93. Cascante M, Boros LG, Comin-Anduix B, Atauri P, Centelles JJ, Lee PWN (2002) Metabolic control analysis in drug discovery and disease. Nat Biotechnol 20:243–249
94. Wu L, Wang W, van Winden WA, van Gulik WM, Heinjen JJ (2004) A new framework for the estimation of control parameters in metabolic pathways using lin-log kinetics. FEBS J 271:3348–3359
95. Bergmann F, Sauro HM (2006) SBW – a modular framework for systems biology. In: Proceedings of the 38th conference on Winter simulation, Monterey, CA, USA, 1637–1645
96. Olivier BG, Rohwer JM, Hofmeyr HS (2005) Modeling cellular systems with PySCeS. Bioinformatics 21:560–561
97. Hoops S, Sahle S, Gauges R, Lee C, Pahle J et al (2006) COPASI – a COmplex PAthway SImulator. Bioinformatics 22:3067–3074
98. Hucka M, Finney A, Sauro HM, Bolouri H, Doyle JC et al (2003) The systems biology markup language (SBML): a medium for representation and exchange of biochemical network models. Bioinformatics 19(4):524–531
99. Bergmann F, Sauro HM (2008) Comparing simulation results of SBML capable simulators. Bioinformatics 24:1963–1965
100. Vallabhajosyula RR, Sauro HM (2006) Complexity reduction in biochemical networks. In: Proceedings of the 38th conference on Winter simulation, Monterey, CA, USA, 1690–1697
101. Surovtsova I, Sable S, Pahle J, Kummer U (2006) Approaches to complexity reduction in a Systems Biology Research Environment. In: Proceedings of the 38th conference on Winter simulation, Monterey, CA, USA, 1683–1689
102. Sauro HM, Ingalls B (2004) Conservation analysis in biochemical networks: computational issues for software writers. Biophys Chem 109(1):1–15
103. Vallabhajosyula RR, Chickarmane V, Sauro HM (2006) Conservation analysis of large biochemical networks. Bioinformatics 22:346–353
104. Chickarmane V, Paladugu SR, Bergmann F, Sauro HM (2005) Bifurcation discovery tool. Bioinformatics 21(18):3688–3690
105. Bergmann F, Vallabhajosyula RR, Sauro HM (2006) Computational tools for modeling protein networks. Curr Proteomics 3(3):181–197
106. Varma A, Palsson BO (1994) Metabolic flux balancing: basic concepts, scientific and practical use. Nat Biotechnol 12:994–998
107. Raman K, Chandra N (2009) Flux balance analysis of biological systems: applications and challenges. Brief Bioinform 10:435–449
108. Segrè D, Vitkup D, Church GM (2002) Analysis of optimality in natural and perturbed metabolic networks. Proc Natl Acad Sci USA 99:15112–15117
109. Shlomi T, Berkman O, Ruppin E (2005) Regulatory on/off minimization of metabolic flux changes after genetic perturbations. Proc Natl Acad Sci USA 102:7695–7700
110. Kell D (2006) Systems biology, metabolic modeling and metabolomics in drug discovery and development. Drug Discov Today 11:1085–1092
111. Sweetlove LJ, Last RL, Fernie AR (2003) Predictive metabolic engineering: a goal for systems biology. Plant Physiol 132:420–425
112. Ballarini P, Guido R, Mazza T, Prandi D (2009) Taming the complexity of biological pathways through parallel computing. Brief Bioinform 10(3):278–288
113. Arkin AP, Ross J, McAdams HH (1998) Stochastic kinetic analysis of developmental pathway bifurcation in Phage λ-infected *Escherichia coli* cells. Genetics 149:1633–1648

114. Gillespie DT (1976) A general method for numerically simulating the stochastic time evolution of coupled chemical species. J Comput Phys 22:403–434
115. Gillespie DT (1977) Exact stochastic simulation of coupled chemical reactions. J Phys Chem 81:2340–2361
116. van Kampen NG (1992) Stochastic processes in physics and chemistry. NHPL, Elsevier Science
117. Wilkinson DJ (2006) Stochastic modelling for systems biology. Chapman and Hall, CRC Press, Boca Raton, Florida, USA
118. Meng TC, Somani S, Dhar P (2004) Modeling and simulation of biological systems with stochasticity. In Silico Biol 4:293–309
119. Adalsteinsson D, McMillen D, Elston TC (2004) Biochemical network stochastic simulator (BioNetS): software for stochastic modeling of biochemical networks. BMC Bioinform 5:24
120. Li H, Cao Y, Petzold LR, Gillespie DT (2008) Algorithms and software for stochastic simulation of biochemical reacting systems. Biotechnol Prog 24(1):56–61
121. Ramsey S, Orrell D, Bolouri H (2005) Dizzy: stochastic simulation of large-scale genetic regulatory networks. J Bioinform Comput Biol 13:49
122. Ullah M, Schmidt H, Cho K-H, Wolkenhauer O (2006) Deterministic modeling and stochastic simulation of biochemical pathways using MATLAB. Syst Biol 153:53–60
123. Vallabhajosyula RR, Sauro HM (2007) A stochastic simulation GUI for biochemical networks. Bioinformatics 23:1859–1861

Chapter 6

An Introduction to Gaussian Bayesian Networks

Marco Grzegorczyk

Abstract

The extraction of regulatory networks and pathways from postgenomic data is important for drug discovery and development, as the extracted pathways reveal how genes or proteins regulate each other. Following up on the seminal paper of Friedman et al. (J Comput Biol 7:601–620, 2000), Bayesian networks have been widely applied as a popular tool to this end in systems biology research. Their popularity stems from the tractability of the marginal likelihood of the network structure, which is a consistent scoring scheme in the Bayesian context. This score is based on an integration over the entire parameter space, for which highly expensive computational procedures have to be applied when using more complex models based on differential equations; for example, see (Bioinformatics 24:833–839, 2008). This chapter gives an introduction to reverse engineering regulatory networks and pathways with Gaussian Bayesian networks, that is Bayesian networks with the probabilistic BGe scoring metric [see (Geiger and Heckerman 235–243, 1995)]. In the BGe model, the data are assumed to stem from a Gaussian distribution and a normal-Wishart prior is assigned to the unknown parameters. Gaussian Bayesian network methodology for analysing static observational, static interventional as well as dynamic (observational) time series data will be described in detail in this chapter. Finally, we apply these Bayesian network inference methods (1) to observational and interventional flow cytometry (protein) data from the well-known RAF pathway to evaluate the global network reconstruction accuracy of Bayesian network inference and (2) to dynamic gene expression time series data of nine circadian genes in *Arabidopsis thaliana* to reverse engineer the unknown regulatory network topology for this domain.

Key words: Systems biology, Regulatory networks, Signalling pathways, Bayesian networks, Gaussian networks, Bayesian inference

1. Introduction

In systems biology, there has been increased interest in learning regulatory networks and signalling pathways from postgenomic data. Following up on the seminal paper of Friedman et al. (*1*), Bayesian networks have been widely applied as a popular tool to this end. Their popularity stems from the tractability of the marginal likelihood of the network structure, which is a consistent scoring

Qing Yan (ed.), *Systems Biology in Drug Discovery and Development: Methods and Protocols*, Methods in Molecular Biology, vol. 662, DOI 10.1007/978-1-60761-800-3_6, © Springer Science+Business Media, LLC 2010

scheme in the Bayesian context. This score is based on an integration over the entire parameter space, for which highly expensive computational procedures have to be applied when using more complex models based on differential equations; for example, see (*2*). To obtain the closed-form expression of the marginal likelihood referred to above, two probabilistic models with their respective conjugate prior distributions have been employed in the past: the multinomial distribution with the Dirichlet prior, leading to the so-called BDe score (*3*), and the linear Gaussian distribution with the normal-Wishart prior, leading to the BGe score (*4*). These approaches are restricted in that they either require the data to be discretised (BDe) or can only capture linear regulatory relationships (BGe). In this chapter, we focus on the Gaussian BGe score and the inference follows the Bayesian paradigm, that is, the topology of the unknown network is inferred by a Markov Chain Monte Carlo (MCMC) sampling scheme for network structures. Such a "Bayesian model averaging" approach is preferable to alternative approaches that search for one single best network with the highest posterior probability, for example, by a greedy-search algorithm. Usually, the data sets are sparse so that there is intrinsic uncertainty about the interactions. Consequently, there is a large amount of valid networks which do explain the data (approximately) equally well. A "Bayesian model averaging" approach aims to learn only those interactions that can be found in most of the high-scoring networks, that is, it aims to extract only those interactions (edge connections) that have a high marginal posterior probability.

The extraction of regulatory networks and pathways is important for drug discovery and development, as the extracted pathways reveal how genes or proteins regulate each other. That is, it can be seen from the extracted pathway which genes interact with each other, and so which ones may be "good" candidates as drug targets, as their down- or up-regulation by a drug may also have an effect on many other agents (which are regulated by them). On the other hand, it can be seen which genes do not interact with the others or can be found downstream in the network without any regulating effect on others. Loosely speaking, genes or proteins which do not have any effect on other agents are less likely to be "good" candidates as drug targets.

2. Bayesian Network Methodology

In this section, an introduction to the principles of Bayesian network methodology is given. Subheading 2.1 deals with the fundamentals of static and dynamic Bayesian networks. Markov Chain Monte Carlo (MCMC) inference using the structure MCMC sampler of

Madigan and York (*5*) is discussed in Subheading 2.2. How to compute edge posterior probabilities and AUC values from the MCMC sample is explained in Subheading 2.3. The Gaussian BGe scoring metric (Bayesian metric for Gaussian networks having scoring equivalence) of Geiger and Heckerman (*4*) for (1) static observational, (2) static interventional, and (3) dynamic time series data is presented in detail in Subheading 3.

2.1. Static and Dynamic Bayesian Networks

Static Bayesian networks (BNs) are interpretable and flexible models for representing probabilistic relationships among interacting variables (e.g., genes or proteins). At a qualitative level, the graph of a BN describes the relationships between the variables $X_1, \ldots, X_N$ in the form of conditional (in-)dependence relations. At a quantitative level, local relationships among variables are described by conditional probability distributions. Formally, a BN is defined by a graph *G*, a family of conditional probability distributions, and their parameters q, which together specify the joint probability distribution over the variables $P(X_1,\ldots,X_N|G,q)$.

The graph *G* of a BN consists of *N* nodes, representing the variables $X_1, \ldots, X_N$, and a set of directed edges connecting the nodes. The set of directed edges indicates conditional (in-)dependence relations. If there is a directed edge pointing from node X_i to node X_j, symbolically: $X_i \rightarrow X_j$, then X_i is called a *parent (node)* of X_j, and X_j is called a *child* (node) of X_i. The *parent (nodes) set* of a node X_n, symbolically π_n, is defined as the set of all parent nodes of X_n, that is the set of all nodes from which an edge points to X_n in *G*. We say that a node X_n is *orphaned* if it has an empty parent set: $\pi^n = \{\ \}$. If a node X_k can be reached by following a path of directed edges starting at node X_i, then X_k is called a *descendant (node)* of X_i, and X_i is called an *ancestor (node)* of X_k. The topology of a *static* Bayesian network is defined to be a DAG, that is, a directed graph in which no node can be its own descendant or ancestor. Graphically, this means that there are no cycles of directed edges (loops) in DAGs. For example, in the DAG shown in the left panel of Fig. 1, there are five domain variables,

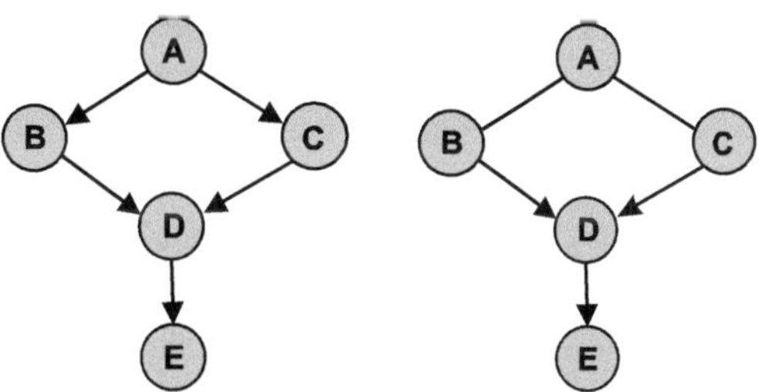

Fig. 1. Example of a Bayesian network with five nodes. *Left panel*: The DAG with five nodes and five directed edges. *Right panel*: The completed partially directed acyclic graph (CPDAG) representation of the DAG shown in the left panel. Two edges of the graph are reversible. These edges are replaced by undirected edges in the CPDAG representation.

symbolically A, …, E. Node A is a parent (node) of both nodes B and C. Nodes B and C are child nodes of A. Since there are paths of directed edges leading from node A to node , for example: $A \to B \to D \to E$, node A is an ancestor (node) of E. The parent set of node A is the empty set, the parent set of D is given by $\pi_D = \{B,C\}$, since the graph possesses the two edges: $B \to D$ and $C \to D$ and there is no other edge pointing to node D. The joint probability distribution in static BNs factorises as follows:

$$P(X_1,.,X_N \mid G,q) = \prod_{i=1}^{N} P(X_i \mid \pi_i, q_i) \tag{1}$$

where $q = (q_1,\cdots,q_N)$ is a vector of unknown parameters, and the parent node sets π_i are implied by the graph G, symbolically: $\pi_i = \pi_i(G)$. Thus, each DAG implies a set of conditional (in-)dependence relations for static BNs. These relations give a unique factorisation of the joint probability distribution. In the factorisation, each node X_i depends on its parent nodes π_i only, and the parameter vector q consists of N sub-vectors, such that each sub-vector q_i specifies the local probability distribution $P(X_i \mid \pi_i, q_i)$. For example, the graph (DAG) shown in the left panel of Fig. 1 implies the following factorisation

$$\begin{aligned} P(A,\cdots,E \mid G,q) = {} & P(A \mid q_A)\cdot P(B \mid A,q_B)\cdot P(C \mid A,q_C) \\ & \cdot P(D \mid \{A,B\},q_D)\cdot P(E \mid D,q_E) \end{aligned}$$

where $q = (q_A,q_B,q_C,q_D,q_E)$. More than one DAG can imply the same set of conditional independencies and if two DAGs assign the same set of conditional independencies assumptions, those DAGs are said to be *equivalent*. This relation of graph equivalence imposes a set of *equivalence classes* over DAGs. The DAGs within an equivalence class have the same underlying undirected graph, but may disagree on the direction of some of the edges. Verma and Pearl (*6*) prove that two DAGs are equivalent if and only if they have the same *skeleton* and the same set of *v-structures*. The skeleton of a DAG is defined as the undirected graph which results from ignoring all edge directions. And a v-structure denotes a configuration $X_j \to X_i \leftarrow X_k$ of two directed edges converging on the same node X_i without an edge between X_j and X_k (*7*). Chickering (*8*) shows that equivalence classes of DAGs can be uniquely represented *using (completed) partially directed acyclic graphs* (CPDAGs). A CPDAG contains the same skeleton as the original DAG, but possesses both directed (compelled) and undirected (reversible) edges. An edge $X_i \to X_j$ is compelled in a CPDAG when all DAGs of this equivalence class contain this *directed* edge, while every reversible (undirected) edge $X_i - X_j$ in the CPDAG representation denotes that some DAGs in the equivalence class contain the directed edge $X_i \to X_j$ while others contain the oppositely orientated edge $X_i \leftarrow X_j$. A directed edge in a DAG is compelled in the CPDAG if it is participating in a v-structure, otherwise it may be either compelled or reversible. An algorithm that takes as input a DAG and outputs the corresponding CPDAG representation can be found in (*7*).

For example, in the DAG in the left panel of Fig. 1, the edges $B \rightarrow D$ and $C \rightarrow D$ are both compelled, because reversing one of these two edges would delete the v-structure: $B \rightarrow D \leftarrow C$. The edge $D \rightarrow E$ is also compelled, as its reversal would give two novel v-structures, symbolically: $B \rightarrow D \leftarrow E$ and $C \rightarrow D \leftarrow E$. The CPDAG representation of the DAG shown in the left panel of Fig. 1 can be found in the right panel. The CPDAG was extracted using the "DAG-to-CPDAG" algorithm from Chickering (*7*).

Although Bayesian networks (BNs) are based on DAGs, that is *directed* acyclic graphs, it is important to note that not all directed edges in a BN can be interpreted causally. Like a BN, a *causal network* is mathematically represented by a directed graph. However, the edges in a causal network have a stricter interpretation: the parents of a variable are its immediate causes. In the presentation of a causal network, it is meaningful to make the *causal Markov assumption* (*9*): Given the value of a variable's immediate causes, it is independent of its earlier causes. Under this assumption, a causal network can be interpreted as a BN in that it satisfies the corresponding Markov independencies. However, the reverse does not hold.

The Gaussian BGe scoring metric of Geiger and Heckerman, on which we will focus in this chapter, specifies the distributional form and the parameters q_i of the local probability distributions $P(X_i \mid \pi_i, q_i)\,(i = 1,\ldots, N)$ in Eq. (1). That is, the BGe score assigns a *local* conditional probability distribution to each node, namely: the probability distribution of X_i conditional on the variables in its parent set X_i. The parent sets are implied by the underlying directed acyclic graph G. According to Eq. (1), the local conditional probability distributions together specify the joint probability distribution $P(X_1,.., X_N \mid G, q)$.

Given a data set D, the Gaussian BGe score can be used to compute the posterior probability $P(G|D)$ of a graph G given the data. For later considerations, we assume that the data matrix D is of size N-by-m and that each of the m columns corresponds to an *independent* realisation of the N variables $X_1, \ldots X_N$. $D_{i,j}$ is the jth realisation of X_i and $D_{\pi_i,j}$ is the jth realisation of the parent node set π_i of X_i. For the posterior probability, we have:

$$P(G \mid D) = \frac{P(G,D)}{P(D)} = \frac{P(D \mid G) \cdot P(G)}{\sum_{G^* \in \Omega} P(D \mid G^*) \cdot P(G^*)} \tag{2}$$

where $P(G)$ $(G \in \Omega)$ is the prior probability distribution over the space Ω of all possible DAGs for the variables $X_1, \ldots X_N$, and $P(D|G)$ is the *marginal likelihood*, that is the probability of the data D given the graph topology G. The *marginal likelihood* of the data G is the integral over the parameter space:

$$P(D \mid G) = \int P(D, q \mid G) dq = \int P(D \mid G, q) \cdot P(q \mid G) dq \tag{3}$$

where $X(q|G)$ is the prior distribution of the parameter vector q.

For a fixed parameter vector q, the likelihood term $P(D|G,q)$ in Eq. (3) can be factorised according to the factorisation of the joint probability distribution in Eq. (1):

$$P(D \mid G, q) = \prod_{i=1}^{N} \prod_{j=1}^{m} P(X_i = D_{i,j} \mid \pi_i = D_{\pi_i, j}, q_i) \tag{4}$$

The denominator in Eq. (2) can be seen as a normalisation constant G:

$$P(G \mid D) = \frac{P(D \mid G) \cdot P(G)}{Z_1} \tag{5}$$

In principle, Z_1 can be computed as the sum over all possible directed acyclic graphs (DAGs). But the number of valid DAGs, that is the cardinality of the set Ω, grows super-exponentially in the number of nodes N. Consequently, the sum is intractable for domains with more than $N = 6$ nodes. Therefore, we concentrate on the numerator $P(D|G)$ $P(G)$ which is proportional to the posterior probability. A commonly employed graph prior $P(G)$ ($G \in \Omega$) is the uniform distribution over Ω. Another graph prior, which we will employ in our data examples, is given by:

$$P(G) = \frac{1}{Z_2} \prod_{i=1}^{N} \binom{N-1}{|\pi_i|}^{-1} \tag{6}$$

where Z_2 is another normalisation constant, and $|\pi_i|$ is the cardinality of the parent set π_i of node G. We note that this graph prior $P(G)$ consists of N local factors: one for each network node. This graph prior assumes for each node X_i that all cardinalities of its parent set π_i are equally likely. For example, the probability of an empty parent set (of cardinality 0) is the same as the probability of a parent set of cardinality 1. As there are $N-1$ possible parent sets of cardinality 1 for node X_i but only one empty parent set, the parent set of cardinality 0 must be $(N-1)$-times as likely as each of the parent sets of cardinality 1.

For the Gaussian BGe model, a closed-form solution of the integral in Eq. (3) can be derived under fairly weak assumptions. *Parameter independence* means that the prior distribution $P(q|G)$ of the unknown parameters N local prior distributions: one for each parameter sub-vector q_i:

$$P(q \mid G) = \prod_{i=1}^{N} P(q_i \mid G)$$

where q_i is the parameter sub-vector required for the ith local conditional probability distribution $P(X_i \mid \pi_i, q_i)$. *Parameter modularity* means that the probability of the parameter sub-vector q_i

in $P(q_i|G)$ depends on the parent node set $\pi_i = \pi_i(G)$ of X_i only. That is, for $i = 1, \ldots, n$ we have: $P(q_i \mid G) = P(q_i \mid \pi_i)$ so that:

$$P(q \mid G) = \prod_{i=1}^{N} P(q_i \mid \pi_i) \tag{7}$$

Inserting Eqs. (4) and (7) into Eq. (3) yields:

$$P(D \mid G) = \int \left(\prod_{i=1}^{N} \prod_{j=1}^{m} P(X_i = D_{i,j} \mid \pi_i = D_{\pi_i,j}, q_i) \right) \cdot \left(\prod_{i=1}^{N} P(q_i \mid \pi_i) \right) dq$$

and it was derived by Geiger and Heckerman(*4*):

$$P(D \mid G) = \prod_{i=1}^{N} \left(\int P(q_i \mid \pi_i) \cdot \left(\prod_{j=1}^{m} P(X_i = D_{i,j} \mid \pi_i = D_{\pi_i,j}, q_i) \right) dq_i \right)$$

For notational convenience we set:

$$\Psi[D_i^{\pi_i}] = \int P(q_i \mid \pi_i) \cdot \left(\prod_{j=1}^{m} P(X_i = D_{i,j} \mid \pi_i = D_{\pi_i,j}, q_i) \right) dq_i \tag{8}$$

where $D_i^{\pi_i} = \{D_{i,j}, D_{\pi_i,j}) : 1 \leq j \leq m\}$ is the subset of the data D pertaining to the m realisations of node X_i and the nodes in its parent set π_i. We refer to the $\Psi[D_i^{\pi_i}]$ terms as *local scores*, and we note that $\Psi[D_i^{\pi_i}]$ can be computed from the data subsets $D_i^{\pi_i}$, that is from the m realisations of X_i and the corresponding m realisations of X_i's parent nodes; the realisations of the other variables are not required. Employing Eq. (8), the marginal likelihood can be re-written as follows:

$$P(D \mid G) = \prod_{i=1}^{N} \Psi[D_i^{\pi_i}] \tag{9}$$

Finally, inserting Eqs. (6) and (9) into Eq. (5) we obtain for the posterior probability $P(G|D)$:

$$P(G \mid D) = \frac{\left(\prod_{i=1}^{N} \Psi[D_i^{\pi_i}] \right) \cdot \left(\frac{1}{Z_2} \prod_{i=1}^{N} \binom{N-1}{|\pi_i|} \right)}{Z_1} = \frac{1}{Z} \cdot \prod_{i=1}^{N} \Psi[D_i^{\pi_i}] \cdot \binom{N-1}{|\pi_i|}$$

where $Z = Z_1 \cdot Z_2$ is a normalisation factor.

We note that the functional form of the local scores $\Psi[D_i^{\pi_i}]$ depends on the stochastic model that is employed. Two widely applied stochastic models are (1) the linear Gaussian model with a normal-Wishart distribution as prior (BGe-model), and (2) the multinomial distribution with a Dirichlet prior (BDe-model). Details about the BDe-model can be found in (*3*) and (*10*). The linear Gaussian BGe scoring metric of Geiger and Heckerman (*4*) will be discussed in more detail in Subheading 3.

When instead of m independent (steady state) observations for the domain variables $X_1, \ldots X_N$ time series data $(X_1(t), \cdots X_N(t))_{t=1,\cdots,m}$ have been collected, ***dynamic Bayesian networks*** (DBNs) can be employed. In DBNs, each edge corresponds to an interaction with a time delay τ; for example, for $\tau = 1$ an edge pointing from X_i to X_j means that the realisation $D_{j,t}$ of X_j at time point t is influenced by the realisation $D_{i,t-1}$ of X_i at time point $\tau - 1$. In DBNs, parameters are tied such that the transition probabilities between time slices $\tau - 1$ and t are the same for all t, that is, DBNs are homogeneous Markov models. Because of the time delay of interactions, there is a bipartite graph structure between two time points $\tau - 1$ and t and the acyclicity-constraint is guaranteed to be satisfied. An illustration is given in Fig.2. The recurrent state space graph in the left panel is cyclic, as it contains a *self-loop* $X \rightarrow X$, that is, node X is its own parent node. When unfolding the state space graph in time, one obtains the *dynamic* graph shown in the left panel which does *not* contain any cycles. Similar to static Bayesian networks DBNs are based on the following homogeneous Markov chain expansion:

$$P(D \mid G, q) = \prod_{i=1}^{N} \prod_{t=2}^{m} P(X_i = D_{i,t} \mid \pi_i = D_{\pi_i,t-1}, q_i) \tag{10}$$

where π_i denotes the parent set of X_i. Accordingly, the DBN counterpart of Eq. (9) is given by:

$$P(D \mid G) = \prod_{i=1}^{N} \Psi[D_i^{\pi_i(t-1)}] \tag{11}$$

where $D_i^{\pi_i(t-1)} = \{D_{i,t}, D_{\quad i,t-1} : 2 \leq t \leq m\}$ is the subset of the data D pertaining to the last $m - 1$ realisations of node X_i and the realisations of its parent nodes in X_i at the corresponding $m - 1$ previous time points. More precisely, the dynamic counterparts of the local scores are given by:

$$\Psi[D_i^{\pi_i(t-1)}] = \int P(q_i \mid \pi_i) \cdot \prod_{t=2}^{m} P(X_i = D_{i,t} \mid \pi_i = D_{\pi_i,t-1}, q_i) dq_i$$

Since no realisations for parent nodes at time point $t = 1$ are available, the first observations $D_{1,1}, \cdots D_{N,1}$ at time point $t = 1$ cannot be included when computing local scores for dynamic Bayesian

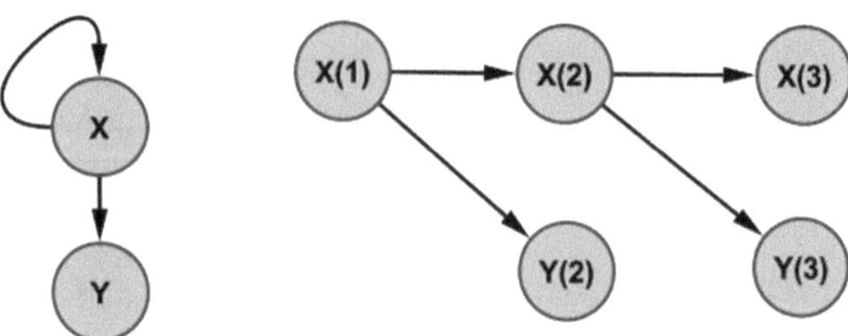

Fig. 2. Recurrent network and unfolded dynamic network. *Left panel*: A recurrent state space graph containing two nodes X and Y. Node X has a recurrent self-loop and acts as a regulator of node Y. *Right panel*: Unfolding the graph shown in the left panel in time gives the dynamic Bayesian network (DBN) with a bipartite graph structure between two adjacent time points t and $t+1$.

networks (DBNs). Therefore, for time series of length m the "effective" sample size, that can be used for the computation of local scores, is equal to $m - 1$.

Finally, we note that the time delay $\tau = 1$ of interactions in DBNs yields that each graph represents a unique factorisation of the joint probability distribution of the variables. There are no equivalence classes as for the DAGs in static Bayesian networks (BNs).

2.2. Structure MCMC Sampling of Bayesian Networks

In the context of static BNs, different Markov Chain Monte Carlo (MCMC) methods have been proposed for sampling DAGs G from the posterior distribution $P(G|D)$ (e.g., see (*5*), (*11*), or (*12*)). The structure MCMC approach of Madigan and York (*5*) generates a sample of DAGs $G_1,\cdots G_T$ from the posterior distribution by a Metropolis Hastings sampler in the space of DAGs. Given a graph (DAG) X_i, in a first step a new DAG G_{i+1} is proposed with the following proposal probability $Q(G_{i+1} \mid G_i)$

$$Q(G_{i+1} \mid G_i) = \begin{cases} \mid N(G_i) \mid^{-1}, & G_{i+1} \in N(G_i) \\ 0, & G_{i+1} \notin N(G_i) \end{cases}$$

where $N(G_i)$ denotes the ***neighbourhood*** of G_i, that is the set of all DAGs that can be reached from G_i by deletion, addition or reversal of ***one single edge*** of the current graph G_i. $|N(G_i)|$ is the cardinality of the set $N(G_i)$. We note that the graphs in $N(G_i)$ have to be acyclic. Therefore, it has to be checked which edges can be added to G_i and which edges can be reversed in G_i ***without*** violating the acyclicity-constraint. Edge deletions are always valid, since the deletion of an edge cannot violate the acyclicity-constraint.

In the Metropolis Hastings algorithm, the proposed graph G_{i+1} is accepted with the acceptance probability: $A(G_{i+1} \mid G_i) = \min\{1, R(G_{i+1}|G_i)\}$ where

$$R(G_{i+1} \mid G_i) = \frac{P(G_{i+1} \mid D)}{P(G_i \mid D)} \frac{Q(G_i \mid G_{i+1})}{Q(G_{i+1} \mid G_i)} = \frac{P(D \mid G_{i+1}) \cdot P(G_{i+1})}{P(D \mid G_i) \cdot P(G_i)} \frac{\mid N(G_i) \mid}{\mid N(G_{i+1}) \mid}$$

and the Markov chain is left unchanged, symbolically $G_{i+1} = G_i$ if the new graph G_{i+1} is not accepted. The stochastic process $\{G_i\}_{i=1,2,3,.}$ is a Markov chain in the space of DAGs, whose Markov transition kernel $T(\tilde{G} \mid G)$ for a move from G to $\tilde{G}$ is given by the product of the proposal probability and the acceptance probability: For $G \neq \tilde{G}$:

$$T(\tilde{G} \mid G) = Q(\tilde{G} \mid G) \cdot A(\tilde{G} \mid G)$$

and

$$T(G \mid G) = 1 - \sum_{\tilde{G} \in N(G)} Q(\tilde{G} \mid G) \cdot A(\tilde{G} \mid G)$$

Per construction, it is guaranteed that the Markov transition kernel satisfies the equation of detailed balance:

$$\frac{P(\tilde{G} \mid D)}{P(G \mid D)} = \frac{T(\tilde{G} \mid G)}{T(G \mid \tilde{G})}$$

and, thus, it converges to the posterior distribution $P(G|D)$ as its stationary distribution:

$$P(\tilde{G} \mid D) = \sum_{G} T(\tilde{G} \mid G) \cdot P(G \mid D)$$

The structure MCMC sampling scheme for static BNs can be straightforwardly modified in order to sample DBNs. For static BNs, the ***neighbourhood*** of a DAG G is defined as the set of all DAGs that can be reached from G by deletion, addition or reversal of one single edge. For DBNs, we define that the neighbourhood of a (not-necessarily acyclic) directed graph is the set of all (not-necessarily acyclic) directed graphs that can be reached from G either by deletion or by addition of one single edge. Thereby, in principle, a node can become its own parent node (like node X in Fig. 2). We refer to the corresponding edge as a *self-loop*. If this appears to be implausible for the domain, then such self-loops can be excluded as invalid edges, and graphs possessing self-loops, such as $X \rightarrow X$, are removed from the respective graph neighbourhoods.

A reasonable approach adopted in most Bayesian network applications is to impose a limit on the cardinality of the parent sets. This limit is referred to as the *fan-in*. The practical advantage of the restriction on the maximum number of edges converging on a node is a reduction of the computational complexity, which improves the convergence. Fan-in restrictions can be justified in the context of biological expression data, as many experimental results have shown that the expression of a gene is usually controlled by a comparatively small number of active regulator genes, while on the other hand regulator-genes seem to be nearly unrestricted in the number of genes they regulate. The imputation of a fan-in restriction leads to a further reduction of the graph's neighbourhoods: Graphs that contain nodes with too many parents, that is, more than the fan-in value, have to be removed from the respective neighbourhoods.

2.3. Posterior Probability of Edges and AUC Values

The structure MCMC algorithm (described in Subheading 2.2) can be used to generate a graph sample $G_1, \ldots G_T$ and usually the next step is to compute the marginal posterior probabilities of *edge relation features*. For static Bayesian networks, we extract the CPDAG of each DAG in the sample and we distinguish between *undirected* and *directed edge relation features*. There is an undirected edge relation feature between X_i and X_j ($i < j$) in G if there is an edge connection between X_i and X_j, that is, G may possess either the edge $X_i \rightarrow X_j$ or the oppositely oriented edge $X_i \leftarrow X_j$. That is, undirected edge relation features refer to edge connections without taking the edge directions into account. For directed edge relation features, the CPDAG representations of the graphs in the sample are required. We interpret undirected edges in CPDAGs as superpositions of two (directed) edges pointing in opposite direction. Thus, there is a directed edge relation feature from X_i to X_j ($i \neq j$) in G if there is either (1) an undirected edge between X_i and X_j (i.e., an edge pointing in both directions in our interpretation) in its CPDAG or (2) a directed edge pointing from X_i to X_j in its CPDAG. That is, in our interpretation directed edge relation features refer to edges that are *not* oppositely oriented in the CPDAG representation.

There are no equivalence classes for dynamic Bayesian networks (DBNs), and we can interpret all edges causally. Consequently, in DBNs our focus is on directed edge relation features only. We define for DBNs that there is a directed edge relation feature from X_i to X_j ($i \neq j$) in the graph G if G possesses the directed edge $X_i \rightarrow X_j$.

An estimator for the marginal posterior probabilities of an edge relation feature F is given by the fraction of graphs in the sample that possess the edge relation feature of interest:

$$\hat{P}(F \mid D) = \frac{1}{T}\sum_{t=1}^{T} I_F(G_t)$$

where $I_F(.)$ is a binary indicator variable over the space of graphs, which is 1 if the edge relation feature F is present in the graph, and 0 otherwise.

When the true graph or at least a gold-standard graph for the domain is known, the concept of receiver-operator-characteristic *(ROC) curves* can be used to evaluate the network reconstruction accuracy of the Bayesian network inference. We assume that $e_{i,j} = 1$ indicates that there is an (directed or undirected) edge relation feature between X_i and X_j in the true graph, while $e_{i,j} = 0$ indicates that this edge relation feature is not given. Bayesian networks infer a posterior probability estimate $\hat{P}(F_{i,j} \mid D)$ for each edge relation feature $e_{i,j}$.

Let $\varepsilon(\theta) = \{e_{i,j} \mid \hat{P}(F_{i,j} \mid D) > \theta\}$ denote the set of all edge relation features whose estimated posterior probabilities exceed a given threshold θ. Given the threshold θ, the number of true positive (TP), false positive (FP), and false negative (FN) edge relation feature findings can be counted, and the *sensitivity* $S = \frac{TP}{TP + FN}$ and the *inverse specificity* $I = \frac{FP}{TN + FP}$ can be computed. But rather than selecting an arbitrary value θ for the threshold, this procedure can be repeated for several values of θ and the ensuing sensitivities can be plotted against the corresponding inverse specificities. This gives the *receiver-operator-characteristic* (ROC) curve.

A quantitative measure for the learning performance can be obtained by integrating the ROC curve so as to obtain the *area under the ROC curve*, which is usually referred to as AUC value. We note that larger AUC values indicate a better learning performance, whereby 1 is an upper limit and corresponds to a perfect estimator, while 0.5 corresponds to a random estimator.

An alternative and more intuitive criteria is given by $(TP|FP=5)$ counts: A threshold ψ is imposed on the estimated edge relation feature posterior probabilities such that five false positive (FP) edges are extracted and the corresponding number of true positive (TP) edge relation features, symbolically $(TP|FP=5)$, exceeding the threshold θ, is counted (*13*).

3. The Gaussian BGe Scoring Metric for Bayesian Networks

In the first Subheading 3.1, we describe the standard linear Gaussian BGe scoring metric (Bayesian metric for Gaussian networks having score equivalence) for Bayesian networks as developed by Geiger and Heckerman (*4*). The required modifications of this BGe scoring metric when analysing interventional or dynamic data will be described in Subheadings 3.2 and 3.3. Subheading 3.1 focuses on BGe for observational static data, that is, passively observed independent (steady-state) observations, Subheading 3.2 describes how interventional data can be included. The last Subheading 3.3 explains how the BGe score can be modified when dynamic data, that is, time series observations, are available. Closed-form solution formulae of the marginal likelihood are given for all three cases.

3.1. The BGe Scoring Metric for Static Observational Data

Given a data set D with m observations of the variables $X_1, \ldots, X_N$:

$$D = \begin{pmatrix} D_{1,1} & D_{1,2} & \cdots & D_{1,m-1} & D_{1,m} \\ D_{2,1} & D_{2,2} & \cdots & D_{2,m-1} & D_{2,m} \\ \vdots & \vdots & \ddots & \vdots & \vdots \\ D_{N-1,1} & D_{N-1,2} & \cdots & D_{N-1,m-1} & D_{N-1,m} \\ D_{N,1} & D_{N,2} & \cdots & D_{N,m-1} & D_{N,m} \end{pmatrix} \quad (12)$$

so that $D_{i,j}$ denotes the jth realisation of the ith node X_i, and the jth column of D: $D_{.,j} = (D_{1,j}, \cdots D_{N,j})'$ is the jth realisation vector of the variables. The Gaussian BGe model assumes that the set of observation vectors $D_{.,j}$ $(j = 1, \ldots, m)$ is a random sample from a multivariate Gaussian distribution $N(\mu, \Sigma)$ with an unknown mean vector μ and an unknown covariance matrix Σ. The prior joint distribution of the mean vector μ and the precision matrix $W = \Sigma^{-1}$ is supposed to be the normal-Wishart distribution. That is, the conditional distribution of μ given W is the $N(\mu_0, (\upsilon \cdot W)^{-1})$ with

$\upsilon > 0$, and the marginal distribution of W is a Wishart distribution with $\alpha > N + 1$ degrees of freedom and covariance matrix T_0, denoted $W(\alpha, T_0)$. The condition $\alpha > N + 1$ ensures that the second moments of the posterior distribution are finite (see also Eq. (26) in (*4*)). Geiger and Heckerman (*4*) show that the marginal likelihood $P(D|G)$ can then – under fairly weak conditions of parameter independence and parameter modularity – be computed in closed form. We define:

$$T_{D,m} = T_0 + S_{D,m} + \frac{\upsilon \cdot m}{\upsilon + m}(\mu_0 - \bar{D}_m)\cdot(\mu_0 - \bar{D}_m)^T \tag{13}$$

where

$$\bar{D}_m = \frac{1}{m}\sum_{j=1}^{m} D_{.,j}$$

is the mean of the m observation vectors and

$$S_{D,m} = \sum_{j=1}^{m}(D_{.,j} - \bar{D}_m)\cdot(D_{.,j} - \bar{D}_m)^T$$

is the (empirical) covariance matrix of the observation vectors multiplied by the factor $(m - 1)$.

T_0, π_0, α, and υ are hyperparameters of the normal-Wishart prior and have to be specified in advance. T_0 is an N-by-N matrix, μ_0 is a N-by-1 column vector, and υ and α are 1-dimensional scalars and usually referred to *as total prior precision parameters.* It can be seen from Eq. (13) that the hyperparameters T_0 and μ_0 refer to the terms $S_{D,m}$ and $\bar{D}_m$. That is, the matrix T_0 and the vector μ_0 reflect the user's prior belief about the unknown covariance matrix Σ and the unknown expectation vector μ of the joint probability distribution of the domain variables $X_1, \ldots, X_N$. If the user's prior belief is that the unknown covariance matrix Σ may be given by Σ_p, for example, $\Sigma_P = I_{N,N}$, where $I_{N,N}$ is the N-dimensional identity matrix, then we recommend setting:

$$T_0 = \frac{\upsilon \cdot (\alpha - N - 1)}{\upsilon + 1} \cdot \Sigma_P.$$

Accordingly, if the user's prior belief is that the unknown expectation vector μ may be given by μ_p, for example, $\mu_P = (0, \cdots, 0)^T$ then we recommend setting: $\mu_0 = \mu_P$. The total prior precision parameters υ and α reflect the user's certainty about the hyperparameters T_0 and μ_0. The higher the hyperparameters α and υ are set, the stronger the influence of the hyperparameter vector μ_0 and the hyperparameter matrix T_0 on $T_{D,m}$. Setting the total prior precision parameters to their minimal values $\alpha = N + 2$ and $\upsilon = 1$ gives an "uninformative" prior with a weak effect on $T_{D,m}$. Theoretical considerations and more details on how to specify the hyperparameters can be found in (*4*)

In the BGe model, the marginal likelihood is given by (*4*):

$$P(D \mid G) = \prod_{i=1}^{N} \Psi[D_i^{\pi_i}] = \prod_{i=1}^{N} \frac{P(D^{\{X_i,\pi_i\}} \mid G_F(\{X_i,\pi_i\}))}{P(D^{\{\pi_i\}} \mid G_F(\{\pi_i\}))} \tag{14}$$

where X_i is the *i*th variable and π_i is the parent set of X_i in the graph *G*. $D^{\{X_i,\pi_i\}}$ and $D^{\{\pi_i\}}$ are sub-matrices of the data matrix *D* consisting only of those rows that correspond to the variables in the subsets $S_1 = \{X_i,\pi_i\}$ and $S_2 = \{\pi_i\}$. $G_F(\{X_i,\pi_i\})$ and $G_F(\{\pi_i\})$ correspond to *full (sub-)graphs* for the variable subsets $S_1 = \{X_i,\pi_i\}$ and $S_2 = \{\pi_i\}$, that is, to subgraphs with the maximal number of edges. Full graphs do not impose any independency relations on the variables in the subsets $S_1 = \{X_i,\pi_i\}$ and $S_2 = \{\pi_i\}$.

The marginal likelihood of a data subset $D^S \subseteq D$, which consists of the *m* realisations of the $N^\Diamond$-dimensional subset $S \subseteq \{X_1,\cdots,X_N\}$, can be computed when a full graph $G_F(S)$ for the variables in the subset *G* is given:

$$P(D^S \mid G_F(S)) = (2\pi)^{-N^\Diamond \cdot m/2} \cdot \left\{\frac{\upsilon}{\upsilon+m}\right\}^{N^\Diamond/2} \cdot \frac{c(N^\Diamond,\alpha)}{c(N^\Diamond,\alpha+m)} \cdot \det(T_0^S)^{\alpha/2} \cdot \det(T_{D,m}^S)^{-(\alpha+m)/2} \tag{15}$$

where $\det(T_0^S)$ and $\det(T_{D,m}^S)$ denote the determinants of the submatrices T_0^S and $T_{D,m}^S$ of T_0 and $T_{D,m}$ consisting only of those $N^\Diamond$ rows and columns that correspond to variables in the subset *S*. $T_{D,m}$ was defined in Eq. (13), and the factors $c(N^\Diamond,\alpha)$ and $c(N^\Diamond,\alpha+m)$ can be computed with the following formula:

$$c(N,\alpha) = \left\{2^{\alpha \cdot N/2} \cdot \pi^{N \cdot (N-1)/4} \cdot \prod_{i=1}^{N} \Gamma\left(\frac{\alpha+1-i}{2}\right)\right\}^{-1} \tag{16}$$

Finally, we note that the probability $P(D^S \mid G_F(S))$ in Eq. (14) is equal to 1 for empty sets of variables, symbolically: $S = \{\ \}$.

3.2. The BGe Metric for Static Interventional Data

Although most of the available biological expression and pathway data are passively observed (so called *observational* data), sometimes experimenters can actively intervene and externally set certain domain variables, using for example, gene knock-outs or over-expressions (*interventional* data). If these interventions are *ideal*, then the intervened variables are set deterministically using forces outside the studied domain; so their values no longer depend on the other domain variables. However, their assigned values can influence the values of other variables in the studied domain; consequently, the intervened data points are useful for discovering causal relationships (directed edges). Under fairly weak conditions, a combination of observational and ideal interventional data can be analysed using Bayesian networks. These conditions are described in detail in (*14*) for the BGe scoring metric. Only some small modifications are required. Each local score $\Psi[D_i^{\pi_i}]$ in the marginal likelihood (see Eq. (9)) is computed only from those data points in which the variable X_i was not intervened. We define the m-by-1 column vector I as follows: $I(j) = i$ indicates that the ith variable X_i was intervened in the jth observation, and we replace Eq. (9) by the following expression:

$$P(G \mid D) = \prod_{i=1}^{N} \Psi[D_{i,I}^{\pi_i}] \tag{17}$$

where $D_{i,I}^{\pi_i} = \{D_{i,j}, D_{\pi_i,j} \mid j \in \{1,\cdots,m\} : I(j) \neq i\}$ is the subset of the data D pertaining to those $m_i \leq m$ realisations of node X_i and its parent set nodes in π_i where node X_i was *NOT* intervened, symbolically $I(j) \neq i$.

The observations that can be used for computing the local scores vary from node to node. For example, for node X_i, we have to replace the data matrix shown in Eq. (12) by the sub-matrix $D(i)$ of size N-by-m_i. The matrix $D(i)$ can be extracted from the matrix D by deleting (removing) all columns (observations) where X_i was intervened. That is, for $i = 1, \ldots, n$, we have to remove all those columns $j \in \{1,\cdots,m\}$ with $I(j) = i$ from the data matrix D to obtain $D(i)$. Afterwards, N different matrices $T_{D(i)}, m_i$ – one for each variable X_i – have to be computed from the matrices $D(i)$:

$$T_{D(i),m_i} = T_0 + S_{D(i),m_i} + \frac{\upsilon \cdot m_i}{\upsilon + m_i}(\mu_0 - \bar{D}(i)_{m_i}) \cdot (\mu_0 - \bar{D}(i)_{m_i})^T$$

and Eq. (14) has to be replaced by:

$$P(D \mid G) = \prod_{i=1}^{N} \Psi[D_{i,I}^{\pi_i}] = \prod_{i=1}^{N} \frac{P(D(i)^{\{X_i,\pi_i\}} \mid G_F(\{X_i,\pi_i\}))}{P(D(i)^{\{\pi_i\}} \mid G_F(\{\pi_i\}))} \tag{18}$$

The counterpart of Eq. (15) for interventional data is given by:

$$P(D(i)^S \mid G_F(S)) = (2\pi)^{-N^{\diamond} \cdot m_i/2} \cdot \left\{\frac{\upsilon}{\upsilon + m_i}\right\}^{N^{\diamond}/2} \cdot \frac{c(N^{\diamond},\alpha)}{c(N^{\diamond},\alpha + m_i)} \cdot \det(T_0^S)^{\alpha/2} \cdot \det(T_{D(i),m_i}^S)^{-(\alpha+m_i)/2} \tag{19}$$

where $G_F(S)$ is a full graph for the subset S of variables of cardinality $N^{\diamond}$, and T_0^S and $T_{D(i),m_i}^S$ of T_0^S and $T_{D(i),m_i}^S$ consisting only of those $N^{\diamond}$ rows and columns that correspond to variables in the subset S.

Finally, we note that the definition of equivalence classes has to be generalised. For pure observational data, two DAGs assert the same set of independence assumptions among the domain variables if and only if they have the same skeleton and the same set of v-structures. This definition of equivalence has to be modified when considering a mixture of observational and ideal interventional measurements. Loosely speaking, all edges being connected to an intervened node become automatically directed (compelled) in the CPDAG representation. Details on how to extract CPDAG representations when interventional data are given can be found in (*14*).

3.3. The BGe Score for Dynamic Observational Data

When (instead of independent observations) time series data have been collected for the domain: $(X_1(t),\cdots X_N(t))_{t=1,\cdots,i}$, DBNs can be employed. In DBNs, each edge corresponds to an interaction with a time delay τ; for example, for $\tau = 1$ an edge pointing from X_i to X_j means that the realisation $D_{j,t}$ of X_j at time point t is influenced by the realisation $D_{i,t-1}$ of X_i at the previous time point $t-1$. To take this time delay into account, new data matrices – one for each domain variable – have to be extracted from the original data matrix D shown in Eq. (12). For dynamic data, the columns do not represent independent (steady-state) observations: the tth column $D_{.,t}$ of D is the realisation of the variables at time point t $(t = 1,\cdots,m)$.

In principle, there are two alternatives, and it depends on whether it should be allowed for self-loops, that is edges having the same node as starting and end point. For example, in Fig. 2, node X has a self-loop, that is, X is its own parent node.

3.3.1. Dynamic BGe Score Without Self-Loops

If self-loops, such as $X_i \rightarrow X_i$, are interpreted as invalid edges, we build the following N matrices of size N-by-(m – 1) from the (time series) data matrix given in Eq. (12):

$$D(i) = \begin{pmatrix} D_{1,1} & D_{1,2} & \cdots & D_{1,m-2} & D_{1,m-1} \\ \vdots & \vdots & & \vdots & \vdots \\ D_{i-1,1} & D_{i-1,2} & \cdots & D_{i-1,m-2} & D_{i-1,m-1} \\ D_{i,2} & D_{i,3} & \cdots & D_{i,m-1} & D_{i,m} \\ D_{i+1,1} & D_{i+1,2} & \cdots & D_{i+1,2,m-2} & D_{i+1,2,m-1} \\ \vdots & \vdots & \ddots & \vdots & \vdots \\ D_{n,1} & D_{N,2} & \cdots & D_{N,m-2} & D_{N,m-1} \end{pmatrix} \qquad (20)$$

$(i = 1, ..., N)$. That is, we obtain $D(i)$ by deleting the last column of D and replacing the ith row by $(D_{i,2},\cdots D_{i,m})$afterwards. In other words – loosely speaking – we shift the ith row of D leftwards by 1 and remove all incomplete columns afterwards. For each data set $D(i)$, we then compute the matrix $T_{D(i),m-1}$

$$T_{D(i),m-1} = T_0 + S_{D(i),m-1} + \frac{\upsilon\cdot(m-1)}{\upsilon+(m-1)}(\mu_0 - \bar{D}(i)_{m-1})\cdot(\mu_0 - \bar{D}(i)_{m-1})^T$$

and we replace Eq. (14) by:

$$P(D\,|\,G) = \prod_{i=1}^{N} \Psi[D_i^{\pi_i(t-1)}] = \prod_{i=1}^{N} \frac{P(D(i)^{\{X_i,\pi_i\}}\,|\,G_F(\{X_i,\pi_i\}))}{P(D(i)^{\{\pi_i\}}\,|\,G_F(\{\pi_i\}))} \qquad (21)$$

The counterpart of Eq. (15) for dynamic data is given by:

$$P(D(i)^S \mid G_F(S)) = (2\pi)^{-N^{\diamond}\cdot(m-1)/2} \cdot \left\{\frac{\upsilon}{\upsilon+(m-1)}\right\}^{N^{\diamond}/2} \cdot \frac{c(N^{\diamond},\alpha)}{c(N^{\diamond},\alpha+(m-1))} \cdot \det(T_0^S)^{\alpha/2} \cdot \det(T_{D(i),m-1}^S)^{-(\alpha+(m-1))/2} \quad (22)$$

where $G_F S$ is a full graph for the subset G of variables of cardinality $N^{\diamond}$, and T_0^S and $T_{D(i),m-1}^S$ are sub-matrices of T_0^S and $T_{D(i),m_i}^S$ consisting only of those $N^{\diamond}$ rows and columns that correspond to variables in the subset G.

3.3.2. Dynamic BGe Score with Self-Loops

Alternatively, if self-loops, such as $X_i \rightarrow X_i$, are interpreted as valid edges, we build the following N matrices of size $(N+1)$-by-$(m-1)$ from the (time series) data matrix given in Eq. (12):

$$D(i) = \begin{pmatrix} D_{1,1} & D_{1,2} & \cdots & D_{1,m-1} \\ D_{2,1} & D_{2,2} & \cdots & D_{2,m-1} \\ \vdots & \vdots & \ddots & \vdots \\ D_{N,1} & D_{N,2} & \cdots & D_{N,m-1} \\ D_{i,2} & D_{i,3} & \cdots & D_{i,m} \end{pmatrix} \quad (23)$$

$i = 1, \ldots, n$. That is, $D(i)$ is extracted by deleting the last column of D and adding a novel row $(D_{i,2}, \cdots D_{i,m})$, that is the ith row of D shifted leftwards by 1, as the $(N+1)$th row. We can identify the $(N+1)$th row in $D(i)$ with a new variable X_{N+1}. This new variable is the ith domain variable with a time shift of size $\tau = 1$, and we note that from that perspective, the novel data matrices $D(i)$ consist of observations for $N + 1$ variables. Accordingly, the hyperparameters T_0 and μ_0 have to be an $(N+1)$-by-$(N+1)$ matrix and an $(N+1)$-by-1 column vector, respectively, here. As before, we can compute the matrix $T_{D(i),m-1}$ for each data set $D(i)$,

$$T_{D(i),m-1} = T_0 + S_{D(i),m-1} + \frac{\upsilon\cdot(m-1)}{\upsilon+(m-1)}(\mu_0 - \bar{D}(i)_{m-1})\cdot(\mu_0 - \bar{D}(i)_{m-1})^T$$

and we replace Eq. (14) by:

$$P(D \mid G) = \prod_{i=1}^{N} \Psi[D_i^{\pi_i(t-1)}] = \prod_{i=1}^{N} \frac{P(D(i)^{\{X_{N+1},\pi_i\}} \mid G_F(\{X_{N+1},\pi_i\}))}{P(D(i)^{\{\pi_i\}} \mid G_F(\{\pi_i\}))} \quad (24)$$

The counterpart of Eq. (15) for dynamic data is again given by:

$$P(D(i)^S \mid G_F(S)) = (2\pi)^{-N^{\diamond}\cdot(m-1)/2} \cdot \left\{\frac{\upsilon}{\upsilon+(m-1)}\right\}^{N^{\diamond}/2} \cdot \frac{c(N^{\diamond},\alpha)}{c(N^{\diamond},\alpha+(m-1))} \cdot \det(T_0^S)^{\alpha/2} \cdot \det(T_{D(i),m-1}^S)^{-(\upsilon+(m-1))/2} \quad (25)$$

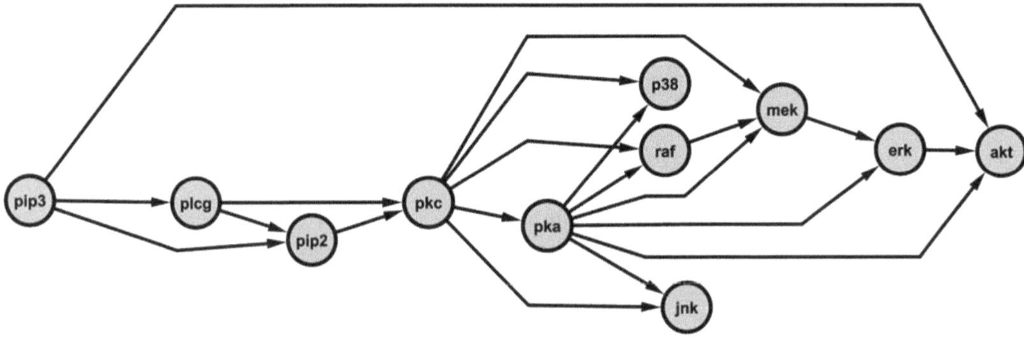

Fig. 3. The RAF-pathway as presented by Sachs et al. (*15*).

where $G_F(S)$ is a full graph for the subset S of variables of cardinality $N^{\blacklozenge}$, and T_0^S and $T_{D(i),m-1}^S$ are sub-matrices of T_0^S and $T_{D(i),m_i}^S$ consisting only of those $N^{\blacklozenge}$ rows and columns that correspond to variables in the subset S.

4. Data

4.1. The RAF Signalling Pathway

The RAF signalling pathway, shown in Fig. 3, is a biologically well-known regulatory network, which describes the intracellular relationships among different molecules involved in *signal transduction*. In the cascade of cellular protein signalling events special enzymes (protein kinases) modify target proteins (substrates) by adding phosphate groups to them (phosphorylation). This leads to a functional change of the targets so that further chemical reactions follow in the signalling cascade. As protein kinases are known to regulate the majority of cellular pathways and cell growth, deregulated kinase activity can lead to diseases, such as cancer. Sachs et al. (*15*) measured the expression levels of $N = 11$ phosphorylated proteins and phospholipids of the RAF signalling pathway in thousands of human immune system cells with flow cytometry experiments. In addition to about 1,200 pure observational measurements, the $N = 11$ molecules in the cascade were also measured after nine different molecular cues. More precisely, the molecules were profiled 15 min after nine different stimulations of the network. For each of these molecular interventions, more than 600 measurements were made, whereby an effect on the molecules in the cascade could be observed for six of these perturbations. As it is known that these interventions predominantly influence only a single molecule in the cascade, they can be considered as ideal interventions. A brief summary of the effects of the six molecular interventions on the measured molecules activities can be found in Table 1. Three molecular interventions

Table 1
Experimental interventions in the RAF signalling pathway. The table shows the effects of the ideal interventions in the flow cytometry experiment on the RAF network shown in Fig. 3

Reagent	Effect
Akt inhibitor	Inhibits AKT
G06976	Inhibits PKC
Psitectorigenin	Inhibits PIP2
U0126	Inhibits MEK
Phorbol myristate acetate	Activates PKC
8-Bromo adenosine 3′5′-cyclic monophospahte	Activates PKA

having no observable effect on the cascade were discarded from the analysis. More details on the probe preparations, the exact experimental conditions as well as more information about the stimulatory agents can be found in (*15*). From this flow cytometry data set, Werhli et al. (*13*) sampled data subsets of size m = 100 each for a comparative evaluation study of different graphical model. That is, five pure observational data sets of size m = 100 were sampled from the 1,200 observational measurements, and five interventional data sets of size m = 100 were composed by sampling 14 measurements for each type of intervention, and including a further set of 16 unperturbed pure observational measurements.

As a gold-standard network for the RAF-pathway is known, we will use the flow cytometry protein data to evaluate the network reconstruction accuracy of the Bayesian network methodology presented in Subheading 3.

4.2. Circadian Genes in Arabidopsis thaliana

We also consider two gene expression time series from *Arabidopsis thaliana* cells, which were sampled at 13 × 2 h time intervals with Affymetrix microarray chips, and Robust Multi-Array (RMA) normalized. The expressions were measured twice independently under experimentally generated constant light condition, but differed with respect to the pre-histories. In the first experiment, E_{20}, the plant was entrained in a 10 h:10 h light/dark-cycle, while the plant in the second experiment, E_{28}, was entrained in 14 h:14 h light/dark-cycle. Our analysis focuses on N = 9 genes, namely LHY, CCA1, TOC1, ELF4, ELF3, GI, PRR9, PRR5, and PRR3, which are known to be involved in circadian regulation (*16*).

As the true regulatory relationships are unknown, we will infer the unknown regulatory networks for both experiments E_{20} and E_{28} in Subheading 6.

5. Simulations

In all our observational static and dynamic structure MCMC simulations, data were standardised to zero mean and marginal variance of 1 for all dimensions. The static interventional measurements could not be analysed without a more sophisticated pre-processing. Occasionally, there was a discrepancy between expected and observed concentrations for intervened nodes, for example, some inhibitions had not led to low concentrations while some activations had not led to high concentrations. This is because the measured concentrations do not reflect the true activities of the corresponding protein. It was therefore decided to replace in each interventional data set the values of the activated (inhibited) nodes by the maximal (minimal) concentration of that node measured under a general perturbation of the system. Afterwards, quantile-normalisation was used to normalise each interventional data set. That is, for each of the $N = 11$ molecules its $m = 100$ realisations were replaced by quantiles of the standard Gaussian distribution, symbolically: $N(0,1)$, as follows: For each of the $N = 11$ molecules, the jth highest realization was replaced by the $(j/m)-$quantile of the standard normal distribution, and the ranks of identical realisations were averaged.

The hyperparameters of the normal-Wishart prior of the BGe model were chosen as uninformative as possible subject to certain regulatory conditions discussed in (*4*): $\upsilon = 1$, $\alpha = N + 2$, $\mu_0 = (0,\ldots,0)^T$, and $T_0 = 0.5 \cdot I_{N,N}$ where $I_{N,N}$ is the N-by-N identity matrix. We note that N has to be replaced by $N+1$ for the dynamic BGe model which allows for self-loops (see Subheading 3.3). For all data sets, we set both the burn-in and the sampling-phase lengths of our MCMC simulations to 500,000 each and sampled every 1,000 iterations during the sampling-phase. We ran the structure MCMC algorithm several times (independently) with different graph initialisations, and we applied the standard diagnostic based on the potential scale reduction factor (see (*17*)) to ensure that in this way a sufficient degree of convergence had been reached. Here, we report the results of the MCMC runs which were seeded by an empty graph without any edges. For the static flow cytometry protein data from the RAF-pathway, the true (gold-standard) network is known so that we can assess the global network reconstruction accuracy in terms of AUC and ($TP|FP = 5$) values as explained in Subheading 2.3. For the circadian genes in *Arabidopsis thaliana*, we do not have a true or a gold-standard network. Therefore, we can only reverse engineer the network by applying dynamic Bayesian networks with the BGe scoring metric, but we cannot evaluate the accuracy of our network predictions.

6. Empirical Results

6.1. RAF-Pathway

The five pure observational flow cytometry protein data sets for the RAF-pathway can be analysed using the standard Gaussian BGe scoring metric for static data presented in Subheading 3.1, the five interventional data sets must be analysed with the modified BGe model for interventional data presented in Subheading 3.2. Since a gold-standard network for the RAF-pathway is known (see Fig. 3), the global network reconstruction accuracy can be evaluated in terms of AUC and (TP/FP = 5) values. We distinguish for both network reconstruction accuracy criteria between undirected and directed edge relation features. Histograms of the average AUC values (left panel) and the average (TP/FP = 5) values (right panel) for the observational and interventional data are shown in Fig. 4. The overall impression from the histograms in

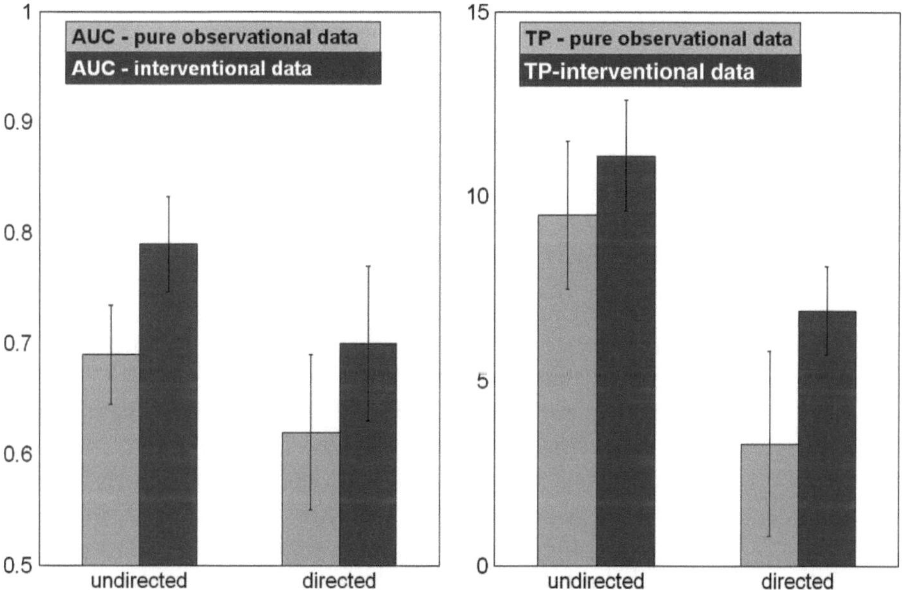

Fig. 4. AUC and TP count histograms. Summary of average AUC (*left panel*) and (TP/FP = 5) values (*right panel*) for the observational and interventional flow cytometry data from the RAF-pathway. The network shown Fig. 3 was used to evaluate the correctness of the extracted *undirected* and *directed* edge relation features. The bars in the histograms represent average AUC values and (TP/FP = 5) counts across five independent data sets each. The error bars of the histograms correspond to one standard deviation.

both panels is that the network reconstruction accuracy is consistently worse for the observational data (light grey bars on the left) than for the interventional data (dark grey bars on the right). Moreover, another trend can be seen: for both criteria, the highest bar refers to learning the undirected edge connections from interventional data sets, and the lowest bars refers to learning the directed edges from pure observational data. Both trends are plausible: First, there is more information in interventional data than in pure observational data so that a higher network reconstruction accuracy can be expected. Second, the task of learning undirected edge connections is less difficult than reverse-engineering the edge connections along with their correct edge orientations. The left panel of Fig. 4 shows histograms of AUC values, and – as a first sanity check – it can be seen that all average AUC values are higher than 0.5 so that the Bayesian network inference leads to a better global network reconstruction accuracy than a random predictor. The highest AUC value of about 0.8 is reached when learning the undirected edge connections from the interventional data sets. For the observational data, the average AUC value for undirected edges is lower than 0.7 and so even lower than the AUC value of 0.71 for learning directed edges from interventional data. The AUC value for directed edges from observational data is 0.63. As AUC values are hard to interpret, we now consider the $(TP|FP = 5)$ criteria in the left panel of Fig. 4. The $(TP|FP = 5)$ value is the number of true positive (TP) edges that can be learned when setting the threshold θ on the marginal posterior probabilities such that five false positive (FP) edges are extracted. As a consequence of the interventions, the number of correctly predicted undirected edges increases slightly from 9.6 to 11.2. The number of correctly predicted directed edges shows a more substantial increase from 3.7 to 7.1. We note that 7.1 correct edge findings are not impressive when on the other hand five false positive edges are extracted. As there are 20 true directed edges in the RAF-pathway shown in Fig. 3, 7.1 true positive correct edges correspond to a sensitivity of 35 percent only. On the other hand, there are 90 false positive directed *non-edges* in the RAF-pathway so that five false positives (FPs) correspond to a specificity of nearly 95 percent. However, this example demonstrates that Bayesian network methodology can help elucidating the regulatory relationships between proteins from flow cytometry data. We note that Sachs et al. (*15*) obtained a much higher network reconstruction accuracy when inferring the RAF-pathway from all $m = 5400$ (interventional) flow cytometry data points with the multinomial BDe scoring metric (*3*) for Bayesian networks.

6.2. Circadian Genes in Arabidopsis

The two dynamic gene expression time series E_{20} and E_{28} of nine circadian genes from *Arabidopsis thaliana* can be analysed with the dynamic BGe score presented in Subheading 3.3. The formulae for the dynamic BGe score depend on whether it is allowed for self-loops or not. Self-loops, such as $X \to X$, describe autocorrelations and can be ruled out altogether to focus on a gene's interaction with other genes. Inferring spurious self-loops may hamper the extraction of the true regulatory mechanisms among genes. On the other hand, ruling out feedback loops altogether, will not provide a sufficient remedy for all applications, as some genes might actually exhibit regulatory feedback loops (e.g., in molecular biology: transcription factors regulating their own transcription), and it is generally not known in advance where these nodes are.

For illustrative purposes, we apply both variants and we cross-compare the results, that is, the reverse-engineered gene regulatory networks. Figure 5 shows the extracted regulatory networks for both dynamic gene expression data sets E_{20} and E_{28}. In all four panels, only those edges are shown that possess a marginal posterior probability higher than $\theta = 0.75$. The top row in Fig. 5 shows the extracted networks for the E_{20} and the E_{28} data when it is allowed for self-loops. For E_{20} three self-loops: *PRR9* $\to$ *PRR9*, *GI* $\to$ *GI*, *LHY* $\to$ *LHY* and further eight edges among genes have been extracted. These eight edges can also be found in the network in bottom left panel, which was extracted from E_{20} by ruling out self-loops. But in the latter case, three additional edges are extracted (excess the posterior probability threshold $\theta = 0.75$): *CCA1* $\to$ *LHY*, *PRR5* $\to$ *GI* and *GI* $\to$ *PRR5*. It appears that these three edges are related to two of the three self-loops in the top left panel. Instead of the two self-loops *GI* $\to$ *GI* and *LHY* $\to$ *LHY* in the top left panel of Fig. 5, the feedback paths *GI* $\to$ *PRR5* $\to$ *GI* and *LHY* $\to$ *CCA1* $\to$ *LHY* appear in the bottom left panel. For the third self-loop *PRR9* $\to$ *PRR9*, there appears to be no compensation by a feedback path. However, we conclude that the inclusion/exclusion of self-loops does not have a strong effect on the extracted interactions among genes for the E_{20} experiment.

For the E_{28} data, the situation is comparable. Allowing for direct feedback loops (see top right panel) yields a network consisting of one self-loop: *GI* $\to$ *GI* and nine further edges. These nine edges are also extracted when ruling self-loops out as invalid edges (see bottom right panel). Here, it seems that the self-loop *GI* $\to$ *GI* is compensated by the feedback path *GI* $\to$ *PRR5* $\to$ *GI*. However, there are three additional edges, symbolically: *PRR5* $\to$ *TOC1*, *CCA1* $\to$ *LHY* and *ELF3* $\to$ *PRR9* in the bottom right panel which do not seem to be related to self-loops.

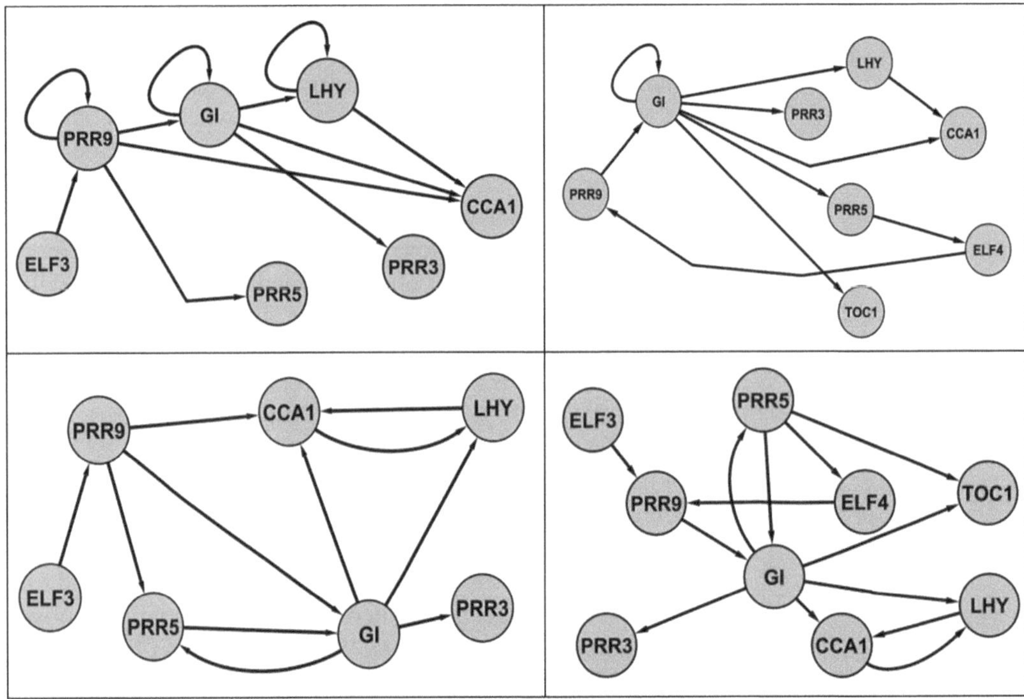

Fig. 5. Extracted gene networks for the circadian genes in *Arabidopsis thaliana*. *Top left panel*: E_{20} with self-loops, *top right panel*: E_{28} with self-loops, *bottom left panel*: E_{20} without self-loops, *bottom right panel*: E_{28} without self-loops. In all four cases, the network topologies have been extracted with the structure MCMC algorithm (*see* Subheading 2.2) and the corresponding two different dynamic BGe scoring metrics. In the top panels, it was allowed for self-loops as valid edges and in the bottom panels self-loops were ruled out as invalid edges; *see* Subheading 3.3 for details. The panels show all directed edges with a marginal posterior probability higher than $\theta = 0.75$. For clarity, genes not being connected to other genes have been omitted in the network representations.

Overall, it can be seen from the extracted networks that the node *GI* (gene name: 'GIGANTEA') seems to play a fundamental role in the circadian clock of *Arabidopsis thaliana*. In all four panels of Fig. 5 node, *GI* is highly connected to other genes and regulates between three and five other genes. It is known that *GI* belongs to the group of 'evening genes' in the circadian clock of *Arabidopsis thaliana*, since its regulating activity tends to peak at the *subjective* night, that is during absence of light. In the first instance, this may appear a little counter-intuitive, as the gene expressions were measured under experimentally generated constant light condition in both experiments E_{20} and E_{28}. But the entrainment of the plants in dark:light cycles (before data collection) may have led to a subjective daytime-phase-shift of the circadian regulation. A review of the most important genes in the circadian clock in *Arabidopsis thaliana* can be found in (*16*).

7. Discussions

This chapter has given an introduction to the fundamentals of reverse-engineering gene regulatory networks and protein signalling pathways with Bayesian networks. After an introduction to the principles of Bayesian networks, the structure MCMC algorithm for inferring the (unknown) topology of a regulatory network has been presented. Afterwards, the Gaussian BGe scoring metric has been described in detail for three different cases (1) BGe score for pure observational static (steady-state) data, (2) BGe score for interventional static (steady state) data, and (3) BGe score for observational dynamic (time series) data. The three scoring metrics have been applied to real biological data. In the first application, pure observational as well as interventional *static* flow cytometry protein data from the RAF-pathway have been analysed. The global network reconstruction accuracy could be assessed in terms of AUC values, as the regulatory protein interactions of the RAF signalling pathway are widely known. It turned out that interventional data, that is, data where experimentalists can activate or inhibit certain genes or proteins by experimental conditions, lead to a superior learning performance. This finding is consistent with (*13*), (*17*), and (*18*) where the same RAF-pathway protein data sets have been analysed in a comparative evaluation study of different graphical models. The dynamic BGe scoring metric has been employed for analysing gene expression time series of nine circadian genes in *Arabidopsis thaliana*. Two different variants of the *dynamic* BGe scoring metric were applied: First, self-loops were ruled out as invalid edges, and second, it was allowed for self-loops, that is genes regulating their own transcription. It could be seen that the inference results (extracted networks) depend on whether self-loops are ruled out or not.

Recently, it has been shown in (*12*) that the convergence of the structure MCMC sampler for Bayesian networks can be substantially improved by introducing a new and more extensive edge reversal move, which allows for much larger steps in the graph space than the single-edge-operations (edge deletions, additions, and reversals) of the classical structure MCMC algorithm. Henceforth, for domains with lots of variables – where the convergence of the MCMC inference may tend to be poor – it should be considered to employ the upgraded structure MCMC algorithm for Bayesian inference. A disadvantage of the Gaussian BGe scoring metric is that it models only *linear* relationships in the data. Non-linear regulatory relationships among genes or proteins cannot be inferred. The BDe scoring metric for Bayesian networks has a higher modelling flexibility but requires the data to be discretised. This always incurs an information loss and so cannot be seen as remedy for this problem. Recently, some mixture model approaches based on the Gaussian BGe scoring metric for inferring non-stationary gene regulatory networks have been proposed (e.g., see (*19*) and (*20*)). These Gaussian Bayesian Mixture (BGM) Bayesian network models can be seen as a consensus in between the flexible but discrete multinomial BDe metric and the continuous but linear Gaussian BGe metric for Bayesian networks. A brief comparison of the network reconstruction accuracy of the BDe metric and the BGe metric for reconstructing the RAF-pathway can be found in (*21*).

Acknowledgements

Marco Grzegorczyk is supported by the Graduate school "Statistische Modellbildung" of the Department of Statistics at TU Dortmund University.

References

1. Friedman N, Linial M, Nachman I, Pe'er D (2000) Using Bayesian networks to analyze expression data. J Comput Biol 7:601–620
2. Vyshemirsky V, Girolami MA (2008) Bayesian ranking of biochemical system models. Bioinformatics 24:833–839
3. Cooper GF, Herskovits E (1992) A Bayesian method for the induction of probabilistic networks from data. Mach Learn 9:309–347
4. Geiger D, Heckerman D (1995) Learning Gaussian networks. In: Proceedings of the tenth conference on uncertainty in artificial intelligence, 235–243, Seattle, Washington, USA, 29–31 July 1994
5. Madigan D, York J (1995) Bayesian graphical models for discrete data. Int Stat Rev 63:215–232
6. Verma T, Pearl J (1990) Equivalence and synthesis of causal models. In: Proceedings of the 6th conference on uncertainty in artificial intelligence, 6, 220–227
7. Chickering DM (2002) Learning equivalence classes of Bayesian network structures. J Mach Learn Res 2:445–498
8. Chickering DM (1995) A transformational characterization of equivalent Bayesian network structures. In: International conference on uncertainty in artificial intelligence (UAI), 11, 87–98
9. Pearl J (2000) Causality: models, reasoning and intelligent systems. Cambridge University Press, London, UK
10. Heckerman D (1999) A tutorial on learning with Bayesian networks, Learning in Graphical Models. In: Jordan MI (ed) Adaptive computation and machine Learning. MIT Press, Cambridge, pp 301–354
11. Friedman N, Koller D (2003) Being Bayesian about network structure. Mach Learn 50:95–126
12. Grzegorczyk M, Husmeier D (2008) Improving the structure MCMC sampler for Bayesian networks by introducing a new edge reversal move. Mach Learn 71:265–305
13. Werhli AV, Grzegorczyk M, Husmeier D (2006) Comparative evaluation of reverse engineering gene regulatory networks with relevance networks, graphical Gaussian models and Bayesian networks. Bioinformatics 22:2523–2531
14. Wernisch L, Pournara I (2004) Reconstruction of gene networks using Bayesian learning and manipulation experiments. Bioinformatics 20:2934–2942
15. Sachs K, Perez O, Pe'er DA, Lauffenburger DA, Nolan GP (2005) Causal protein-signaling networks derived from multiparameter single-cell data. Science 308(5721):523–529
16. Salome P, McClung C (2004) The *Arabidopsis thaliana* clock. J Biol Rhythms 19:425–435
17. Grzegorczyk, M (2006) Comparative evaluation of different Graphical Models for the Analysis of Gene Expression Data. Doctoral Thesis, Department of Statistics, Dortmund University
18. Grzegorczyk M, Husmeier D, Werhli AV (2008) Reverse engineering gene regulatory networks with various machine learning methods. In: Emmert-Streib F, Dehmer M (eds) Analysis of microarray data: a network-based approach. Wiley-VCH, Weinheim

19. Grzegorczyk M, Husmeier D, Edwards KD, Ghazal P, Millar AJ (2008) Modelling non-stationary gene regulatory processes with a non-homogeneous Bayesian network and the allocation sampler. Bioinformatics 24:2071–2078
20. Grzegorczyk M, Husmeier D (2009) Modelling non-stationary gene regulatoy processes with a non-homogeneous Bayesian network and the change point process. In: Manninen et al (eds) Proceedings of the 6th international workshop on computational systems biology (WCSB 2009), TICSP series 48
21. Grzegorczyk M (2008) Comparison of two different stochastic models for extracting protein regulatory pathways with Bayesian networks. J Toxicol Environ Health A 71: 780–787

Chapter 7

Derivation of Large-Scale Cellular Regulatory Networks from Biological Time Series Data

Benjamin L. de Bivort

Abstract

Pharmacological agents and other perturbants of cellular homeostasis appear to nearly universally affect the activity of many genes, proteins, and signaling pathways. While this is due in part to nonspecificity of action of the drug or cellular stress, the large-scale self-regulatory behavior of the cell may also be responsible, as this typically means that when a cell switches states, dozens or hundreds of genes will respond in concert. If many genes act collectively in the cell during state transitions, rather than every gene acting independently, models of the cell can be created that are comprehensive of the action of all genes, using existing data, provided that the functional units in the model are collections of genes. Techniques to develop these large-scale cellular-level models are provided in detail, along with methods of analyzing them, and a brief summary of major conclusions about large-scale cellular networks to date.

Key words: Large-scale cellular-level networks, Gene transcription, Module, Regulatory influence, Modeling, Attractor, Dynamics

1. Background

An ideal drug will alter the function of only a single target protein, signaling pathway or cellular function, thereby reducing the chance of harmful side effects. However, there is considerable evidence that this is not the case in nearly all drugs that make it to the market. For example, chronic treatment with the antidepressants moclobemide, clorgiline, amitriptyline, each alter the expression of up to 10% of all genes in the rat brain (1). Pharmacological treatments (including, e.g., aspirin, ibuprofen, and sildenafil) typically alter the expression levels of 1,600 genes at a time in liver cells (2, 3). Application of any of 33 different chemokines to cultured murine b-lymphocytes (4) alters the expression of between ~50 and ~600 genes (5).

Qing Yan (ed.), *Systems Biology in Drug Discovery and Development: Methods and Protocols*, Methods in Molecular Biology, vol. 662, DOI 10.1007/978-1-60761-800-3_7, © Springer Science+Business Media, LLC 2010

Even the cancer therapeutic agent imatinib, noted for its specificity of action and minimal side effects (6), alters the expression levels of 46 different genes (7).

The problem of off-target pharmacological activity is likely confounded by the screening methods that are frequently used in the early rounds of drug development. By screening collections of potential drugs in vitro for their effect on the enzymatic activity of a single protein or the activity of a single signaling pathway (8), the effects of the pharmacological agents across a broad number of targets are not considered. To do otherwise would involve testing potential drugs in vivo, or in simplified systems by testing the drugs against a panel of functional assays that could reveal off-target effects. These are impractical alternatives at the early stages of drug development. However, in addition to preliminary screens not necessarily selecting for drugs with high specificity, there may be fundamental reasons that drugs, or for that matter, any agent sufficiently potent to perturb cellular activity, will induce broad effects across many targets. This would occur if the networks of regulation underlying the molecular behavior of the cell (transcription network, metabolic network, etc.) employ very high degrees of self-regulation.

If many or all molecular components are involved in their own regulation, this leads to attractor states in the behavioral dynamics of the cell, configurations which are self-stabilizing and require concerted change in many components before the cell will switch into a new state. These attractor states correspond to the distinct cell-types giving rise to different tissues, as well as nontransient pathological states associated with disease. Because these states are self-sustaining due to the action of many genes or pathways, any drug that is capable of switching the cell between these states will of necessity be altering the activity of many genes or pathways. The view that biological systems act this way has historical precedent (9, 10), largely rooted in the observation that self-regulating systems will, by their very nature, support homeostasis.

Complete self-regulation of the cell is one extreme view. The other end of this spectrum could be characterized as "feed-forward" regulation in which cellular components are either regulators, or targets of regulators, and no components are regulated, either directly or indirectly, by their own activity. While the true behavior of the cell surely lies between these viewpoints (11), the assumption of feed-forward regulation is often implicit in the molecular biology literature in the guise of genes being identified as "master-regulators," mutations and phenotypes being linked in a one-to-one fashion, and the hope that drugs will be developed targeting only a single protein or signaling pathway.

In addition to systems biological arguments that support a considerable degree of self-regulation in molecular networks based on the first principle argument of the need for homeostatic response in the face of random external fluctuations and intrinsic noise (12–14), there is recent direct experimental evidence of attractor states in cell regulatory networks. One prediction of the attractor view of the cell is that there will exist multiple parallel trajectories by which a cell can switch between states (15, 16). That is, once the cell attains a state within the boundary of the basin of attraction of a particular attractor, it will follow an energetically minimal trajectory to that attractor, with this trajectory being dependent on the cell state when the basin is entered. This has been observed in cultured neutrophils, in which differentiating cues applied in separate cultures ultimately yield cell populations with the same expression profiles, but with differing intermediate transcriptional states (16). Equivalently, the application of a number of distinct combinations of transcription factors is sufficient to reprogram differentiated cells into pleuripotent stem cells (17). Most recently, populations of hematopoietic stem cells were observed to have normally distributed concentrations of the cell-surface marker Sca-1. Cultures seeded with those cells with either the highest or lowest concentrations of Sca-1 were allowed to grow for two weeks, and in this time restored the original concentration distribution of the gene (18).

Bioinformatic analyses corroborate the attractor view. The number of genes required to distinguish cell-states is proportional to the number of genes involved in the regulatory maintenance of those states. If a handful of master regulator genes are sufficient to maintain cell-states by directly or indirectly regulating numerous target genes, then the activity levels of only those master regulators should be sufficient to identify the cell states. Instead, approximately 200 genes (19) are required to convey the diversity liver cell states in response to pharmacological perturbation (2) as well as the diversity of gene expression levels in healthy and cancerous tissue types (20). The convergence of experimental, statistical, and theoretical evidence has led to several calls for an attractor model to be the default view of cellular regulatory networks (19, 21, 22).

2. Introduction to Global Models of Cellular Regulation

In the face of compelling evidence that the concerted action of numerous cellular components in a nonhierarchical regulatory network underlies cellular regulation, new models are needed. In order to be predictive of the collective behavior of the entire regulatory network, these models must be global, that is, every regulatory component in the cell must have a corresponding component in the model. Cataloguing global regulation would seem to require assessing the pair-wise regulatory relationships between all components of the cell. But even this would be insufficient, as the relevant unit of regulation can very well be a binary (23), ternary (24), or higher order complex of genes or gene products. Thus, the amount of data available required for the inference of a global regulatory network with gene or gene-product-level resolution is going to be unavailable for the foreseeable future.

The methods explored here take an alternative approach to this problem. Rather than scaling up the number of observations until sufficiently many have been recorded to infer a high-resolution global network, the resolution of the network is scaled down (while still retaining its global scale) to match the amount of available data. This is accomplished by aggregating numerous small-scale cellular components and their regulatory influences into large-scale components, a move that is statistically justified provided the genes within a group behave more like each other than genes outside the group. This approach is contrary to conventional approaches of applying large, comprehensive cellular datasets to modeling the mutual regulation of small networks of individual genes (25–27), and has some philosophical support in the observation that even a single gene's behavior is an aggregated behavior composed of the combined action of two loci, hundreds of base pairs with distinct chemical properties, and temporal averaging, etc. For the sake of simplicity, from now on, the components of these networks will be referred to as genes (being aggregated into gene groups), though the method is transferable to other cellular components such as organelles, metabolites, proteins, etc., as will be discussed below.

Once the number of components has been reduced sufficiently, that is, by reducing the resolution of the model, so that the number of potential regulatory interactions between groups of genes is less than the number of data observations available, the specific regulatory relationships can be determined numerically, given appropriate data. Time series data has been used for this purpose (28–30), as it allows the inference of effective regulatory relationships between gene groups of the form "activation of gene group A predicts increased/decreased activity in gene group B, over time interval *t*." The remainder of this chapter will be dedicated to the numerical and statistical methods used to aggregate biologically and pharmacologically relevant large-scale gene groups, derive their regulatory influences using time series data, verify the predictive power of the regulatory model, visualize the models, and interpret the models along several different conceptual lines. Specific conclusions about large-scale global transcription networks in murine b-lymphocytes (4, 28, 29) will then be briefly discussed.

3. Data Sources

To first approximation, to solve for the mutual influences of n gene groups requires n distinct time interval observations of the system. This assumes only first order interactions, where the relevant regulatory units are groups of genes, rather than pairs (or triplets, etc.) of gene groups. Multiple ($s-1$) time intervals can be derived from a single time series $t_1, t_2, t_3 \ldots, t_s$ provided the intervals between all observations are identical. Since the relevant data for the regulatory effects of gene (transcript) groups are derived from microarrays (31) and are comparatively costly and time-consuming to obtain, there are currently few sources of data that enable the construction of even medium-scale cellular networks.

Embryonic time course data from *Drosophila* has been sufficient to solve for the mutual interaction of 17 gene groups (30). Perhaps the best current source, which is also publicly available, is the Alliance for Cellular Signaling (AfCS) data set which records the transcriptional activity of ~15,000 genes in murine b-lymphocytes at 0.5, 1, 2, and 4 h after the administration of 32 different chemokine perturbants (plus an untreated control series) (4). Although the time intervals for each treatment in this dataset are not suitable for sequential use, the 33 distinct time series nominally allow the derivation of the mutual regulatory influences of 33 different gene groups over distinct time steps of 0.5, 1, 1.5, 2, 3, and 3.5 h. In practice, derived networks should be composed of fewer than 33 gene groups because of noise in microarray analysis (32, 33), except that by invoking a statistical assumption of sparseness in the gene group regulatory network, it is possible to identify the strong regulatory influences in cellular networks with up to 72 gene groups (discussed below).

Processed microarray data can be used without further manipulation, resulting in models that predict the absolute level of gene group expression after a fixed time interval, given the initial levels. However, absolute gene expression levels do not necessarily correspond to levels of activity, as the relevant dose of a gene required to alter cellular physiology varies by orders of magnitude across genes. Thus, it is reasonable to normalize gene (and thus gene group) activity levels to a baseline, and replace the raw expression level data with change in expression values. These are typically log-transformed so that a 50% decrease in gene expression is treated as having the same magnitude as a doubling. Networks generated from data of this sort predict changes in gene group expression following changes in gene group expression.

While this chapter focuses on construction of large-scale transcription regulatory networks using microarray data, the methods described here are applicable to numerous other types of data. In fact, any collection of time series data that measures regulatory influence-exerting components and is comprehensive of all said components is sufficient. Other examples include metabolites, proteins (translation products or proteins in their final covalently modified forms), methylation sites, hormone secretion, organelle content in the cell, neuronal activity, fMRI data, etc. For now, microarrays are the most economical source of pharmacologically relevant data.

4. Data Aggregation

The goal of aggregating genes into gene groups to reduce the number of mutual regulatory influences to be modeled is essentially a problem of dimensional reduction (34, 35). In the case of time series data, this means starting with n genes, each of which has a value at a number of time points t. At the end of the reduction process, the original n genes will be reduced to n' gene groups, each of which has a value at each t that represents the values of all its component genes at that time point. Many statistical methods are applicable to this problem, and three will be presented here (principal component analysis, clustering, and a priori definition), though many other methods are appropriate.

Principal component analysis (PCA) (36) rotates multidimensional data in order to minimize the correlation between any two dimensions. Applying this to time series data, where genes are distinct observations in a t-dimensional space, results in n "principal components" which are mutually orthogonal linear combinations of the original n data vectors. These principal components will be ranked by the amount of statistical variation in the original data that each one reflects. If data are available to solve for the mutual regulatory influences of k gene groups, then one should consider the first k principal components. If in total they reflect a considerable portion of the original data variation (e.g., more than 90%), then these principal components can be used directly as large-scale gene groupings, by identifying those genes contributing to each principle component with large magnitude coefficients (each principal component being simply a linear combination of all gene values). Applying a coefficient threshold will result in a list of the primary contributing genes for each principle component. Choice of this threshold depends on the distribution of coefficients, but should be low enough to closely approximate all principle components once the remaining low-coefficient genes are eliminated, but at the same time minimize the inclusion of genes in multiple groups. Alternatively, it has been argued that allowing individual genes to fall into multiple groups can reflect context-dependent coexpression of genes (modularity) (37).

Several methods under the broad heading of "clustering" can be used to place genes into groups according to similarity in their behavior across the time series (38–40). Since clustering approaches are by their nature hierarchical, several studies have examined the scaling of gene membership in groups based on similar expression (41–43). This tuning of scale can be applied to generate a set of gene groups of the appropriate size to infer a mutual regulatory model.

Tree building clustering methods assign genes to the terminals of a binary tree such that the average behavior of terminals on adjacent branches will be as similar as possible. Because there is no evolutionary relationship of common descent underlying the collection of genes varying across time points, cladistic methods to build the cluster tree are not appropriate. Instead "distance methods" such as Unweighted Pair Group Method with Arithmetic mean (UPGMA) (44) or the Fitch-Margoliash algorithm (45), in which branches are united by similarity of adjacent gene terminals, are appropriate. The latter method is preferable, as it performs local branch swapping to maximize similarity and does not assume equal amounts of dissimilarity across equivalent branch distances (ultrametricity), but can be computationally prohibitive for more than 100 genes (almost certainly the case in the construction of global cellular networks).

These distance methods require an $n \times n$ distance matrix measuring the similarity between genes, and the choice of distance metric is important. Euclidean distance metrics, typically employed, will identify genes that behave similarly across all conditions. By contrast, a "1-correlation coefficient metric" will identify genes that are proportionally coregulated across conditions, without distinguishing genes that are inhibited from genes that are activated at the same times.

Once a clustering tree has been constructed, one need only identify the level of the tree at which precisely k branches exist (where k is the number of components to be modeled in the cellular network model), and divide the genes into groups based on the branches found at or below that level. Depending on the distribution of similarities between the genes, this can result in gene groups with radically varying numbers of constituent genes. This is, not a problem per se, though it can increase statistical noise in very small gene groups.

However, if one desires groups of more similar size, it can help to employ an algorithm that uses the original t-dimensional data, rather than a distance metric to partition the n genes. These methods include the k-means algorithm (here k is part of the name of the algorithm) (46) and the Self Organizing Map (SOM) (47) which identify an arbitrary number of "characteristic gene profiles" whose centroids span the diversity of gene profiles present in the time series data. These "characteristic gene profiles" can be directly used as the representative value of the group of genes, rather than the average behavior of the genes within the group. SOMs additionally have the advantage of spatially arraying groups of genes (called modules in SOM terminology) so that adjacent groups are likely to have related behavior across time. This can help with visualizing the model once mutual regulatory influences have been calculated.

Finally, genes can be grouped together on the basis of an a priori categorization. For example, the Gene Ontology (GO) database (48) compiles information about the biological process a gene is involved in, its biochemical mechanism, and its localization within the cell, as determined by direct experimental evidence or computationally inferred homology to genes with experimental evidence. Thus, genes can be grouped on the basis of shared GO labels, yielding intuitive groupings whose identities are known from the beginning of the analysis. This method requires statistical assessment to guarantee that the groups defined a priori are more meaningful than random groups. Conversely, using the other methods to group genes by similarity in behavior demands a subsequent statistical analysis to determine the functional identity of each group.

5. Group Assignment Statistical Control

Before proceeding to derive the mutual regulatory influences between gene groups, it is necessary to confirm that the groups represent intrinsically meaningful biological units. This is required because nothing in the network-inferring methodology would prevent the derivation of a regulatory network from randomly generated or otherwise nonsensical gene groups. Instead, it must first be shown that the diversity of gene behaviors within the group can be reasonably replaced with a single representative behavior. This can be done by either of two statistical methods. First, if the variance (or standard deviation) of gene values within a gene group at each time point is less than the variance (or standard deviation) across all genes at that time point, the group is enriched for an intrinsically meaningful behavior. Comparably, if the average pair-wise correlation between gene time courses within a group is less than the average pair-wise correlation between all gene time courses, the group is meaningful. The p-values associated with the observation of these variances (or standard deviation) or correlations can be determined empirically from the data by generating a large number of random groupings (shuffling the gene-to-group assignments will preserve the gene group size distribution during this test) and determining the frequency with which lower variances or higher correlations are observed.

One's preferred threshold of statistical significance (e.g., $p<0.05$) can then be applied to these frequencies.

6. Inferring Group Function

If groups were defined a priori by sharing a functional classification, such as a GO label, then the identity of the group is already know. Groups defined by this method will, by definition, have lower intragroup correlations and higher variance between gene time courses than groups defined by similarity in behavior, and if the functional identity of groups defined by similar behavior can be inferred, they are consequently statistically preferable. To infer gene group functional identity, some functional classification of genes, such as the GO dataset, is required. This dataset is nearly ideal, as it is hierarchical with general cellular functions composed of multiple more specific functions. The extent to which one of these functions is associated with a gene group can be assessed using the chi-squared statistic, by calculating the number of genes within a group expected to have a particular GO label, given that label's frequency across all genes and the size of the particular gene group being considered. Applying one's statistical significance threshold of choice identifies which GO labels are more abundantly associated with genes in each group than expected by chance alone.

Because the chi-squared statistic works best when the observed values are normally distributed around the expected mean (as a rule of thumb, this occurs when at least six genes are expected to have a particular label), this statistic will not reliably identify the function of small groups, or assign to groups uncommon GO labels. Additionally, since a gene can belong to several GO labels, multiple correspondences between a gene group and GO labels cannot be considered independent events and are therefore recalcitrant to statistical analysis. If necessary, this problem can be addressed by resampling GO label assignments so that each gene is only assigned a single GO label, and repeating the GO label-gene group correspondence.

Assignment of functional GO identities to similarity-defined gene groups can lead to gene groups with many (even dozens) of associated functions (28, 29). This leads to some difficulty in interpreting the meaning of each group, a problem that can be superficially mitigated by imposing a more strict significance threshold in the statistical assignment of GO functions to gene groups. Moreover, selecting related experimental perturbations would likely enrich for certain gene group-to-function associations. For example, if the original time series data comprises only observations of transcription following cellular shocks such as heat, cold, osmotic change, and oxidation, GO label GO:0006950 (response to stress) should be highly associated with one or more groups, and more highly so than other GO labels whose signals are not enriched in the data. With respect to drug discovery, time courses reflecting the broadest spectrum of physiological response to drug treatment (e.g., all stages of tumorigenesis in several different tissue types) are ideal for mapping function onto similarly behaving genes, as well as identifying relevant regulatory influences.

7. Reducing Group Data to Representative Data

Having defined groups, the collective behavior of all the genes within the group must be identified. If *k*-means or SOM was used to assign the groups, then representative time series for each group are already present as the centroid of *k*-means partition or SOM module. These are not necessarily equal to the centroid (average) of all the values of the genes in each group at each time point, but will be very close to it for the large *n* used in generating cellular networks. The median value of all genes with the groups at each time point can also be used as the representative time series. In the author's experience, which of these methods is used to reduce the many behaviors of the genes within a group to a single behavior makes no discernible qualitative difference in the final models or analyses.

8. Deriving the Regulatory Model

The number of classes of model that can be derived from time series data is exceedingly large (49). Here, the focus will be on first-order linear models that have been shown to predict the activity of large groups of genes over hours-long timescales (28–30). In this model, the activity of a gene group i at time $t+1$ equals the activity of gene group 1 at time t times the extent to which gene group 1 predicts activity in gene group i, plus the activity of gene group 2 at time t times the extent to which gene group 2 predicts activity in gene group i, etc.: $X_{i,t+1} = a_{1i}X_{1,t} + a_{2i}X_{2,t} + \ldots + a_{ki}X_{k,t}$. Here $X_{i,t}$ is the activity value of gene i at time t, k is the number of gene groups, and a_{lm} is the extent to which activity in gene group l predicts activity in gene group m. In matrix notation, $\mathbf{X}_{t+1} = \mathbf{A}\mathbf{X}_t$, where $\mathbf{X}_t$ is the vector of gene group activities at time t, and $\mathbf{A}$ is the $k \times k$ transition matrix of pair-wise effective regulatory influences.

The k^2 entries of $\mathbf{A}$ are the models parameters that must be solved for using our gene group data. To solve for this many parameters directly requires at least k^2 equations. Every time interval in the data provides k equations, one for each gene group. As mentioned above, microarray data contains considerable noise. Thus k should be significantly smaller than the number of time intervals observed, and the regulatory influence values in the matrix **A** should be calculated using a least-squares fit that minimizes the discrepancy between the model predictions and all observations. For AfCS data, in which 33 time intervals are available, the values in a 12×12 transition matrix converge when 28 or more time interval observations are used in their calculation (29), suggesting that the baseline ratio of microarray observations to gene groups should exceed 2:1.

Because the regulatory influences will be calculated as a least-squares fit to the available data, the resulting values will each represent the average regulatory effect of a gene group on another. This average value will always be defined, even if the value that would be inferred for this regulatory influence were to vary significantly between the time intervals chosen for its derivation. This can be assessed by a bootstrapping (50) method in which many random subsets of the original time intervals (sufficient in number to derive a converged transition matrix) are chosen, used to derive a transition matrix, and each regulatory influence is evaluated across these replicates for consistency in its value (28, 29). Given the estimates of each regulatory influence across these replicates, it is possible to assign statistical confidence to their consistency across regulatory contexts (i.e., different observations). In particular, if the magnitude of the mean of the estimates is more than twice the standard deviation of the estimates, there is a >95% chance the influence is nonzero with sign corresponding to that of the mean of its estimates. Applying a statistical significance threshold such as this one also enables the construction of regulatory networks, which are not trivially complete (containing all pair-wise regulatory influences), from those regulatory influences in the transition matrix found to have significantly consistent values across regulatory contexts. These networks will form the basis of some of the visualizations discussed below.

The value of the estimates of the regulatory influence between gene groups for which no effective regulation exists will be highly variable as a function of noise in the various time interval observations. Thus, the bootstrapping method described above will eliminate not only regulatory influences that exist but are inconsistent across different contexts, but also those influences that do not really exist at all. It takes less information to specify that a regulatory influence does or does not exist than to specify its value (a total of $-(k^2)(p\log(p) + (1-p)\log(1-p))$ bits, where p is the fraction of regulatory influences that actually exist, according to Shannon information theory) (51). By allowing the bootstrapping algorithm to eliminate inconsistent regulatory influences, we can use this liberated information to solve for those regulatory influences that do exist between larger numbers of gene groups (29). As the information required to indicate which regulatory influences are present or absent grows by k^2, and the max number of regulatory influences that can be numerically solved grows as $33k$ (in the AfCS case), even given the absence of noise, and arbitrarily high sparseness in the regulatory network, there is an upper bound on the number of gene groups whose sparse mutual influences can be determined. Practically, with the AfCS data, the mutual regulation of 72 gene groups was derived, with 337/5,184 possible regulatory influences receiving statistical support (29).

This framework can be generalized to nonlinear and higher order regulatory models. For example, a system of equations of the form $X_{i,t+1} = a + (b_{1i}X_{1,t} + c_{1i}X_{1,t}^{\ 2}) + (b_{2i}X_{2,t} + c_{2i}X_{2,t}^{\ 2}) + \ldots + (b_{ki}X_{k,t} + c_{ki}X_{k,t}^{\ 2})$ could be solved for the quadratic regulatory influences on gene group X_i from all other gene groups. This model has $k+2k^2$ parameters, restricting the total number of interacting gene groups which could be modeled given a dataset of fixed size. Similarly, the model $X_{i,t+1} = (a_{11i}X_{1,t}X_{1,t} + a_{21i}X_{2,t}X_{1,t} + \ldots + a_{k1i}X_{k,t}X_{1,t}) + (a_{22i}X_{2,t}X_{2,t} + \ldots + a_{k2i}X_{k,t}X_{2,t}) + \ldots + a_{kki}X_{k,t}X_{k,t}$ identifies the linear second-order influences on X_i from all other pairs of gene groups, but has $(k^3 - k^2)/2$ parameters. These examples show the versatility of the modeling framework, but are likely unnecessary for most purposes, as the linear first-order model is capable of predicting the large-scale behavior of cells over all time intervals solvable using the AfCS data with $r > 0.95$ (28).

9. Verifying Model Predictivity

The predictive value of the large-scale model can be tested by applying the inferred transition matrix to data from the original time series and assessing the correlation between the prediction of the model ($\mathbf{AX}_t$) and the observed cell-state ($\mathbf{X}_{t+1}$). However, this approach can yield inappropriately high correlations because the $\mathbf{X}_t$ to $\mathbf{X}_{t+1}$ transition may have been used to determine the values of $\mathbf{A}$. Therefore, a cross-validation technique should be used, in which subsets of the original time series data are set aside, the model parameters are inferred using the remaining data, and then the correlation of the model's predictions and actual observation is tested using the data that were set aside (52). This should be done so that the dataset retained for calculation of the model parameters is large enough to assure convergence of the model (see above). In order to obtain a full assessment of the model performance across a variety of removed data, random subsets should be removed over many replicates, and the mean correlation between the model prediction and observed data assessed.

10. Model Visualization

Even though these cellular models radically reduce the number of regulatory influences compared to the cell at its highest resolution, they still generate copious numerical values in need of interpretation. This task can be somewhat simplified by visualizing aspects of the data. This section cannot be exhaustive of the useful methods to display large-scale models, but will seek to highlight a few techniques that have been of particular use to the author.

The aggregated dataset composed of gene groups by time interval observations can be highly informative in the context of drug development. Displaying this matrix as a colored 2-dimensional array, with gene groups clustered (using one of the distance methods discussed above) by similarity in their responses across time observations, shows critical information. For example, in the case of the AfCS data, the gene group enriched for involvement in cation homeostasis, small molecule transport, and oxygen metabolism, is strongly and specifically activated 0.5 and 1 h after b-lymphocytes are incubated in the presence of CGS-21680 hydrochloride (28), a selective agonist of the adenosine receptor (53). If the time series data are composed of several independent treatments, rather than a few long time courses with many time points, these treatments can also be clustered to reveal which treatments induce the same changes in large-scale gene groups, a useful technique considering the correlations between gene expression changes, drug chemical structure, and drug target (54).

The matrix of regulatory influences (**A**) can be similarly treated, by coloring influences according to their sign and magnitude, sorting rows (which correspond to gene groups) according to their similarity in regulatory inputs, and columns by their similarity in regulatory output. In large-scale transcription networks, this view reveals a striking pattern – columns of constant color – e.g., if a gene group is an activator of one gene group, it will tend to be an activator of all other gene groups. This target nonspecificity occurs over all time steps analyzed, especially those with high magnitudes of influence, and appears to be a property exclusive to networks composed of the largest gene groups (29). The phenomenon likely reflects the gating effect of a rate-limiting process in the cell (such as metabolite synthesis) onto which each gene group has a specific activating or inhibiting effect.

Once bootstrapping has been used to identify regulatory influences with the greatest statistical support, the gene groups and those regulatory influences can be displayed as a network in which gene groups are nodes and directional edges (arrows) connecting two gene groups indicate a regulatory influence from one to the other. Numerous programs are available to draw networks in a visually pleasing manner, a problem for which there is no clear solution, but the following approach works well in the context of cellular networks: use the regulatory outputs of the gene groups (from the transition matrix **A**) to generate a pair-wise distance matrix between them, and multidimensional scaling (MDS) to map the gene groups in this distance space onto the plane (55). Depending on the configuration of the MDS, the ranking of pair-wise distances between the gene groups will be preserved, or an approximation of the proportional distances will be imposed. Thus, gene groups that activate or inhibit the same targets will be near one another. The specific regulatory influences can then be added as arrows to the network. A similar effect can be accomplished by treating the regulatory outputs of each gene group as a point in k-dimensional space, performing PCA on the gene groups in this space, and plotting them along axes of the first and second principal components.

Lastly, if SOM was used to generate the gene groups, the spatial SOM array can be used to visualize similarity in gene groups targeted by activators or inhibitors. This method has been used to show that while gene groups in adjacent SOM modules are not particularly likely to exert similar regulatory outputs, they are very likely to receive similar regulatory inputs. This approach also has revealed that gene groups undergoing constitutive responses to chemokine perturbation are less regulatory of other gene groups than gene groups with transiently fluctuating responses (29).

11. Interpretation of Regulatory Networks

Results from large-scale regulatory networks have been interpreted on numerous conceptual axes. In closing, a few of the reported observations will be briefly discussed to illustrate the diversity of analyses possible with large-scale cellular networks. With the AfCS data, it is possible to distinguish the effects of one gene group on another over 0.5, 1, 1.5, 2, 3, and 3.5 h. The specific regulatory influences between gene groups often fluctuate dramatically across these intervals, indicating that the regulation between large gene groups is brought about by multiple mechanisms with different characteristic kinetics. Additionally, the average magnitude of regulatory influences fluctuates over time, suggesting the cell can perform a fixed number (~12) of major transcriptional shifts during the cell cycle. The speed of this "transcriptional impulse" goes up as finer-scale networks are considered (i.e., the 72 gene group network) (29).

Numerous other network attributes change as a function of the scale of the network. For example, the largest-scale networks (composed of 12 gene groups) have normally distributed regulatory outputs across gene groups (i.e., all gene groups exert some influences). In finer-scale networks (composed of up to 72 gene groups), the distribution is power-law dependent (i.e., a few groups exert many influences and most groups exert none), consistent with the edge distribution in fine-scale genetic interaction and protein physical interaction networks (56). The largest-scale networks are almost free of target-specific regulation. This begins to appear in networks of intermediate scale, and is prevalent by the $k=72$ network, suggesting that rate-limiting cellular functions dominate the largest cellular processes without gating smaller-scale behaviors. Additionally, positive feedback is nearly absent at the largest scale, but found at smaller scales, consistent with the largest cellular activities operating under strong homeostatic constraints (28, 29). Lastly, this multiscale analysis can reveal how large-scale cellular functions are brought about by the combined action of small-scale functions, by coupling the GO identity of smaller gene groups to the GO identity of the larger groups they compose (29). All of these observations are potentially relevant to drug development as they imply scale-specific constraints on drug effect.

Acknowledgments

The author would like to thank Jay Strader for helpful comments on the manuscript.

References

1. Drigues N, Poltyrev T, Bejar C, Weinstock M, Youdim MB (2003) cDNA gene expression profile of rat hippocampus after chronic treatment with antidepressant drugs. J Neural Transm 110:1413–1436
2. Natsoulis G, El Ghaoui L, Lanckriet GR, Tolley AM, Leroy F, Dunlea S, Eynon BP, Pearson CI, Tugendreich S, Jarnagin K (2005) Classification of a large microarray data set: algorithm comparison and analysis of drug signatures. Genome Res 15:724–736
3. Slatter JG, Templeton IE, Castle JC, Kulkarni A, Rushmore TH, Richards K, He Y, Dai X, Cheng OJ, Caguyong M, Ulrich RG (2006) Compendium of gene expression profiles comprising a baseline model of the human liver drug metabolism transcriptome. Xenobiotica 36:938–962
4. Gilman AG, Simon MI, Bourne HR, Harris BA, Long R, Ross EM, Stull JT, Taussig R, Arkin AP, Cobb MH, Cyster JG, Devreotes PN, Ferrell JE, Fruman D, Gold M, Weiss A, Berridge MJ, Cantley LC, Catterall WA, Coughlin SR, Olson EN, Smith TF, Brugge JS, Botstein D, Dixon JE, Hunter T, Lefkowitz RJ, Pawson AJ, Sternberg PW, Varmus H, Subramaniam S, Sinkovits RS, Li J, Mock D, Ning Y, Saunders B, Sternweis PC, Hilgemann D, Scheuermann RH, DeCamp D, Hsueh R, Lin KM, Ni Y, Seaman WE, Simpson PC, O'Connell TD, Roach T, Choi S, Eversole-Cire P, Fraser I, Mumby MC, Zhao Y, Brekken D, Shu H, Meyer T, Chandy G, Heo WD, Liou J, O'Rourke N, Verghese M, Mumby SM, Han H, Brown HA, Forrester JS, Ivanova P, Milne SB, Casey PJ, Harden TK, Doyle J,

Gray ML, Michnick S, Schmidt MA, Toner M, Tsien RY, Natarajan M, Ranganathan R, Sambrano GR (2002) Overview of the alliance for cellular signaling. Nature 420:703–706

5. Zhu X, Hart R, Chang MS, Kim JW, Lee SY, Cao YA, Mock D, Ke E, Saunders B, Alexander A, Grossoehme J, Lin KM, Yan Z, Hsueh R, Lee J, Scheuermann RH, Fruman DA, Seaman W, Subramaniam S, Sternweis P, Simon MI, Choi S (2004) Analysis of the major patterns of B cell gene expression changes in response to short-term stimulation with 33 single ligands. J Immunol 173:7141–7149
6. Deininger MW, Druker BJ (2003) Specific targeted therapy of chronic myelogenous leukemia with imatinib. Pharmacol Rev 55:401–423
7. Balabanov S, Bartolovic K, Komor M, Kanz L, Hofmann WK, Brümmendorf TH (2005) Gene expression profiling of normal hematopoietic progenitor cells under treatment with imatinib in vitro. Leukemia 19:1483–1485
8. Kerr DJ, Workman P (1994) New molecular targets for cancer chemotherapy. Salem, MA USA: CRC Press: 1–194
9. Kauffman SA (1969) Metabolic stability and epigenesis in randomly constructed genetic nets. J Theor Biol 22:437–467
10. Waddington CH (1956) Principles of embryology. Allen and Unwin Ltd., London
11. Davidson EH, Rast JP, Oliveri P, Ransick A, Calestani C, Yuh CH, Minokawa T, Amore G, Hinman V, Arenas-Mena C, Otim O, Brown CT, Livi CB, Lee PY, Revilla R, Rust AG, Pan Z, Schilstra MJ, Clarke PJ, Arnone MI, Rowen L, Cameron RA, McClay DR, Hood L, Bolouri H (2002) A genomic regulatory network for development. Science 295:1669–1678
12. Bar-Yam Y, Epstein IR (2004) Response of complex networks to stimuli. Proc Natl Acad Sci USA 101:4341–4345
13. de Visser JA, Hermisson J, Wagner GP, Ancel Meyers L, Bagheri-Chaichian H, Blanchard JL, Chao L, Cheverud JM, Elena SF, Fontana W, Gibson G, Hansen TF, Krakauer D, Lewontin RC, Ofria C, Rice SH, von Dassow G, Wagner A, Whitlock MC (2003) Perspective: evolution and detection of genetic robustness. Evolution 57:1959–1972
14. Huang S (1999) Gene expression profiling, genetic networks, and cellular states: an integrating concept for tumorigenesis and drug discovery. J Mol Med 77:469–480
15. Ao P, Galas D, Hood L, Zhu X (2008) Cancer as robust intrinsic state of endogenous molecular-cellular network shaped by evolution. Med Hypotheses 70:678–684
16. Huang S, Eichler G, Bar-Yam Y, Ingber DE (2005) Cell fates as high-dimensional attractor states of a complex gene regulatory network. Phys Rev Lett 94:128701
17. Bang AG, Carpenter MK (2008) Deconstructing pluripotency. Science 320:320–321
18. Chang HH, Hemberg M, Barahona M, Ingber DE, Huang S (2008) Transcriptome-wide noise controls lineage choice in mammalian progenitor cells. Nature 453:544–547
19. Bar-Yam Y, Harmon D, de Bivort B (2009) Systems biology. Attractors and democratic dynamics. Science 323:1016–1017
20. Sui AI, Wiltshire T, Bataloc S, Lapp H, Ching KA, Block D, Zhang J, Soden R, Hayakawa M, Kreiman G, Cooke MP, Walker W, Hogenesch JB (2004) A gene atlas of the mouse and human protein-encoding transcriptomes. Proc Natl Acad Sci USA 101:6062–6067
21. Tyler AL, Asselbergs FW, Williams SM, Moore JH (2009) Shadows of complexity: what biological networks reveal about epistasis and pleiotropy. Bioessays 31:220–227
22. Kestler HA, Wawra C, Kracher B, Kühl M (2008) Network modeling of signal transduction: establishing the global view. Bioessays 30:1110–1125
23. Halazonetis TD, Georgopoulos K, Greenberg ME, Leder P (1988) c-Jun dimerizes with itself and with c-Fos, forming complexes of different DNA binding affinities. Cell 55:917–924
24. Cortes P, Flores O, Reinberg D (1992) Factors involved in specific transcription by mammalian RNA polymerase II: purification and analysis of transcription factor IIA and identification of transcription factor IIJ. Mol Cell Biol 12:413–421
25. D'Haeseleer P, Wen X, Fuhrman S, Somogyi R (1999) Linear modeling of mRNA expression levels during CNS development and injury. Pac Symp Biocomput, 41–52
26. Gardner TS, di Bernardo D, Lorenz D, Collins JJ (2003) Inferring genetic networks and identifying compound mode of action via expression profiling. Science 301:102–105
27. Tegner J, Yeung MK, Hasty J, Collins JJ (2003) Reverse engineering gene networks: integrating genetic perturbations with dynamical modeling. Proc Natl Acad Sci USA 100:5944–5949
28. de Bivort B, Huang S, Bar-Yam Y (2004) Dynamics of cellular level function and regulation derived from murine expression array data. Proc Natl Acad Sci USA 101:17687–17692

29. de Bivort B, Huang S, Bar-Yam Y (2007) Empirical multiscale networks of cellular regulation. PLoS Comput Biol 3:1968–1978
30. Haye A, Dehouck Y, Kwasigroch JM, Bogaerts P, Rooman M (2009) Modeling the temporal evolution of the *Drosophila* gene expression from DNA microarray time series. Phys Biol 6:016004
31. Lockhart DJ, Winzeler EA (2000) Genomics, gene expression and DNA arrays. Nature 405:827–836
32. Chen Y, Kamat V, Dougherty ER, Bittner ML, Meltzer PS, Trent JM (2002) Ratio statistics of gene expression levels and applications to microarray data analysis. Bioinformatics 18:1207–1215
33. Koren A, Tirosh I, Barkai N (2007) Autocorrelation analysis reveals widespread spatial biases in microarray experiments. BMC Genomics 8:164
34. Gorban AN, Kegl B, Wunsch DC, Zinovyev A (2008) Principle manifolds for data visualization and dimension reduction. Springer-Verlag, Berlin
35. Lee JA, Verleysen M (2007) Nonlinear dimensionality reduction. Springer, New York, NY, USA
36. Jolliffe IT (1986) Principal component analysis. Springer-Verlag, New York, NY, USA
37. Ihmels J, Friedlander G, Bergmann S, Sarig O, Ziv Y, Barkai N (2002) Revealing modular organization in the yeast transcriptional network. Nat Genet 31:370–377
38. Eisen MB, Spellman PT, Brown PO, Botstein D (1998) Cluster analysis and display of genome-wide expression patterns. Proc Natl Acad Sci USA 95:14863–14868
39. Iyer VR, Eisen MB, Ross DT, Schuler G, Moore T, Lee JC, Trent JM, Staudt LM, Hudson JJ, Boguski MS, Lashkari D, Shalon D, Botstein D, Brown PO (1999) The transcriptional program in the response of human fibroblasts to serum. Science 283:83–87
40. Tamayo P, Slonim D, Mesirov J, Zhu Q, Kitareewan S, Dmitrovsky E, Lander ES, Golub TR (1999) Interpreting patterns of gene expression with self-organizing maps: methods and application to hematopoietic differentiation. Proc Natl Acad Sci USA 96:2907–2912
41. Bergmann S, Ihmels J, Barkai N (2004) Similarities and differences in genome-wide expression data of six organisms. PLoS Biol 2:E9
42. Ihmels J, Levy R, Barkai N (2004) Principles of transcriptional control in the metabolic network of *Saccharomyces cerevisiae*. Nat Biotechnol 22:86–92
43. Ihmels J, Bergmann S, Berman J, Barkai N (2005) Comparative gene expression analysis by differential clustering approach: application to the *Candida albicans* transcription program. PLoS Genet 1:e39
44. Sneath PHA, Sokal RR (1973) Numerical taxonomy. Freeman, San Francisco, CA, USA
45. Fitch WM, Margoliash E (1967) Construction of phylogenetic trees. Science 155:279–284
46. MacQueen JB (1967) Proceedings of the 5th Berkeley symposium on mathematical statistics and probability 1, 281–297
47. Kohonen T (2001) Self-organizing maps. Springer, Berlin, Heidelberg, Germany
48. Gene Ontology Consortium (2001) Creating the gene ontology resource: design and implementation. Genome Res 11:1425–1433
49. Rao TS (1993) Developments in time series analysis. Chapman & Hall, London, UK
50. Efron B (1979) Bootstrap methods: another look at the jackknife. Ann Stat 7:1–26
51. Weaver W, Shannon CE (1949) The mathematical theory of communication. University of Illinois Press, Urbana, pp 3–28
52. Stone M (1974) Cross-validatory choice and assessment of statistical predictions. J R Stat Soc Series B 36:111–147
53. Kull B, Svenningsson P, Fredholm BB (2000) Adenosine A(2A) receptors are colocalized with and activate g(olf) in rat striatum. Mol Pharmacol 58:771–777
54. Scherf U, Ross DT, Waltham M, Smith LH, Lee JK, Tanabe L, Kohn KW, Reinhold WC, Myers TG, Andrews DT, Scudiero DA, Eisen MB, Sausville EA, Pommier Y, Botstein D, Brown PO, Weinstein JN (2000) A gene expression database for the molecular pharmacology of cancer. Nat Genet 24:236–244
55. Torgerson WS (1958) Theory and methods of scaling. Wiley, New York, NY, USA
56. Albert R (2005) Scale-free networks in cell biology. J Cell Sci 118:4947–4957

Chapter 8

Translational Bioinformatics and Systems Biology Approaches for Personalized Medicine

Qing Yan

Abstract

Systems biology and pharmacogenomics are emerging and promising fields that will provide a thorough understanding of diseases and enable personalized therapy. However, one of the most significant obstacles in the practice of personalized medicine is the translation of scientific discoveries into better therapeutic outcomes. Translational bioinformatics is a powerful method to bridge the gap between systems biology research and clinical practice. This goal can be achieved through providing integrative methods to enable predictive models for therapeutic responses. As a media between bench and bedside, translational bioinformatics has the mission to meet challenges in the development of personalized medicine. On the biomedical side, translational bioinformatics would enable the identification of biomarkers based on systemic analyses. It can improve the understanding of the correlations between genotypes and phenotypes. It would enable novel insights of interactions and interrelationships among different parts in a whole system. On the informatics side, methods based on data integration, data mining, and knowledge representation can provide decision support for both researchers and clinicians. Data integration is not only for better data access, but also for knowledge discovery. Decision support based on translational bioinformatics means better information and workflow management, efficient literature and resource retrieval, and communication improvement. These approaches are crucial for understanding diseases and applying personalized therapeutics at systems levels.

Key words: Systems biology, Pharmacogenomics, Personalized medicine, Translational, Bioinformatics, Outcomes, Biomarkers, Genotypes, Phenotypes, Data integration, Data mining, Knowledge representation, Decision support, Interactions, Workflow

1. Translational Bioinformatics: The Bridge Between the Gap of Systems Biology and Personalized Medicine

Systems biology and pharmacogenomics are emerging and promising fields that will provide a thorough understanding of diseases and enable personalized therapy. Pharmacogenomics studies the genetic basis of individual variations in response to drug therapies (1). The goal of pharmacogenomics is to achieve personalized

Qing Yan (ed.), *Systems Biology in Drug Discovery and Development: Methods and Protocols*, Methods in Molecular Biology, vol. 662, DOI 10.1007/978-1-60761-800-3_8, © Springer Science+Business Media, LLC 2010

therapy by predicting the susceptibility to diseases and response to drugs and vaccines (2). Such study should not be limited to single genes or single nucleotide polymorphisms (SNPs), as genes interact with each other. Systems biology is needed to study the interactions among biological elements toward the understanding of diseases at the system level (1). As a new approach in analyzing biological systems at all levels of information, systems biology may offer novel strategies for drug discovery and development.

However, one of the most significant obstacles in the practice of personalized medicine is the translation of scientific discoveries into better therapeutic outcomes. A critical factor in the successful translation from bench to bedside is the access and analysis of integrated data within and across functional domains (3). For example, most of the clinical and basic research data are currently stored in disparate and separate databases, it is inefficient for researchers to access these data (4). Although there has been an overwhelming demand for data management, few tools are available that meet the demand.

With the plethora of various technologies, the exponential growth of medical information far surpasses the ability to digest or apply it in clinics. However, traditional clinical information systems have not focused on knowledge management or decision support functions (5). With the strong demand for personalized medicine based on systems biology and pharmacogenomics, the need for novel informatics support to improve communication between basic scientists and clinicians becomes increasingly urgent.

Translational bioinformatics is a powerful method to bridge the gap between systems biology research and the practice of personalized medicine. Bioinformatics uses computational approaches to solve problems and improve the communication, understanding, analysis, and management of biomedical information (6). As defined by the American Medical Informatics Association (AMIA), translational bioinformatics is a new field "to optimize the transformation of increasingly voluminous biomedical data, and genomic data in particular, into proactive, predictive, preventive, and participatory health (7)."

For instance, the most difficult and critical parts in clinical outcome assessment include inefficient management of clinical workflow, ineffective communication, and the lack of a centralized data management. Biomedical informatics applications in translational medicine would enable effective management of the workflow in both clinical and research environments (8). Translational bioinformatics would improve the integration of clinical and laboratory data streams. Applications such as electronic record architecture and concept representation would facilitate the share of information, as well as the establishment, implementation, and the compliance of the standards. In addition, translational bioinformatics can help the reduction of clinical risks and the efficient use of healthcare resources. For example, computer-based information systems are the most cost-effective and promising strategy for preventing adverse therapeutic events (9).

Most importantly, translational bioinformatics can promote the practice of personalized medicine. It will empower scientists and clinicians to design personalized strategies to bring the right drugs with the right dosages to the right people (see Fig. 1). Such approaches will help overcome therapeutic resistance and adverse effects, and improve communications among multidisciplinary groups.

Such goals can be achieved through providing integrative methods to enable predictive models for therapeutic responses. Informatics studies in systems biology may enable the simulation of networks of interacting components (10). Combined with high-throughput studies, bioinformatics can help identify patient's genetic profiles and patient subgroups in order to develop the optimal therapy.

As a media between bench and bedside, translational bioinformatics has the mission to meet challenges in the development of personalized medicine. These challenges come from two sides of the area, the informatics side and the biomedical side, as illustrated in Fig. 1. They define the vision that drives the way by which translational bioinformatics will be studied, developed, and applied.

On the biomedical side, translational bioinformatics would enable the identification of systems-based biomarkers, understanding of genotype–phenotype correlations, and modeling of systemic interactions and interrelationships. These cannot be achieved without concrete informatics basis. Therefore, on the

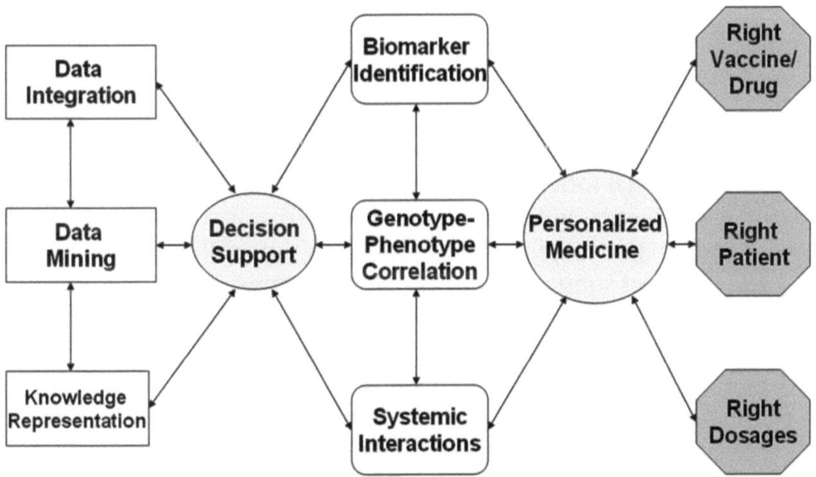

Fig. 1. The roadmap of translational bioinformatics and systems biology for the development of personalized medicine.

informatics side, methods based on data integration, data mining, and knowledge representation can provide decision support for both researchers and clinicians.

2. Applying Translational Bioinformatics in Analyses of Biomarkers and Systemic Interactions

An important objective for translational bioinformatics is to help the identification of biomarkers. Biomarkers are indicators of biologic states. They are usually analyzed in a variety of functional pathways at different systems levels (11). They are objectively measured and represent the responses to therapeutic interventions. These include genetic markers at the molecular level, and imaging markers at the system level. As the main elements in predicting efficacy and safety from animal to man, biomarkers are key signs in a translational process that can be accountable for 80–90% of translational success (12).

Biomarkers have great potentials for improving disease diagnosis, treatment selection and the prevention of side effects. Predictive and prognostic biomarkers for characterizing different subsets of patients have been used for outcome prediction and assessment in a variety of diseases. These diseases include cancer, cardiovascular diseases, respiratory diseases, rheumatoid arthritis, and neurological diseases (11, 13–18).

For instance, patient's genetic profiles would provide the basis for the prediction of a patient's response to particular therapeutics, and empower health practitioners to make the right decisions for the treatment (10). Biomarkers would enable the identification of specific patient groups at risks, such as in atherosclerotic disease (19). The establishment of the association between biomarkers and certain patients may lay the ground for the objective and accurate practice of personalized medicine.

Translational bioinformatics would enable the incorporation of biomarkers into clinical trial design and outcome assessment. Informatics applications can facilitate pattern recognition, and expedite and validate the discovery of biomarkers. These are crucial for understanding the mechanisms of patient responses to diseases and therapeutics (20). To achieve the goal of personalized therapy, informatics tools are especially useful for the prediction of substantial disease subpopulations using biomarkers.

For example, a bioinformatics tool was used to illustrate the correlation between biomarkers and treatment outcomes in breast cancer patients (21). The software graphed the relationship between the protein expression of p53 and patient survival. It showed two distinct subpopulations of tumors with different levels of expressers. In addition, the tool enabled the recognition of the association between different biomarkers, such as Human

Epidermal growth factor Receptor 2 (HER2) and Estrogen Receptor (ER) expression in the cohort of breast cancer patients, in correlation with patient survival outcomes.

Another mission of translational bioinformatics is to help elucidate genotype–phenotype correlations (see Fig. 1). Understanding of the correlations is one of the key issues that needs to be solved in pharmacogenomics and personalized medicine (1). Such understanding may help change the focus of medical practice from diseases to humans and bring hope for the transformation from disease treatment to prevention (22). The exploration of genotype–phenotype correlations may lead to novel insights in disease pathogenesis and treatments at various systems levels. For such purposes, translational bioinformatics is important for correlating genetic structural variations such as polymorphisms with phenotypic response data.

Furthermore and most importantly, as shown in Fig. 1, translational bioinformatics can improve the understanding of systemic interactions. For example, in the case of influenza prevention and treatment, the systemic interactions include those among humans, influenza viruses, vaccines, drugs, and the environment (see Chapter 14). The core value of systems biology is that it considers a whole system as more than the simple sum of its parts. This is because the behavior of a whole system does not merely come from its separated building blocks, but rather through the complex interrelationships and interactions among them. As an important tool in systems biology studies, translational bioinformatics would enable such a holistic view.

This view is opposite to the reductionist approach, which percepts a system as the sum of its parts and the dissection of these parts would just be enough to understand the whole system's behavior. Through integrating and analyzing complex data, translational bioinformatics can help establish predictive models for interrelationships and interactions among the components of the system. Such models can in turn help predict the behavior of the whole system, such as patient responses to vaccines or drugs.

3. Data Integration Methods in Translational Bioinformatics for Decision Support

3.1. Decision Support Based on Translational Bioinformatics

On the informatics side, translational bioinformatics can provide knowledge management and decision support functions for applying systems biology in the practice of personalized medicine. The decision-making process in both clinical and laboratory settings is becoming more and more complicated with the growing amount of data combined with factors such as time, cost, and individualized treatment. For translational studies, a variety of data sources are involved, from high-throughput data to clinical data. The understanding and interpretation of the data are difficult because of knowledge domain barriers. Translational bioinformatics can correlate the relevant information to promote data sharing across domains, and link networks of researchers, clinicians, and patients.

Decision support systems (DSS) based on translational bioinformatics can be incorporated into electronic medical record systems and other clinical information systems. Here decision support includes information management and providing patient-specific recommendations by health practitioners (23). In addition, effective decision support means better documentation and workflow, efficient literature and resource retrieval, and communication improvement.

The goal of such decision support is to bring the "right knowledge to the right people in the right form at the right time (24)." Workflow integration is a critical component toward the success of decision support (23). A widely used methodology for modeling the workflow and decision support processes is the Unified Modeling Language (UML) (23, 25). With a specific methodology designed for biomedical sciences, UML can be used for modeling of biomedical knowledge to facilitate concept extraction, elicitation, representation, sharing, and delivery of knowledge to both researchers and clinicians (25).

As shown in Fig. 1, an essential component in translational bioinformatics for decision support is data integration, which may enable effective workflow management in both laboratory and clinical environments. Data integration is the fundamental method for supplying answers to decision-making questions. Such questions include the identification of gene expression profile with diseases and therapeutic responses.

A direct benefit of data integration is the promotion of time and economic efficiency, as well as clinical outcomes. For example, the integration of information from multiple layers of biological regulation from multiple genome-wide data sources has been suggested to improve the prediction of cancer outcome (26). Combined with data mining techniques, data integration is the basis for finding correlations and knowledge discovery (27), which is crucial for systems biology studies.

3.2. Data Integration Approaches in Translational Bioinformatics

Data integration is a crucial part in the organization of information for biomarker identification and predictive modeling of interactions. Because systems biology and pharmacogenomics contain knowledge from multiple domains, the data integration process is especially important. Data integration is not only for simple data access, but also for knowledge discovery and decision support (6). Specifically, it is a process that standardizes names and values, resolves inconsistencies in the representation of data, and integrates common values together.

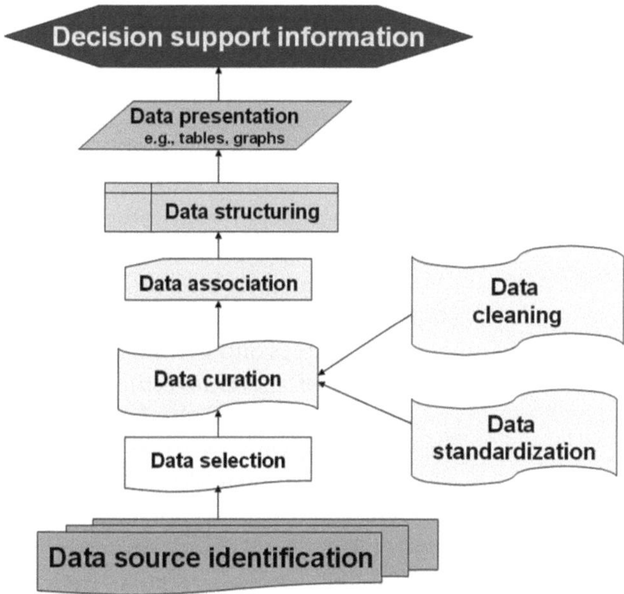

Fig. 2. Data integration processes in translational bioinformatics for decision support.

Figure 2 summarizes the process of data integration for translational bioinformatics. This is a data "evolution" process from "untreated" crude format to the ready-to-use information for decision support. During the data integration process, data are chosen by screening all the available sources and choosing the ones that can best fulfill the requirements. For translational studies, different types of data from various knowledge domains need to be selected and collected, such as nucleotide and amino acid sequence information, expression data, and protein–protein interaction data. As mentioned above, data sources can be from both high-throughput experiments and clinical practice.

For example, for genomic studies, data merging techniques are usually used. That is, different data sets can be concatenated in the database through cross-referencing the sequence identifiers. Another way is to integrate multiple layers of data into one mathematical model, such as a kernel-based integration framework (26). In this approach, data are selected based on the relevant features from all available sources, and combined in a machine learning-based model.

Some tools, such as the one called POINTILLIST, are designed specifically for the integration of data such as those from gene expression arrays, proteomics, and chromatin immunoprecipitation on chip assays (28). The inference of POINTILLIST is based on a weighted statistical method. In this method, each evidence type is assigned a trustworthiness weight. The software package contains programs of Data Manager, Data Normalizer, Significance Calculator, and Evidence-Weighted Inferer (29).

The Data Manager is designed for data subselection. The Data Normalizer can be used for quantile normalization of microarray data. The Significance Calculator is designed to calculate *P*-values for measurements based on observations for negative controls. The Evidence-Weighted Inferer can be used to combine the *P*-values for observations of different evidence types.

The overall data integration approach can be based on the data consolidation or federation approaches. Data consolidation is performed through constructing a central database with a single integrated data model. That is, data are extracted from various data sources and centralized at one place. This approach has the benefit of the enforcement of the standardization of heterogeneous data. It will also help solve interoperability and compatibility issues with other available information systems. However, because it is based on a single data model, it may be hard to evolve and hard to update.

The data federation approach usually links databases together, extracts data on a regular basis, and combines data for queries. This approach has the benefit of providing a single portal and "real time" data access to support customized queries. However, because the data are from different sources, they may not be always in sync.

Some advanced computational techniques can be applied in bioscience. For instance, a data warehouse can be useful to provide a unified platform for data curation (4). Data warehousing is the technique to integrate data from different sources into a common format. It is a collection of subject-oriented databases, and designed exclusively for decision support purposes.

Data are rarely clean. On the basis of data concept extraction and data modeling, such as using UML (25), the selected data need to be cleansed, validated, curated, updated, and structuralized (see Fig. 2). Because biomedical data have a feature of volatility, that is, the relevant knowledge in the database may grow and change over time, updates are needed frequently. Selected data may have different formats from disparate sources. Common values in data need to be integrated together with a consistent and unified format.

Redundancies and inconsistencies are common problems with biomedical data and need to be resolved. For example, one gene may have many different names, such as TAP1, ABC17, ABCB2, APT1, which all refer to the same gene but may have different entries. Such redundancies and inconsistencies need to be solved, such as using genetic nomenclature in Gene Ontology (GO) (see Table 1). Interoperability and links to national databases such as cancer Biomedical Informatics Grid (caBIG) can also be considered.

Standardization efforts are important for the integration of translational bioinformatics into the general health information

Table 1
Resources of biomedical standards for translational bioinformatics studies

Name	URL	Description
Health Level Seven (HL7)	http://www.hl7.org/	Standards for interoperability of health information technology
GELLO	http://www.openclinical.org/gmm_gello.html	An object-oriented guideline query language
GuideLine Interchange Format (GLIF)	http://www.glif.org/glif_main.html	For sharing of clinical practice guidelines
Unified Medical Language System (UMLS)	http://www.nlm.nih.gov/research/umls/	Terminology, classification and coding standards
Systematized Nomenclature of Medicine –Clinical Terms (SNOMED CT)	http://www.nlm.nih.gov/research/umls/Snomed/snomed_main.html	A comprehensive clinical terminology
International Classification of Disease (ICD)	http://www.who.int/classifications/icd/en/	Classifications of diseases
Current Procedural Terminology (CPT-4)	https://catalog.ama-assn.org/Catalog/cpt/cpt_search.jsp	Describes medical, surgical, and diagnostic services
Logical Observation Identifiers Names and Codes (LOINC)	http://loinc.org/	Universal codes and names to identify laboratory and other clinical observations
Clinical Data Interchange Standards Consortium (CDISC)	http://www.cdisc.org/	Standards to support the use of clinical research data and metadata
Digital Imaging and Communication in Medicine (DICOM)	http://medical.nema.org/	A standard for information in medical imaging
Universal Protein Resource (UniProt)	http://www.uniprot.org/	Protein sequence and functional information
dbSNP	http://www.ncbi.nlm.nih.gov/projects/SNP/	Database of single nucleotide polymorphisms (SNPs)
Online Mendelian Inheritance in Man (OMIM)	http://www.ncbi.nlm.nih.gov/omim/	Human genes and genetic phenotypes
Minimum Information About a Microarray Experiment (MIAME)	http://www.mged.org/Workgroups/MIAME/miame.html	For the interpretation of microarray experimental results
Extensible Markup Language (XML)	http://www.w3.org/XML/	A textual data format for encoding documents
Translational Bioinformatics Portal (TBP)	http://bioinformatics.pharmtao.com	A web portal containing resources on translational bioinformatics

system to provide decision support. Standardization methods such as semantic mapping can facilitate communication and data sharing. Table 1 lists some resources of biomedical standards. These standards include Systematized Nomenclature of Medicine – Clinical Terms (SNOMED CT), Digital Imaging and Communications in Medicine (DICOM), and GO. In building genetic pathways and systems biology models, the Systems Biology Markup Language (SBML) can be used (30).

Besides cleaning and standardization, another important step is data structuralizing (see Fig. 2). Lack of uniformly structured data has been considered a significant barrier to translational research (31). Structured information allows rapid and efficient access and retrieval, and makes it easier for further automated processing (32). It would also facilitate the integration with general information systems such as Electronic Health Record (EHR). Structured information such as using ontologies may reduce the complexity of text processing and improve searching performance. For example, documents can be preindexed by a conceptual hierarchy to facilitate concept-based search (33).

With these steps, the data processing results can be presented using tables and graphs for decision support (see Fig. 2). Associated data should be linked, such as linking genetic structural data with functional data. Such data integration approaches are necessary for achieving the goals discussed in Subheading 2, including identifying biomarkers, elucidating genotype–phenotype correlations, and supporting predictive models for interaction networks.

4. Data Mining and Knowledge Representation Methods in Translational Bioinformatics for Decision Support

On the basis of collected and integrated data, data mining and machine learning methods can be used for the discovery of meaningful patterns, relationships, interactions, and clinical rules to build systems biology models. Machine-learning techniques are useful for classifications that have been identified by biomedical experts. Data mining methods also include clustering, decision trees, artificial neural networks (ANN), Bayesian network, and genetic algorithms. Such approaches may lead to the discovery of drug targets. For example, the clustering of expression data at both the gene and protein levels can help identify biomarkers and candidate targets.

Part I of this book introduces many useful computational methods for data mining and modeling in systems biology, such as Bayesian network, clustering, tree building, self-organizing map (SOM), and bootstrapping (see Chapters 5–7). Part II of this book discusses the application of some of these methods in disease modeling and drug development. For instance, many of the immunoinformatics tools have been developed based on ANN methods for the identification of structural and functional patterns (see Chapter 10).

With the discovery of patterns and building of models, knowledge representation becomes important to convey the message. Knowledge representation refers to the expression of knowledge in a format that can be explained and reasoned with by humans and machines (23). For example, the area of ontology studies concept definitions in a domain and the relationships among the concepts. Ontologies have been used to represent clinical guidelines and biomedical facts. Knowledge representation can also be achieved through data modeling and building of multidimensional databases (25). These approaches can help transform biomedical data into useful drug development information, and apply the knowledge for decision support in clinical practice.

5. Conclusion

In summary, translational bioinformatics plays an important role in transforming systems biology and pharmacogenomics into personalized medicine. To achieve this goal, translational bioinformatics can be applied from two aspects: biomedical and informational (see Fig. 1).

On the biomedical side, translational bioinformatics would enable the identification of biomarkers based on systemic analyses. It can improve the understanding of correlations between genotypes and phenotypes. It would also enable novel insights of interactions and interrelationships among different parts in a whole system.

On the informatics side, translational bioinformatics methods based on data integration, data mining, and knowledge representation can provide decision support for both researchers and clinicians. These approaches are crucial for understanding diseases at systems levels, and for the development of personalized and optimal treatment strategies.

References

1. Yan Q (2005) Pharmacogenomics and systems biology of membrane transporters. Mol Biotechnol 29:75–88
2. Meyer UA (2004) Pharmacogenetics – five decades of therapeutic lessons from genetic diversity. Nat Rev Genet 5:669–676
3. Madhavan S, Zenklusen JC et al (2009) Rembrandt: helping personalized medicine become a reality through integrative translational research. Mol Cancer Res 7:157–167
4. Wang X, Liu L, Fackenthal J et al (2009) Translational integrity and continuity: personalized biomedical data integration. J Biomed Inform 42:100–112
5. Greenes RA (2003) Decision support at the point of care: challenges in knowledge representation,

management, and patient-specific access. Adv Dent Res 17:69–73

6. Yan Q (2003) Bioinformatics and data integration in membrane transporter studies. Methods Mol Biol 227:37–60
7. American Medical Informatics Association (AMIA). AMIA strategic plan. Available at: http://www.amia.org/inside/stratplan/. Accessed July 2009
8. Suh KS, Remache YK, Patel JS et al (2009) Informatics-guided procurement of patient samples for biomarker discovery projects in cancer research. Cell Tissue Bank 10:43–48
9. Yan Q et al (2000) Preventing adverse drug events (ADEs): the role of computer information systems. Drug Inf J 34:1247–1260
10. Yan Q (2008) The integration of personalized and systems medicine: bioinformatics support for pharmacogenomics and drug discovery. Methods Mol Biol 448:1–19
11. Aich P, Babiuk LA et al (2009) Biomarkers for prediction of bovine respiratory disease outcome. OMICS 13:199–209
12. Wehling M (2008) Translational medicine: science or wishful thinking? J Transl Med 6:31
13. Emerson JW, Dolled-Filhart M et al (2009) Quantitative assessment of tissue biomarkers and construction of a model to predict outcome in breast cancer using multiple imputation. Cancer Inform 7:29–40
14. Chia S, Senatore F et al (2008) Utility of cardiac biomarkers in predicting infarct size, left ventricular function, and clinical outcome after primary percutaneous coronary intervention for ST-segment elevation myocardial infarction. JACC Cardiovasc Interv 1:415–423
15. Khuseyinova N, Koenig W (2006) Biomarkers of outcome from cardiovascular disease. Curr Opin Crit Care 12:412–419
16. Welsh P, Barber M et al (2009) Associations of inflammatory and haemostatic biomarkers with poor outcome in acute ischaemic stroke. Cerebrovasc Dis 27:247–253
17. Knudsen LS, Klarlund M et al (2008) Biomarkers of inflammation in patients with unclassified polyarthritis and early rheumatoid arthritis. Relationship to disease activity and radiographic outcome. J Rheumatol 35:1277–1287
18. Ozkisacik EA, Discigil B et al (2006) Effects of cyclosporin a on neurological outcome and serum biomarkers in the same setting of spinal cord ischemia model. Ann Vasc Surg 20:243–249
19. Hurks R, Peeters W et al (2009) Biobanks and the search for predictive biomarkers of local and systemic outcome in atherosclerotic disease. Thromb Haemost 101:48–54
20. Radulovic D, Jelveh S et al (2004) Informatics platform for global proteomic profiling and biomarker discovery using liquid chromatography-tandem mass spectrometry. Mol Cell Proteomics 3:984–997
21. Camp RL, Dolled-Filhart M et al (2004) X-tile: a new bio-informatics tool for biomarker assessment and outcome-based cut-point optimization. Clin Cancer Res 10:7252–7259
22. Yan Q (2008) Pharmacogenomics in drug discovery and development. Preface. Methods Mol Biol 448:v–vii
23. Peleg M, Tu S (2006) Decision support, knowledge representation and management in medicine. Yearb Med Inform, 72–80
24. Schreiber G, Akkermans H, Anjewierden A et al (2000) Knowledge engineering and management: the common KADS methodology. The MIT Press, Cambridge, MA
25. Yan Q (2010) Bioinformatics for transporter pharmacogenomics and systems biology: data integration and modeling with UML. Methods Mol Biol 637:23–45
26. Daemen A, Gevaert O, Ojeda F et al (2009) A kernel-based integration of genome-wide data for clinical decision support. Genome Med 1:39
27. Brazhnik O, Jones JF (2007) Anatomy of data integration. J Biomed Inform 40:252–269
28. Hwang D, Rust AG, Ramsey S et al (2005) A data integration methodology for systems biology. Proc Natl Acad Sci USA 102:17296–17301
29. http://magnet.systemsbiology.net/software/Pointillist/. Accessed June 2009
30. Hucka M, Finney A, Bornstein BJ et al (2004) Evolving a lingua franca and associated software infrastructure for computational systems biology: the Systems Biology Markup Language (SBML) project. Syst Biol (Stevenage) 1:41–53
31. Ruttenberg A, Clark T, Bug W et al (2007) Advancing translational research with the Semantic Web. BMC Bioinform 8(Suppl 3):S2
32. Rassinoux AM (2008) Decision support, knowledge representation and management: structuring knowledge for better access. Findings from the yearbook 2008 section on decision support, knowledge representation and management. Yearb Med Inform 80–82
33. Moskovitch R, Martins SB, Behiri E et al (2007) A comparative evaluation of full-text, concept-based, and context-sensitive search. J Am Med Inform Assoc 14:164–174

Part II

Systems Biology Methods for Disease Treatment and Translational Medicine

Chapter 9

Systems Biology and Inflammation

Yoram Vodovotz and Gary An

Abstract

Inflammation is a complex, multiscale biological response to threats – both internal and external – to the body, which is also required for proper healing of injured tissue. In turn, damaged or dysfunctional tissue stimulates further inflammation. Despite continued advances in characterizing the cellular and molecular processes involved in the interactions between inflammation and tissue damage, there exists a significant gap between the knowledge of mechanistic pathophysiology and the development of effective therapies for various inflammatory conditions. We have suggested the concept of translational systems biology, defined as a focused application of computational modeling and engineering principles to pathophysiology primarily in order to revise clinical practice. This chapter reviews the existing, translational applications of computational simulations and related approaches as applied to inflammation.

Key words: Translational research, Inflammation, Mathematical modeling, Sepsis, Trauma, Multidisciplinary research, Systems biology

1. Introduction

Biocomplexity and Translation via Computational Modeling

Biocomplexity refers to distinctive aspects of the structure, organization, and behavior of biological systems, such as nonlinear dynamics due to multiple feedback loops, multiscale emergent properties, robustness to perturbation (often coupled to surprising fragility at point of control), and nonintuitive, paradoxical behavior (1–3). The traditional scientific paradigm of reductionist analysis is often ineffective in fully capturing the behavior of complex systems. This is particularly evident in the biomedical arena, where there is a significant gulf between the volume of mechanistic information regarding underlying cellular and molecular processes versus the ability to translate that information to the level of the entire organism, particularly with respect to the development of effective therapeutics. Translational research aims

Qing Yan (ed.), *Systems Biology in Drug Discovery and Development: Methods and Protocols*, Methods in Molecular Biology, vol. 662, DOI 10.1007/978-1-60761-800-3_9, © Springer Science+Business Media, LLC 2010

to apply scientific discoveries in basic science into clinical practice in order to improve health care (4). Both the United States Food and Drug Administration in its "Critical Path" document (5) and the United States National Institutes of Health in its "Roadmap" statement (6) have explicitly highlighted the need for multidisciplinary teams utilizing computational technology to bridge the gap between mechanistic knowledge obtained from basic science research, as well as the translation of that knowledge via the development of effective clinical regimens. The challenge of managing biocomplexity is most pronounced in attempts to understand and manipulate diseases involving intrinsic systemic regulatory mechanisms, such as inflammation and immunity. Disorders of inflammation include sepsis, trauma, inflammatory bowel diseases, chronic wounds, rheumatologic disorders, and asthma; many other diseases, such as cancer, diabetes, atherosclerosis, Alzheimer's, and obesity are also associated with dysregulated inflammation. Specifically, the NIH Roadmap has recently emphasized the need to apply systems biology methods to the study of inflammation (7).

Overcoming the translational challenges of biocomplexity requires the development of improved multidisciplinary approaches that fall broadly under the umbrella of "systems biology," namely computational and mathematical simulation (in silico methods). Systems biology has been defined in many ways (1, 8–10) and includes approaches that span the multiple scales of organization that characterize biological systems. One end of the spectrum focuses on the basic biology of inflammation at the gene and intracellular levels (11–14). At the other end of the spectrum, pattern recognition has been applied to large, time-series datasets to identify specific genetic and molecular signatures as a means for improved phenotypical characterization of disease (15–24). However, despite progress in establishing a systems approach to biomedical questions, there is a relative paucity of techniques that can mechanistically and dynamically transcend the multiple scales and hierarchies of biological organization. The nonmechanistic nature of many systems biology approaches (such as the "-omic" technologies) limits the ability of these methods to go beyond the phenotypical description of disease toward the translational dynamic representation necessary for the development of effective therapeutics.

Thus, there is a need to modify the way computational analysis is currently implemented in order to best address issues of direct clinical relevance. Mathematical modeling and simulation technologies used in the study of subcellular and cellular processes (1, 2) need to be applied to the translation from the molecule and cell to the organism. We term this approach *translational systems biology* (25–27). Translational systems biology involves using dynamic mathematical modeling based on mechanistic information generated in basic science research to simulate higher-level behaviors at the organ and organism level, thus affecting a means of translating reductionist experimental data to the level of clinically relevant phenomena. This process requires that modeling expertise be combined with domain expertise on both the underlying system mechanisms and the real-world applications. Current mainstays of systems biology, such as various "-omics" technologies, are currently utilized within a pattern analysis framework that relies on statistical methods for data interpretation. Translational systems biology suggests that these "-omics" studies must be combined with computational simulations in order to create a unified framework for mechanistic prediction. Moreover, existing systems approaches stem from and are utilized within the context of the current "fragmented continuum" of health care delivery, in which the domains of preclinical studies, clinical trials, in-hospital care, and eventual long-term care are separate. Translational systems biology calls for the creation of models that are generated with an a priori focus on rapid translational application in areas such as in silico clinical trials, patient diagnostics, rational drug design, and long-term rehabilitative care (26–28).

The inflammation field is the first in which the translational systems biology framework has been applied in a systematic fashion. This concept has been used in the study of the acute inflammatory response manifested clinically in sepsis, trauma, hemorrhagic shock, and wound healing (26–40). This work has been largely carried out under the aegis of the Society of Complexity in Acute Illness (SCAI, website at http://www.scai-med.org). The following sections will discuss the primary computational modeling methods used thus far in studying acute inflammation and give examples of how those models are developed and used with a translational goal.

2. Computational Modeling Methods

The primary methods of dynamic mathematical modeling used thus far in translational systems biology work in acute inflammation are agent-based modeling (ABM) (29, 30, 38, 41–44) and equation-based modeling (EBM) (32–37, 40, 45–52). The two forms of dynamic mathematical modeling have their respective strengths and weaknesses (25–28, 31), but the utilization of both methods in the work described below demonstrates a pragmatic, goal-directed approach not tied to a particular modeling platform (28, 31).

These computational simulations all involve a temporal component that allows them to evolve over time and simulate the behavior of a system based on mechanisms derived from empirical data from the basic science lab. This approach distinguishes dynamic mathematical modeling from statistical mathematical models (such as regression fitting, principal component analysis, and cluster analysis) and classical network models, which are static and not mechanism based. By incorporating both mechanism and behavior, dynamic mathematical modeling allows for the testing of manipulations and modulations of the system, a hallmark of the iterative process that underlies translational systems biology.

ABM consists of viewing a system as an aggregation of components (agents), which can be classified into populations or agent classes based on similar intrinsic rules of behavior (agent-rules). While a particular population of agents will have the same rules for behavior, the behavior of the individual agents is heterogeneous due to agents implementing their rules based on local conditions that may differ considerably. The behavior of the simulation results from the aggregate interactions within the model (29, 41, 53). The advantages of ABM are several. Such models map well to biological phenomena (i.e., cells interacting within tissues and organs) and are therefore fairly intuitive. Agent-rule systems are very often expressed as conditional statements ("if-then"), thereby facilitating the translation of the results of basic science experiments into agent-rules. ABM has an intrinsically spatial component based on its reliance upon local interactions and environmental heterogeneity. The limitations of ABM are (1) they are computationally intensive; and (2) due to the fact that there is often not a direct inferable relationship between the agent-rules and the system's behavior, they can be very difficult to calibrate in a quantitative way (31). An example of an ABM interface, in the setting of a simplified model of the inflammatory response to respiratory syncitial virus infection in the lung, can be seen in Fig. 1 (Mi et al., unpublished observations).

EBMs consist of differential equations that describe how the state variables of a system evolve over time. These equations are often coupled, so that the dynamics of the system depend heavily on the interactions between system variables. Ordinary differential equations (ODEs) use time as the sole independent variable (assuming a well-mixed and homogeneous system), while partial differential equations (PDEs) incorporate spatial variables. Thus far, the primary EBMs used in the TSB community are ODE-based (31, 33, 35–37, 45–47). Like ABMs, EBMs have several advantages. EBMs may be subjected to formal mathematical analysis and therefore can provide insight into underlying "laws" of behavior (54). These models are not as computationally intensive as ABM, and are easier to calibrate quantitatively (31). The limitations of EBM include the fact that they are predicated on the

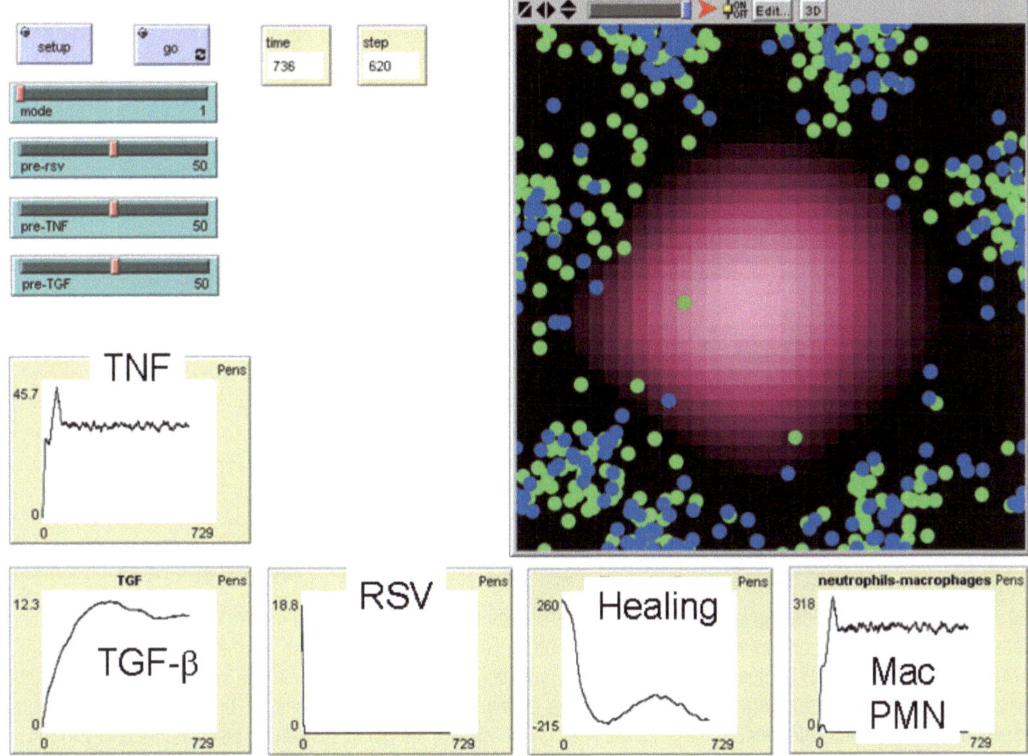

Fig. 1. The Agent-based model (ABM) of inflammation induced by respiratory syncitial virus. An ABM was created in the NetLogo™ platform. This software creates a grid populated with software agents that act and are acted upon by rules. The rules are carried out in a stochastic fashion, creating variability form simulation to simulation. The upper *left panel* depicts the virtual lung grid. Neutrophils are initially *blue* and change to *light blue* when activated. Macrophages start out as *green* and change to *light green* when activated. Respiratory syncitial virus is depicted as *red*. The alveolar epithelial cells start out as *pink* and change to *maroon* when damaged. The graphs show the cumulative behavior of the system. Depicted is a late time point post-RSV infection.

assumption of a homogenously distributed system (for ODEs); thus, they may be less applicable in situations where spatial concerns are present (31, 54, 55). A formal analysis of EBMs may be extremely difficult as the dimensions of differential equations and the number of parameters increase. Perhaps the greatest hurdle to the dissemination of EBM within the general biomedical research community is the fact that dealing with equations is often a daunting task for nonmathematicians. In order to overcome this challenge, graphical tools have been developed to aid in the construction of ODE models. BioNetGen (56) is an example of one of these tools, and a screenshot of a BioNetGen model can be seen in Fig. 2 (57).

The following sections describe translational systems biology modeling efforts at the global/systemic level, at the organ/tissue level, at the subcellular/intracellular level, to simulate clinical

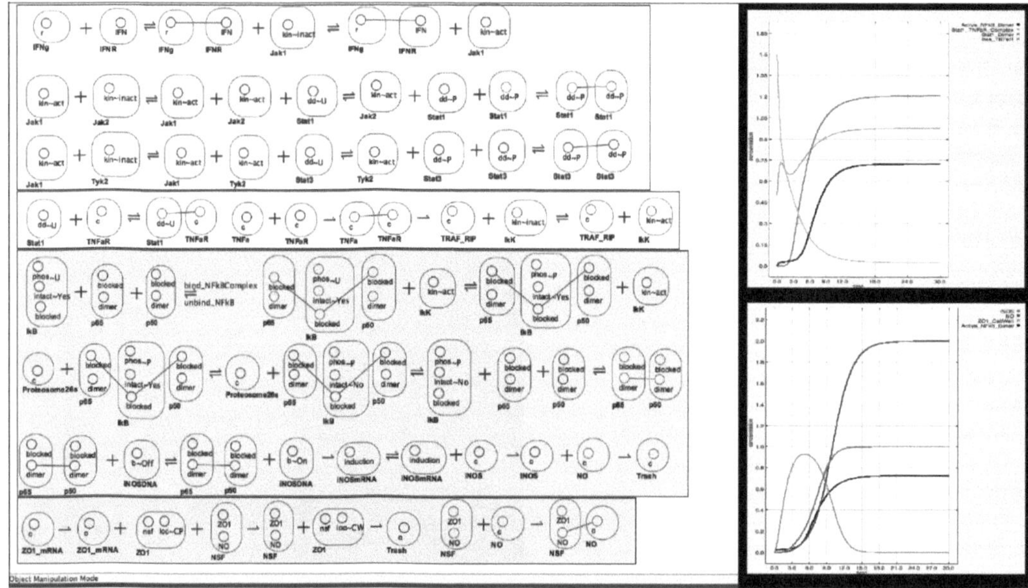

Fig. 2. Example of a BioNetGen graphical interface for an equation-based cell signaling model. This model is of signaling pathways involved in inflammatory signaling within an enterocyte, in which the cytokine interferon-γ (IFN-γ) via the JAK/STAT pathway, activation of Nuclear Factor-κB (NF-κB), induction of the inducible nitric oxide synthase (iNOS), and Tight Junction (TJ) protein metabolism. BioNetGen generates ordinary differential equations for the molecular interactions depicted in the graphical interface. Run results are seen on the *right of the panel*. BioNetGen also has the ability to run in a stochastic mode (not shown).

trials, and to examine wound healing and recovery. We will also address the issue of methodologies used to bring the various levels of models together.

2.1. The Big Picture: Models of Systemic Inflammation

The clinical manifestations of acutely disordered systemic inflammation are seen in sepsis, hemorrhagic shock, and the response to trauma. Initial modeling efforts focused on the general dynamics of systemic inflammation. Because much of the data regarding molecular mediators and the dynamics of cellular populations were derived from the analysis of peripheral blood, we made the approximation that systemic inflammation could be viewed as a well-mixed system, and thus suited the application of ODE/EBM. The initial EBM addressed the body's proinflammatory response to infection as a bactericidal mechanism (45) and was subsequently modified by adding anti-inflammatory mediators to simulate the dynamic control mechanisms involved in maintaining a healthy state (46). The simulations of endotoxemia in this model could account for preconditioning phenomena such as priming or tolerance (47).

While these models reproduced the qualitative dynamics of the inflammatory response to infection, calibration methodologies were necessary if these models were to achieve translational utility. Calibration accomplished in in silico experiments matched, both retrospectively and prospectively, to animal models. These experiments included using in silico simulated data to predict the lethal doses of endotoxin prospectively in mice (33). Furthermore, this model was used to show that the previously overlooked tissue injury necessary to induce hemorrhagic shock in animals was actually the primary influential component of trauma/hemorrhage-induced inflammation (36). Extensions of this EBM were also used to gain insight into the inflammatory complexity of CD14-deficient mice (35) and examine the nonlinearly interacting effects of antibiotics and vaccination in the setting of anthrax (40). More recently, this EBM was expanded to include a more realistic simulation of the process of hemorrhage (duration and intensity), and was validated using a custom-made, computer-controlled, closed-loop apparatus (52). Continuing the translational goal by matching with increasingly complex in vivo models, EBM were used to quantitatively predict circulating cytokine levels in rats, swine, and humans (34). Other investigators utilized EBM to gain insight into the dynamics of bacterial growth in an experimental paradigm of pneumonia (37), as well as to simulate the inflammatory response in the setting of influenza (39). EBMs have also been used to study the various facets of the inflammatory response to burns, including the effects of resuscitation and cell-based therapy (58–65).

A parallel project aimed at modeling systemic inflammation with ABM required the development of a model structure based on interactions among diffusely distributed cell types. For this purpose, inflammation was conceptualized as the interaction between endothelium (the single layer of cells that line blood vessels and delineate the vessel's lumen) and blood-borne inflammatory cells as the basis of the global inflammation ABM (29, 41). This model treated the whole organism as a consisting solely of an endothelial cell surface over which inflammatory cells moved and interacted, and which was able to qualitatively reproduce patterns of diverse clinical outcomes in sepsis (29).

2.2. Utilizing Modularity: Tissue and Organ-Specific Modeling

Despite the appreciation of the global effects of disordered inflammation, clinical management issues and pathophysiology often center on end-organ dysfunction. Patterns of organ failure (such as the interaction between the gut and the lung) often form a pathophysiologic feedback loop, and therefore more comprehensive computational description of the effects of disordered inflammation would require the modeling of specific tissue and organ systems. While it is possible to simulate organs as internally well-mixed systems with compartmentalized EBM (a well-established technique in classical physiology) (66), the majority of more recent tissue/organ-specific modeling done in the translational systems biology community has utilized ABM.

ABM has an intrinsically modular capacity, and this has been utilized to develop an overall modeling architecture. This architecture utilizes a classification scheme of organ structure and organ-to-organ relationship based on cell types and topological organ orientation. Organs are viewed as compilations of layers of epithelial cells directed at maintaining self/nonself integrity via barrier function and the layers of endothelial cells functioning as a means of organ–organ communication. Since the aforementioned ABM of systemic inflammation incorporated the endothelial elements, the next stages of model development focused on epithelial-based models representing the key actors in maintaining organ integrity and function. The first of these ABMs was based on in vitro cell culture data of enterocyte behavior with respect to barrier function. These ABMs were calibrated quantitatively based on their response to inflammatory stimuli (67, 68). This model was then linked to the endothelial ABM to simulate qualitatively both isolated gut response (69, 70) and pulmonary response (71) to inflammation. Finally, these organ-level models are placed in a global topology that organizes organs either in series (such as the Right Heart to Lung to Left Heart) or in parallel (such as the gut and the kidney). This architecture can be seen in Fig. 3 (70).

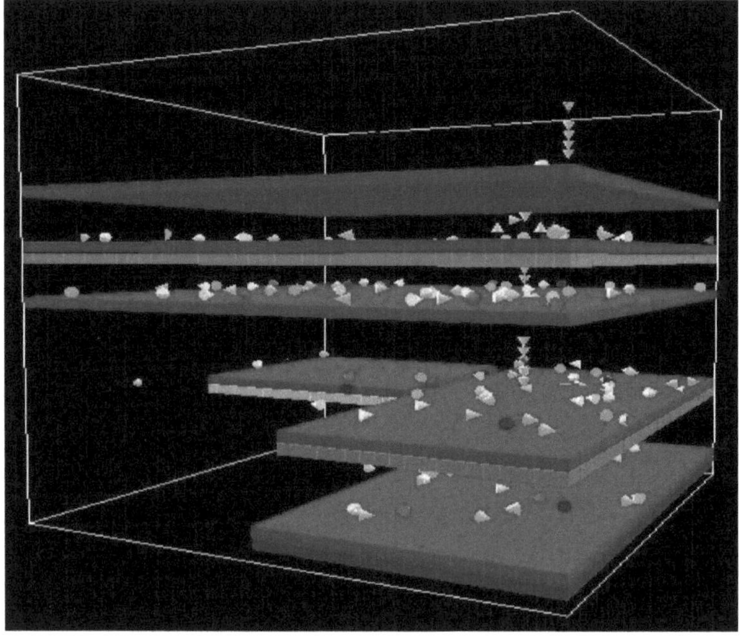

Fig. 3. Example of a multitissue ABM. The layers of cells represent abstracted tissue/organ beds. They are, from the *top*, the *Right Heart*, the Lung, the *Left Heart*, the gut and kidney (*in parallel*), and the liver *below*. Blood flows from *top* to *down* along designated channels, providing means of communication from one tissue bed to the next via secreted mediators and circulating inflammatory cells (multicolored cones).

Linking and relating organs in this fashion allows for the simulation of organ–organ crosstalk, the consequent patterns of multiple organ failure (72) and the effects of organ support.

2.3. Scaling Down: Subcellular and Genomic/Proteomic Modeling

The traditional reductionist scientific paradigm has pushed the boundaries of investigation deep into cells, and as such a great deal of the current central focus of the basic science community is on the intracellular processes of signaling, gene activation, post-transcriptional modification, and signaling/synthetic pathways (73–78). Evaluation of these aspects of cellular behavior adds yet another level of complexity to understand the dynamics of the inflammatory response, and further accentuates the need for the utilization of formal analytic/synthetic methods patterned after tools already developed by the computational and systems biology community (56, 79–83). While these researchers have made significant strides in the understanding and characterization of gene regulation and intracellular signaling, there is a growing recognition within the computational/systems biology community of the need to translate the issues of gene profile information to the clinical setting (15, 17, 84). As another, and parallel, approach to this challenge, translational systems biology emphasizes the development of multiscale methods to integrate and translate the work done at the gene and subcellular levels to higher-scales of biological organization and behavior. In this vein, EBMs have been used to distinguish between the scale of gene expression (magnitude of gene expression) and the scope of gene expression (number of pathways recruited) based on the severity of insult (36). Studies such as this one suggest that existing efforts to address inflammation at the whole-genome expression levels coupled to network/pathway analysis (13, 16) and in vivo validation in gene-deficient mice (13) could be coupled to mathematical modeling to examine and predict the inflammatory biology of these animals (35). The modified mathematical models that account for the role of a given gene could then be examined in the context of in silico clinical trials (30, 32, 40) (see below) to determine if the gene of interest might be a valid therapeutic target. At a more basic level, attempts should be made to address the static nature of "-omic" data by creating a dynamic modeling framework that utilizes expanded cellular automata, dynamic network analysis, multidimensional data structures and discrete-event, mechanism-based, constraining rules (85).

In addition, translational systems biology seeks to aid the integration and unification of knowledge across the entire biomedical research community. Accomplishing this goal requires not only the development of a unified means of representing models, similar to the Systems Biology Markup Language (SBML; http://www.sbml.org) and as can be seen in the introduction of a syntactical modeling grammar (86), but also extends the scope of unification to developing methods of combining and integrating multiple different types of models into hybrid configurations (87). The development of the Functional Unit Representation Method (FURM) (88) proposes a means by which models can be developed in a modular form and used and reused as situations require.

2.4. Gained in the Translation: Simulated Clinical Trials

Translational systems biology is aimed ultimately at improving the clinical care of the patient. Toward this end, a central, translational goal is the use of computational simulations such as those described so far in the design and structure of clinical trials. We suggest that the effective achievement of this goal will fundamentally transform the landscape of clinical trials in sepsis and trauma.

Trial outcome may be analyzed by identifying via simulation the distributions of patients helped, harmed, and not affected by a particular therapy (32). In an effort to utilize this knowledge, there is a great deal of interest in the substratification of patient populations prior to the design of a clinical trial. One of the potential uses of genomic data is to prospectively classify patients and develop targeted groups for interventions (a step in the direction of "personalized medicine.") (89) Translational systems biology groups have developed methodologies to extend gene regulation and cellular control studies to clinically relevant phenomena. EBM have been used to distinguish between the magnitude of gene expression and the scope of gene expression (number of pathways recruited) based on the severity of insult (36). One further direction in the modeling behavior of the inflammatory response is the incorporation of whole-genome expression levels coupled to network/pathway analysis in the model (13, 16), with in vivo validation in appropriate mouse models (such as mice with specific genes deleted (13)). Studies such as this one suggest that the existing efforts to address inflammation at the whole-genome expression levels coupled to network/pathway analysis (13, 16) and in vivo validation in gene-deficient mice (13) could be added to mathematical modeling to examine and predict the observed inflammatory biology (35). The modified mathematical models that account for the role of a given gene could then be examined in the context of in silico clinical trials (30, 32, 40) to determine if the gene of interest might be a valid therapeutic target.

Modeling has been integrated into actual trial design by utilizing the iterative process common to engineering projects, consisting of a knowledge/development loop between the real-world data and the simulation as the information from one source feeds into the next (34). Models thus refined have already been used to assist in the analysis of existing clinical drug trials. In one case, the models were able to produce a "virtual" placebo arm for an open-label Phase IV drug trial, with the advantage of providing an additional frame of reference for the comparison between the Phase IV and its preceding Phase III trial (90). Furthermore, these models have even been able to prospectively predict the outcome of a phase III clinical trial, providing vital information regarding subgroup analysis, and risk stratification within the treatment protocol (Unpublished Data). While limited in their scope of effect (being focused on the efficacy of a particular therapeutic intervention), these accomplishments serve as evidence that the translational systems biology approach does have the capacity to significantly affect the design and implementation of therapies at the point of clinical practice.

2.5. After the Injury: Inflammation, Healing, and Recovery

Wound healing is a process that involves both inflammation and the resolution of the inflammatory response, which culminates in remodeling (91–97). Initial and ongoing work on wound healing has focused on epithelial proliferation and migration, though the insights derived were generally basic rather than translational (55, 98–108). This work was extended with the ultimate goal of predicting the long-term outcome of therapies for sepsis and trauma as well as of chronic inflammatory diseases, wound healing, and even cancer. These ABM were used to examine inflammation and healing in the setting of necrotizing enterocolitis (NEC), a severe inflammatory disease of the newborn intestinal tract (109). Healing of the inflamed intestine and reversal of the proinflammatory cascade occurs through the process of intestinal restitution, which involves the migration of healthy enterocytes to sites of mucosal disruption (110). Initial EBM were focused on modeling enterocyte migration (55). The EBMs were extended to include spatial effects, such as the diffusion of inflammatory agents, chemotaxis, and cell migration in NEC, using PDE (109).

The relationship between inflammation and healing in the chronic wound has also been modeled using an ABM of diabetic foot ulcers (38). An expanded ABM was developed to include the interactions between inflammation and healing in the setting of diabetic foot ulcers. This model was calibrated using values published in the literature regarding normal skin healing, which the ABM was capable of reproducing at baseline. This simulation demonstrated delayed healing in the setting of elevated TNF or reduced TGF-β1 expression (both known aspects of deranged inflammation in diabetes and/or diabetic foot ulcers), recapitulated the beneficial effect of known therapies for diabetic foot ulcers (wound debridement and treatment with platelet-derived growth factor), and was used to suggest novel therapies (38).

Personalized medicine is a longstanding therapeutic goal, and is likewise a central pillar of translational systems biology (89). Toward this end, patient-specific ABMs for vocal fold inflammation have been generated with the goal of identifying individually optimal treatments (111). ABM simulations reproduced trajectories of inflammatory mediators in laryngeal secretions of individuals subjected to experimental phonotrauma up to 4 h postinjury, and predicted the levels of inflammatory mediators 24 h postinjury. Subject-specific simulations also predicted the effects of behavioral treatment regimens to which subjects had not been exposed (111).

Closing the circle with our prior studies on acute inflammation, similar methods were used to create patient-specific EBMs in the setting of polytrauma. Human trauma patients were recruited into an observational study in which blood samples were obtained daily up to 1 week postadmission, then weekly thereafter. Plasma was assessed for TNF, IL-6, IL-10, and NO_2^-/NO_3^-. Trauma was modeled as an exponentially decaying function the EBM developed originally for mice (33). The coefficient of the trauma function was scaled from one to two, with one corresponding the lowest injury severity score (ISS, an established clinical scoring system) and two highest ISS for any patient. The rate constants of the EBM that relate to the generation of TNF, IL-6, IL-10, and NO_2^-/NO_3^- were estimated to fit the time course data of individual patients. Using this methodology, the resultant patient-specific models accurately predicted patient survival when ISS alone could not (Sarkar et al., unpublished observations).

2.6. Finding the Needle in the Haystack: Making Sense of Danger, Damage, and Inflammation Biomarkers

These encouraging translational developments were linked in a stepwise fashion to the developments on the preclinical front. The models have in fact grown from the initial work done to assimilate the current state of knowledge on the inflammatory and wound healing responses. It is becoming clear, however, that continuing to enlarge these simulations will be difficult, given the number of interactions each new component will have with existing model components. Mesarovic et al. have suggested that such a process could ultimately prove intractable and have called instead for a search for organizing principles in complex biological systems (112). Accordingly, we have set about a parallel process of gleaning major integrative insights regarding inflammation, while at the same time attempting to establish a framework within which complex datasets could be used to drive the creation of computational simulations that may be applied for clinical purposes.

The former process has led us to focus on the positive feedback loop of inflammation→damage→inflammation (28). Our overarching hypothesis is that damage-associated molecular patterns (DAMPs, also known as "alarm/danger signals"), which propagate inflammation in both infectious and sterile inflammatory settings using similar recognition systems (28, 113, 114) act as integrators of the inflammatory response and surrogates for an individuals' health status. This property of our simulations allows us to predict both inflammatory trajectories and morbidity/mortality outcomes (28). It is now appreciated that one mechanism by which the host can recognize both pathogen-derived products and DAMPs is via the Toll-like receptor pathway and DAMP ligands such as HMGB1 (115–121). Importantly, our conceptual framework suggests that prior exposure to a given stimulus will modify the response to a subsequent stimulus, a clinically relevant phenomenon known as preconditioning that we have modeled computationally (28, 42, 44, 47, 50) In support of this framework, DAMPs have been recently suggested to play a role in preconditioning (116, 122, 123). While this conceptual framework has allowed us to make translational advances, we fully anticipate that emerging data will suggest changes to this paradigm. For example, DAMPs such as HMGB1 have been suggested to act not only by increasing the production of nominally proinflammatory cytokines such as tumor necrosis factor-α (TNF) (124), but also by the direct or indirect suppression of nominally anti-inflammatory cytokines such as transforming growth factor-β1 (TGF-β1) (125).

Our parallel effort at assimilating experimental data into predictive, mechanistic computational simulations has led us to integrate statistical methods into the process. In so doing, we face a dilemma. Clearly, one factor that can hamper modeling studies is the lack of sufficient data. On the other hand, it is often difficult to make sense of these data, and especially to determine which data should be included in a given model. Our earlier work on modeling the inflammatory response in vivo utilized standard ELISA methodology to assay cytokines (33, 35, 36). The volumes of serum required for this technique, as well as cost considerations, limited our ability to assess inflammatory cytokines. Perhaps fortuitously, these constraints forced us to consider quite carefully which analytes were to be assessed. By trading off having a large number of experimental conditions and time points at which a judiciously considered set of analytes was assessed vs. having a large number of analytes per time point, we were able to gain insights into the acute inflammatory response in mice in diverse settings (33, 35, 36). Nonetheless, we felt that having to make this tradeoff might hamper the capability of our mathematical models, and so deployed a Luminex™ 100 IS apparatus (Luminex, Austin, TX). This device is a continuous, random-access instrument that performs automated chemiluminescent immunoassays on multiple analytes simultaneously in a very small volume of a biofluid (126). Using this apparatus, we have obtained extensive data regarding the acute inflammatory response in both experimental animals (127–129) and humans (see below).

The large datasets obtained in our studies, such as the Luminex™ analytes described above, allow us the unprecedented opportunity to assess the acute inflammatory response with a very high resolution, and in turn create computational simulations to describe and predict features of this response. However, these very same data present the problem of determining the relevance of a given analyte to our mathematical models. Janes et al. described the use of principal component analysis to gain insights into the primary drivers of inflammation and apoptosis in an in vitro setting (130). We sought to determine if similar methodology could be used in vivo, and therefore employed principal component analysis to reduce the number of variables from the larger Luminex™ dataset to a relevant subset, thereby helping define the principal drivers of inflammation in diverse settings. As an example of this approach, we subjected mice to hemorrhagic shock for 1, 2, 3, or 4 h versus sham procedure (surgical cannulation only, followed by obtaining blood samples at the same time points, using a computer-controlled, closed-loop platform (52)). We carried out a principal component analysis of a 20-cytokine Luminex™ panel of plasma samples from these mice. Figure 4 shows that ~98% of the

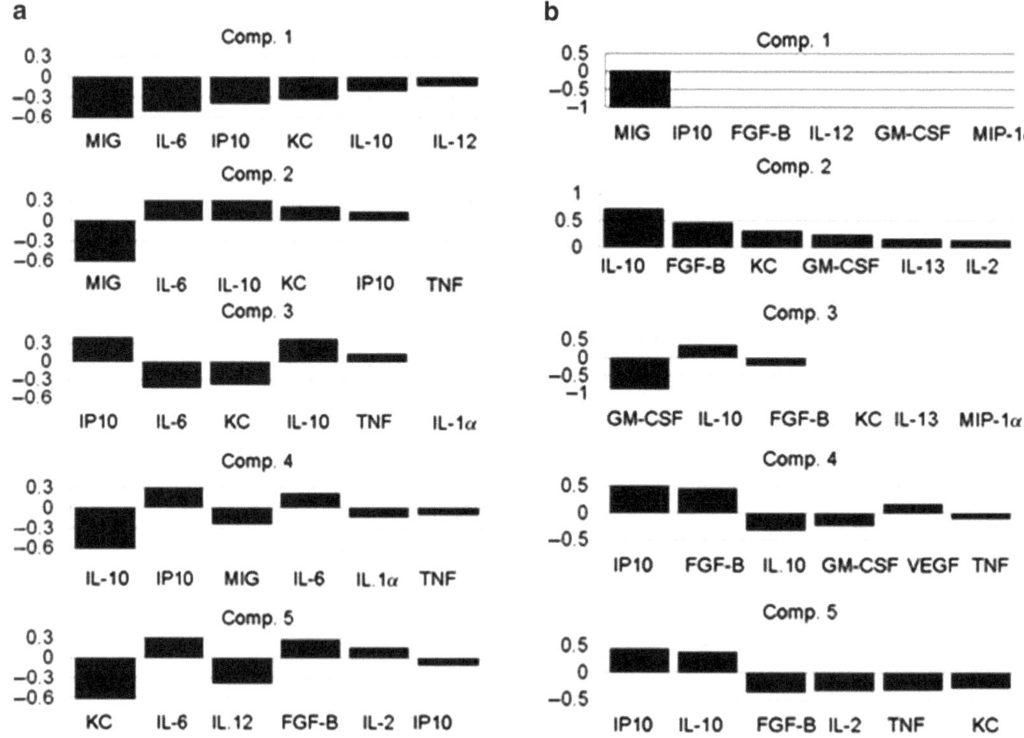

Fig. 4. Principal component analysis in mouse hemorrhagic shock. C57BL/6 mice (6 per group) were subjected either to hemorrhagic shock (25 mmHg using an automated, computer-controlled, closed-loop system (52) for 1, 2, 3, or 4 h (*Panel A*) or to a sham procedure (surgical cannulation only; *Panel B*)) followed by monitoring using the above apparatus for the same time periods. Plasma cytokines were assayed by Luminex™ (BioSource, San Diego, CA). Data were normalized, to emphasize relative rather than absolute changes in the levels of inflammatory analytes. *Bars* indicate relative contribution of a given cytokine. If bars appear absent, it is due to their contributing <1% to the variance observed in the data.

variance in the data obtained in the sham cannulation group could be described by a vector combination consisting predominantly of the chemokines MIG and IP-10, along with the cytokines GM-CSF and IL-10.

Mice subjected to HS were characterized by an inflammatory response dominated by MIG, IL-6, IL-10, IP-10, IL-12, TNF, and KC. These results support our published modeling work that suggests a central role for underlying trauma in hemorrhage-induced inflammation (36), as well as supporting our previous selection of TNF, IL-6, and IL-10 for inclusion in the mathematical model of inflammation in various shock states (33). We have also developed a rational method for translating multiplexed cytokine data obtained from the cerebrospinal fluid into clinically relevant mathematical models in the setting of human traumatic brain injury (Okonkwo et al., manuscript in preparation). These results highlight the capacity of principal component analysis, both for discovering novel tendencies in the data and for defining variables to be included in mechanistic simulations. While additional work is necessary, these latter findings raise the potential for patient-specific simulation and outcome prediction, and suggest how this work can be applied to the human clinical setting.

3. Conclusions and Future Directions

The field of acute inflammation is inundated with literature that describes various aspects of the process but fails to link them in a holistic fashion that has clinical translation as its main goal. Translational systems biology aims to unify mechanisms described in the scientific literature using methods and tools developed by the computational and systems biology communities. By doing so, we hope to suggest novel insights into the pathobiology of inflammation and the intertwined damage/healing response, and add a mechanistic, rational basis to the design and implementation of therapies. The not-too-distant future includes rational, model-driven design and testing of novel therapies; clinical trials that are first run in silico; inpatient care in which diagnosis is aided by mathematical models; and outpatient care plans prepared using model-driven decisions along the fragmented continuum of care.

Areas of active interest in our group include modeling the inflammatory complications of spinal cord injury, the inflammatory response to exercise and rehabilitation form injury, the response of the lung to chronic inflammatory stresses, the inflammatory genesis of cancer, and the cascade of events that lead from arterial injury to intimal hyperplasia. Successful achievement of these objectives will benefit from several advances. Models are currently initiated and modified through a painstaking and time-consuming process of manual extraction of relevant data from the scientific literature. Thus, translational systems biology will benefit from the automated means of searching the literature and mining and extracting data in a form that will support continued updating of the core models (131, 132). Similarly, nonmathematically inclined clinician-investigators often struggle to convert even simple biological interactions into mathematical models using software optimized for mathematicians. Accordingly, translational systems biology would benefit from software designed to facilitate the translation of biological and clinical knowledge into mathematical models.

The ultimate therapeutic utility of these approaches is still in debate within the clinical community (133). We, in the translational systems biology community, hope that the exciting developments outlined herein, and the many more on the way, will build bridges to the larger computational and systems biology communities to aid us in these translational efforts.

Acknowledgments

The authors would like to acknowledge the contributions to this work of the following investigators, students, and postdoctoral fellows: Arie Baratt, Timothy R. Billiar, Frederick D. Busche, David Carney, Carson Chow, Gilles Clermont, Gregory Constantine, Judy Day, Edwin Dietch, Russell Delude, Joyeeta Dutta-Moscato, G. Bard Ermentrout, James Faeder, Rena Feinman, Ali Ghuma, Mitchell P. Fink, David Hackam, Rukmini Kumar, Claudio Lagoa, Ryan M. Levy, Nicole Li, Qi Mi, Maxim Mikheev, Rajaie Namas, Gary Nieman, Patricio Polanco, Jose M. Prince, Juan Carlos Puyana, Angela Reynolds, Beatrice Riviere, Jonathan Rubin, Matthew Rosengart, David L. Steed, Joshua Sullivan, David Swigon, Andres Torres, Jeffrey Upperman, Katherine Verdolini, Ivan Yotov, Ruben Zamora, and Sven Zenker. Additionally, several excellent technicians (Derek Barclay, David Gallo, and Binnie Betten) contributed to this work. We would also like to thank Alan Russell and Clifford Brubaker for their unwavering support. This work was supported in part by the National Institutes of Health grants R01-GM-67240, P50-GM-53789, R33-HL-089082, R01-HL080926, and R01-HL-76157; National Institute on Disability and Rehabilitation Research grant H133E070024; as well as grants from the Commonwealth of Pennsylvania, the Pittsburgh Lifesciences Greenhouse, and the Pittsburgh Tissue Engineering Initiative. Dr. Vodovotz is a cofounder of and consultant to Immunetrics,

Inc., which has licensed from the University of Pittsburgh the rights to commercialize aspects of the mathematical modeling of inflammation. Dr. Vodovotz's involvement with Immunetrics is monitored by the University of Pittsburgh's Entrepreneurial Oversight Committee. Dr. An is also a consultant to Immunetrics, Inc.

References

1. Kitano H (2002) Systems biology: a brief overview. Science 295:1662–1664
2. Csete ME, Doyle JC (2002) Reverse engineering of biological complexity. Science 295:1664–1669
3. Doyle J, Csete M (2007) Rules of engagement. Nature 446:860
4. Keramaris NC, Kanakaris NK, Tzioupis C, Kontakis G, Giannoudis PV (2008) Translational research: from benchside to bedside. Injury 39:643–650
5. Food and Drug Administration (2004) Innovation or stagnation: challenge and opportunity on the critical path to new medical products. 1–38
6. Anonymous (2006) NIH roadmap for medical research: research teams
7. Anonymous (2007) http://nihroadmap.nih.gov/2008initiatives.asp
8. Snoep JL, Westerhoff HV (2005) From isolation to integration, a systems biology approach for building the silicon cell. In: Alberghina L, Westerhoff HV (eds) Systems biology: definitions and perspectives. Springer, Berlin/Heidelberg, p 7
9. Sauer U, Heinemann M, Zamboni N (2007) Genetics. Getting closer to the whole picture. Science 316:550–551
10. Noble D (2006) The music of life. Oxford University Press, Oxford
11. Nguyen A, Yaffe MB (2003) Proteomics and systems biology approaches to signal transduction in sepsis. Crit Care Med 31:S1–S6
12. Yue H, Brown M, Knowles J, Wang H et al (2006) Insights into the behaviour of systems biology models from dynamic sensitivity and identifiability analysis: a case study of an NF-kappaB signalling pathway. Mol Biosyst 2:640–649
13. Gilchrist M, Thorsson V, Li B et al (2006) Systems biology approaches identify ATF3 as a negative regulator of Toll-like receptor 4. Nature 441:173–178
14. Park SG, Lee T, Kang HY et al (2006) The influence of the signal dynamics of activated form of IKK on NF-kappaB and anti-apoptotic gene expressions: a systems biology approach. FEBS Lett 580:822–830
15. Cobb JP, O'Keefe GE (2004) Injury research in the genomic era. Lancet 363:2076–2083
16. Calvano SE, Xiao W, Richards DR et al (2005) A network-based analysis of systemic inflammation in humans. Nature 437:1032–1037
17. Cobb JP, Mindrinos MN, Miller-Graziano C et al (2005) Application of genome-wide expression analysis to human health and disease. Proc Natl Acad Sci USA 102:4801–4806
18. Brownstein BH, Logvinenko T, Lederer JA et al (2006) Commonality and differences in leukocyte gene expression patterns among three models of inflammation and injury. Physiol Genomics 24:298–309
19. Liu T, Qian WJ, Gritsenko MA et al (2006) High dynamic range characterization of the trauma patient plasma proteome. Mol Cell Proteomics 5:1899–1913
20. Omenn GS (2006) Strategies for plasma proteomic profiling of cancers. Proteomics 6:5662–5673
21. Ahrens CH, Wagner U, Rehrauer HK et al (2007) Current challenges and approaches for the synergistic use of systems biology data in the scientific community. EXS 97:277–307
22. Steinfath M, Repsilber D, Scholz M et al (2007) Integrated data analysis for genome-wide research. EXS 97:309–329
23. Tanke HJ (2007) Genomics and proteomics: the potential role of oral diagnostics. Ann N Y Acad Sci 1098:330–334
24. Kourtidis A, Eifert C, Conklin DS (2007) RNAi applications in target validation. Ernst Schering Res Found Workshop 1–21
25. An G, Hunt CA, Clermont G, Neugebauer E, Vodovotz Y (2007) Challenges and rewards on the road to translational systems biology in acute illness: four case reports from interdisciplinary teams. J Crit Care 22:169–175
26. An G, Vodovotz Y (2008) Translational systems biology: introduction of an engineering approach to the pathophysiology of the burn patient. J Burn Care Res 29:277–285

27. Vodovotz Y, Csete M, Bartels J, Chang S, An G (2008) Translational systems biology of inflammation. PLoS Comput Biol 4:1–6
28. Vodovotz Y, Constantine G, Rubin J, Csete M, Voit EO et al (2009) Mechanistic simulations of inflammation: current state and future prospects. Math Biosci 217:1–10
29. An G (2001) Agent-based computer simulation and SIRS: building a bridge between basic science and clinical trials. Shock 16:266–273
30. An G (2004) In-silico experiments of existing and hypothetical cytokine-directed clinical trials using agent based modeling. Crit Care Med 32:2050–2060
31. Vodovotz Y, Clermont G, Chow C, An G (2004) Mathematical models of the acute inflammatory response. Curr Opin Crit Care 10:383–390
32. Clermont G, Bartels J, Kumar R, Constantine G, Vodovotz Y et al (2004) In silico design of clinical trials: a method coming of age. Crit Care Med 32:2061–2070
33. Chow CC, Clermont G, Kumar R et al (2005) The acute inflammatory response in diverse shock states. Shock 24:74–84
34. Vodovotz Y, Chow CC, Bartels J et al (2006) In silico models of acute inflammation in animals. Shock 26:235–244
35. Prince JM, Levy RM, Bartels J et al (2006) In silico and in vivo approach to elucidate the inflammatory complexity of CD14-deficient mice. Mol Med 12:88–96
36. Lagoa CE, Bartels J, Baratt A et al (2006) The role of initial trauma in the host's response to injury and hemorrhage: insights from a comparison of mathematical simulations and hepatic transcriptomic analysis. Shock 26:592–600
37. Ben David I, Price SE, Bortz DM et al (2005) Dynamics of intrapulmonary bacterial growth in a murine model of repeated microaspiration. Am J Respir Cell Mol Biol 33:476–482
38. Mi Q, Rivière B, Clermont G et al (2007) Agent-based model of inflammation and wound healing: insights into diabetic foot ulcer pathology and the role of transforming growth factor-b1. Wound Rep Reg 15:617–682
39. Hancioglu B, Swigon D, Clermont G (2007) A dynamical model of human immune response to influenza A virus infection. J Theor Biol 246:70–86
40. Kumar R, Chow CC, Bartels J et al (2008) A mathematical simulation of the inflammatory response to anthrax infection. Shock 29:104–111
41. An G, Lee I (2000) Complexity, emergence and pathophysiology: using agent based computer simulation to characterize the nonadaptive inflammatory response (Manuscript # 344). InterJournal Complex Systems: http://www.interjournal.org May
42. An G, Faeder JR (2009) Detailed qualitative dynamic knowledge representation using a BioNetGen model of TLR-4 signaling and preconditioning. Math Biosci 217:53–63
43. An G (2008) Introduction of a agent based multi-scale modular architecture for dynamic knowledge representation of acute inflammation. Theor Biol Med Model 5:11
44. An G (2009) A model of TLR4 signaling and tolerance using a qualitative, particle event-based method: introduction of spatially configured stochastic reaction chambers (SCSRC). Math Biosci 217:43–52
45. Kumar R, Clermont G, Vodovotz Y, Chow CC (2004) The dynamics of acute inflammation. J Theor Biol 230:145–155
46. Reynolds A, Rubin J, Clermont G et al (2006) A reduced mathematical model of the acute inflammatory response: I. Derivation of model and analysis of anti-inflammation. J Theor Biol 242:220–236
47. Day J, Rubin J, Vodovotz Y et al (2006) A reduced mathematical model of the acute inflammatory response: II. Capturing scenarios of repeated endotoxin administration. J Theor Biol 242:237–256
48. Bagci EZ, Vodovotz Y, Billiar TR et al (2006) Bistability in apoptosis: roles of bax, Bcl-2 and mitochondrial permeability transition pores. Biophys J 90:1546–1559
49. Bagci EZ, Vodovotz Y, Billiar TR et al (2008) Computational insights on the competing effects of nitric oxide in regulating apoptosis. PLoS One 3:e2249
50. Rivière B, Epshteyn Y, Swigon D, Vodovotz Y (2009) A simple mathematical model of signaling resulting from the binding of lipopolysaccharide with Toll-like receptor 4 demonstrates inherent preconditioning behavior. Math Biosci 217:19–26
51. Daun S, Rubin J, Vodovotz Y et al (2008) An ensemble of models of the acute inflammatory response to bacterial lipopolysaccharide in rats: results from parameter space reduction. J Theor Biol 253:843–853
52. Torres A, Bentley T, Bartels J et al (2008) Mathematical modeling of post-hemorrhage inflammation in mice: studies using a novel, computer-controlled, closed-loop hemorrhage apparatus. Shock 32:172–178
53. Ermentrout GB, Edelstein-Keshet L (1993) Cellular automata approaches to biological modeling. J Theor Biol 160:97–133

54. Seydel R (1994) Practical bifurcation and stability analysis. Springer, New York, NY
55. Mi Q, Riviere B, Cetin S et al (2007) One-dimensional elastic continuum model of epithelial wound healing. Biophys J 93:3745–3752
56. Blinov ML, Faeder JR, Goldstein B, Hlavacek WS (2004) BioNetGen: software for rule-based modeling of signal transduction based on the interactions of molecular domains. Bioinformatics 20:3289–3291
57. Faeder JR, Hlavacek WS, An G (2007) BioNetGen: a tool for formal knowledge representation of intracellular pathways. Shock 27(S1):S31
58. Roa LM, Gomez-Cia T, Cantero A (1988) Analysis of burn injury by digital simulation. Burns Incl Therm Inj 14:201–209
59. Roa L, Gomez-Cia T, Cantero A (1990) Pulmonary capillary dynamics and fluid distribution after burn and inhalation injury. Burns 16:25–35
60. Bert J, Gyenge C, Bowen B et al (1997) Fluid resuscitation following a burn injury: implications of a mathematical model of microvascular exchange. Burns 23:93–105
61. Rosinski M, Yarmush ML, Berthiaume F (2004) Quantitative dynamics of in vivo bone marrow neutrophil production and egress in response to injury and infection. Ann Biomed Eng 32:1108–1119
62. Feng Q, Zhao-Yan H, Zheng-Kang Z, Li-Xing S (2005) The establishment of the mathematical model of the 2nd degree burn injury of human tissues and its application. Conf Proc IEEE Eng Med Biol Soc 3:2918–2921
63. Mercer GN, Sidhu HS (2005) Modeling thermal burns due to airbag deployment. Burns 31:977–980
64. Lv YG, Liu J, Zhang J (2006) Theoretical evaluation of burns to the human respiratory tract due to inhalation of hot gas in the early stage of fires. Burns 32:436–446
65. Denman PK, McElwain DL, Harkin DG, Upton Z (2007) Mathematical modelling of aerosolised skin grafts incorporating keratinocyte clonal subtypes. Bull Math Biol 69:157–179
66. Crampin EJ, Halstead M, Hunter P et al (2004) Computational physiology and the Physiome Project. Exp Physiol 89:1–26
67. An G (2004) Agent based model of cell culture epithelial barrier function: using computer simulation in conjunction with a basic science model. Shock 21(Suppl 2):S13
68. An G (2008) Dynamic knowledge representation using agent based modeling: ontology instantiation and verification of conceptual models. In: Maly I (ed) Systems biology: methods in molecular biology series. Humana, Totowa, NJ
69. An G, Feinman R, Xu D, Deitch E (2004) In-silico unification of different basic science models of gut epithelial barrier function using agent based modeling. Crit Care Med 32:A95
70. An G (2005) Multi-hierarchical agent-based modeling of the inflammatory aspects of the gut. J Crit Care 20:383
71. An G (2006) Agent based models of pulmonary epithelial barrier function. Proc Am Thorac Soc 3:A309
72. An G (2004) Computer simulations of multiple organ failure secondary to shock and sepsis with a multi-tissue, endothelial level agent based model. Shock 21S2:66
73. Lengeler JW (2000) Metabolic networks: a signal-oriented approach to cellular models. Biol Chem 381:911–920
74. Smye SW, Clayton RH (2002) Mathematical modelling for the new millenium: medicine by numbers. Med Eng Phys 24:565–574
75. Goldstein B, Faeder JR et al (2004) Mathematical and computational models of immune-receptor signalling. Nat Rev Immunol 4:445–456
76. Mogilner A, Wollman R et al (2006) Modeling mitosis. Trends Cell Biol 16:88–96
77. Stevens A, Sogaard-Andersen L (2005) Making waves: pattern formation by a cell-surface-associated signal. Trends Microbiol 13:249–252
78. Hlavacek WS, Faeder JR et al (2006) Rules for modeling signal-transduction systems. Sci STKE 2006:re6
79. Batada NN (2004) CNplot: visualizing pre-clustered networks. Bioinformatics 20:1455–1456
80. Iragne F, Nikolski M et al (2005) ProViz: protein interaction visualization and exploration. Bioinformatics 21:272–274
81. Aragues R, Jaeggi D, Oliva B (2006) PIANA: protein interactions and network analysis. Bioinformatics 22:1015–1017
82. Baitaluk M, Qian X et al (2006) PathSys: integrating molecular interaction graphs for systems biology. BMC Bioinform 7:55
83. Lee DY, Yun C et al (2006) WebCell: a web-based environment for kinetic modeling and dynamic simulation of cellular networks. Bioinformatics 22:1150–1151
84. Mathew JP, Taylor BS et al (2007) From bytes to bedside: data integration and

computational biology for translational cancer research. PLoS Comput Biol 3:e12
85. An G (2007) Crossing levels of biological organization: agent based modeling of genomic/proteomic data. Inflamm Res 56(Suppl 2):s208–s209
86. An G (2006) Concepts for developing a collaborative in silico model of the acute inflammatory response using agent-based modeling. J Crit Care 21:105–110
87. Wakeland W, Macovsky L, An G (2007) A hybrid simulation for studying acute inflammatory response. 39–46
88. Ropella GEP, Hunt CA, Sheikh-Bahaei S (2005) Methodological considerations of heuristic modeling of biological systems. Proceedings of the 9th world multi-conference on systemics, cybernetics and informatics
89. Weston AD, Hood L (2004) Systems biology, proteomics, and the future of health care: toward predictive, preventative, and personalized medicine. J Proteome Res 3:179–196
90. Chang S, Baratt A et al (2006) Mathematical model predicting outcomes of sepsis patients treated with Xigris(R): ENHANCE trial. Shock 25:70–71
91. Hart J (2002) Inflammation. 1: its role in the healing of acute wounds. J Wound Care 11:205–209
92. Hart J (2002) Inflammation. 2: its role in the healing of chronic wounds. J Wound Care 11:245–249
93. Goldring SR (2003) Inflammatory mediators as essential elements in bone remodeling. Calcif Tissue Int 73:97–100
94. Guilak F, Fermor B et al (2004) The role of biomechanics and inflammation in cartilage injury and repair. Clin Orthop 423:17–26
95. Ramadori G, Saile B (2004) Inflammation, damage repair, immune cells, and liver fibrosis: specific or nonspecific, this is the question. Gastroenterology 127:997–1000
96. Redd MJ, Cooper L et al (2004) Wound healing and inflammation: embryos reveal the way to perfect repair. Philos Trans R Soc Lond B Biol Sci 359:777–784
97. Diegelmann RF, Evans MC (2004) Wound healing: an overview of acute, fibrotic and delayed healing. Front Biosci 9:283–289
98. Murray JD, Maini PK, Tranquillo R (1988) Mechanochemical models for generating biological pattern and form in development. Phys Rep 171:59–84
99. Murray JD (1989) Mathematical biology. Springer, Heidelberg (Germany)
100. Sherratt JA, Murray JD (1990) Models of epidermal wound healing. Proc Biol Sci 241:29–36
101. Tranquillo RT, Murray JD (1992) Continuum model of fibroblast-driven wound contraction: inflammation-mediation. J Theor Biol 158:135–172
102. Tranquillo RT, Murray JD (1993) Mechanistic model of wound contraction. J Surg Res 55:233–247
103. Cook J (1995) A mathematical model for dermal wound healing: wound contraction and scar formation [dissertation]. University of Washington, Seattle
104. Olsen L, Sherratt JA, Maini PK (1995) A mechanochemical model for adult dermal wound contraction and the permanence of the contracted tissue displacement profile. J Theor Biol 177:113–128
105. Dallon JC, Sherratt JA, Maini PK (2001) Modeling the effects of transforming growth factor-beta on extracellular matrix alignment in dermal wound repair. Wound Repair Regen 9:278–286
106. Sherratt JA, Dallon JC (2002) Theoretical models of wound healing: past successes and future challenges. C R Biol 325:557–564
107. Walker DC, Hill G et al (2004) Agent-based computational modeling of epithelial cell monolayers: predicting the effect of exogenous calcium concentration on the rate of wound closure. IEEE Trans Nanobioscience 3:153–163
108. Walker DC, Southgate J et al (2004) The epitheliome: agent-based modelling of the social behaviour of cells. Biosystems 76:89–100
109. Upperman JS, Lugo B et al (2007) Mathematical modeling in NEC – a new look at an ongoing problem. J Pediatr Surg 42:445–453
110. Hackam DJ, Upperman JS et al (2005) Disordered enterocyte signaling and intestinal barrier dysfunction in the pathogenesis of necrotizing enterocolitis. Semin Pediatr Surg 14:49–57
111. Li NYK, Verdolini K et al (2008) A patient-specific in silico model of inflammation and healing tested in acute vocal fold injury. PLoS ONE 3:e2789
112. Mesarovic MD, Sreenath SN, Keene JD (2004) Search for organising principles: understanding in systems biology. Syst Biol (Stevenage) 1:19–27
113. Matzinger P (2002) The danger model: a renewed sense of self. Science 296:301–305

114. Mollen KP, Anand RJ et al (2006) Emerging paradigm: toll-like receptor 4-sentinel for the detection of tissue damage. Shock 26:430–437
115. Tsung A, Sahai R et al (2005) The nuclear factor HMGB1 mediates hepatic injury after murine liver ischemia-reperfusion. J Exp Med 201:1135–1143
116. Izuishi K, Tsung A et al (2006) Cutting edge: high-mobility group box 1 preconditioning protects against liver ischemia-reperfusion injury. J Immunol 176:7154–7158
117. Yang R, Harada T et al (2006) Anti-HMGB1 neutralizing antibody ameliorates gut barrier dysfunction and improves survival after hemorrhagic shock. Mol Med 12:105–114
118. Tsung A, Zheng N et al (2007) Increasing numbers of hepatic dendritic cells promote HMGB1-mediated ischemia-reperfusion injury. J Leukoc Biol 81:119–128
119. Fan J, Li Y et al (2007) Hemorrhagic shock induces NAD(P)H oxidase activation in neutrophils: role of HMGB1-TLR4 signaling. J Immunol 178:6573–6580
120. Tsung A, Klune JR et al (2007) HMGB1 release induced by liver ischemia involves Toll-like receptor 4 dependent reactive oxygen species production and calcium-mediated signaling. J Exp Med 204:2913–2923
121. Klune JR, Dhupar R et al (2008) HMGB1: endogenous danger signaling. Mol Med 14:476–484
122. Klune JR, Billiar TR, Tsung A (2008) HMGB1 preconditioning: therapeutic application for a danger signal? J Leukoc Biol 83:558–563
123. Aneja RK, Tsung A, Sjodin H, Gefter JV, Delude RL et al (2008) Preconditioning with high mobility group box 1 (HMGB1) induces lipopolysaccharide (LPS) tolerance. J Leukoc Biol 84(5):1326–1334
124. Andersson U, Wang H, Palmblad K, Aveberger AC, Bloom O et al (2000) High mobility group 1 protein (HMG-1) stimulates proinflammatory cytokine synthesis in human monocytes. J Exp Med 192:565–570
125. El GM (2007) HMGB1 modulates inflammatory responses in LPS-activated macrophages. Inflamm Res 56:162–167
126. Dunbar SA (2006) Applications of Luminex xMAP technology for rapid, high-throughput multiplexed nucleic acid detection. Clin Chim Acta 363:71–82
127. Barclay D, Zamora R, Torres A, Namas R, Steed D et al (2008) A simple, rapid, and convenient Luminex™-compatible method of tissue isolation. J Clin Lab Anal 22:278–281
128. Kobbe P, Kaczorowski D, Vodovotz Y, Tzioupis C, Mollen KP et al (2008) Local exposure of bone components to injured soft tissue induces Toll-like-receptor-4 dependent systemic inflammation with acute lung injury. Shock 30:686–691
129. Kobbe P, Vodovotz Y, Kaczorowski D, Mollen KP, Billiar TR et al (2008) Patterns of cytokine release and evolution of remote organ dysfunction after bilateral femur fracture. Shock 30:43–47
130. Janes KA, Albeck JG, Gaudet S, Sorger PK, Lauffenburger DA et al (2005) A systems model of signaling identifies a molecular basis set for cytokine-induced apoptosis. Science 310:1646–1653
131. Cohen AM, Hersh WR (2005) A survey of current work in biomedical text mining. Brief Bioinform 6:57–71
132. Spasic I, Ananiadou S, McNaught J, Kumar A (2005) Text mining and ontologies in biomedicine: making sense of raw text. Brief Bioinform 6:239–251
133. Marshall JC (2004) Through a glass darkly: the brave new world of in silico modeling. Crit Care Med 32:2157–2159

Chapter 10

Immunoinformatics and Systems Biology Methods for Personalized Medicine

Qing Yan

Abstract

The immune system plays an important role in the development of personalized medicine for a variety of diseases including cancer, autoimmune diseases, and infectious diseases. Immunoinformatics, or computational immunology, is an emerging area that provides fundamental methodologies in the study of immunomics, that is, immune-related genomics and proteomics. The integration of immunoinformatics with systems biology approaches may lead to a better understanding of immune-related diseases at various systems levels. Such methods can contribute to translational studies that bring scientific discoveries of the immune system into better clinical practice. One of the most intensely studied areas of the immune system is immune epitopes. Epitopes are important for disease understanding, host–pathogen interaction analyses, antimicrobial target discovery, and vaccine design. The information about genetic diversity of the immune system may help define patient subgroups for individualized vaccine or drug development. Cellular pathways and host immune-pathogen interactions have a crucial impact on disease pathogenesis and immunogen design. Epigenetic studies may help understand how environmental changes influence complex immune diseases such as allergy. High-throughput technologies enable the measurements and catalogs of genes, proteins, interactions, and behavior. Such perception may contribute to the understanding of the interaction network among humans, vaccines, and drugs, to enable new insights of diseases and therapeutic responses. The integration of immunomics information may ultimately lead to the development of optimized vaccines and drugs tailored to personalized prevention and treatment. An immunoinformatics portal containing relevant resources is available at http://immune.pharmtao.com.

Key words: Immunoinformatics, Immune system, Bioinformatics, Software, Databases, Pharmacogenomics, Systems biology, Personalized medicine, Epigenetics, High-throughput, Microarray, Single nucleotide polymorphism (SNP), Pathways, Interactions, Drugs, Vaccines, Cancer, Infectious diseases, Epitopes, Genomics, Proteomics, Structure

1. Immunoinformatics and Systems Biology

The immune system plays an important role in the development of personalized medicine for a variety of diseases including cancer, autoimmune diseases, and infectious diseases. The immune system

Qing Yan (ed.), *Systems Biology in Drug Discovery and Development: Methods and Protocols*, Methods in Molecular Biology, vol. 662, DOI 10.1007/978-1-60761-800-3_10, © Springer Science+Business Media, LLC 2010

has a great diversity in order to fight pathogens and discriminate between self and nonself components. It typically generates more than 10^9 different antibodies and 10^9 different T lymphocytes in one individual (1). Thousands of molecules are involved in the dynamic network of immune responses. Such large number of data require powerful informatics methods for data management and analyses.

Immunoinformatics, or computational immunology, is an emerging area that uses bioinformatics approaches to solve problems and improve the communication, understanding, analysis, and management of immunological information. Bioinformatics plays an indispensible role in designing experiments, such as high-throughput studies, and helping to establish and test hypotheses through data analyses. These essential tasks in drug discovery and development cannot be accomplished with traditional approaches alone.

Immunoinformatics resources include immunological databases, genetic sequence analysis and structural modeling tools, and mathematical models of the immune system (2). Immunoinformatics may provide fundamental methodologies in the study of immunomics, that is, immune-related genomics and proteomics.

Systems biology studies the interactions among biological elements toward the understanding of diseases at the system level (3). The combination of bioinformatics and systems biology approaches can lead to a better understanding of immune-related diseases (1). Complex interactions and feedbacks are involved in the behavior of the immune system, such as the systemic inflammatory reaction. These interactions include those between antigens and the receptors of the immune system, such as B-cell receptors (antibodies), T-cell receptors (TRs or TCRs), and major histocompatibility complex (MHC) receptors.

Systems biology approaches can be applied to modeling the immunological functions for individual persons to enable an optimal interaction between a personalized vaccine or drug and the immune system. Such approaches can establish the basis of personalized medicine and bring the right vaccine or drug to the right person with the right dosages. The integration of immunoinformatics and systems biology approaches may also contribute to bench-to-bedside translational studies that bring scientific discoveries of the immune system into better clinical practice.

Many software tools have been constructed to organize immunological data and information at different systems levels. Here, the methods and resources are briefly reviewed in terms of integrative immunoinformatics and systems biology for the development of personalized medicine. Their applications in vaccine and drug development for the prevention and treatment of disorders such as infectious diseases and cancer are also discussed. An immunoinformatics portal containing relevant information is available at http://immune.pharmtao.com.

2. Epitope Recognition and Prediction for Drug and Vaccine Design

One of the most intensely studied areas of the immune system is immune epitopes. Epitopes are the parts of antigens interacting with receptors of the immune system (4). Receptor recognition can stimulate immune responses. Epitopes are important for disease understanding, host–pathogen interaction analyses, antimicrobial target discovery, and vaccine design. Besides infectious diseases, epitopes have great potentials for the development of cancer vaccines. For example, a naturally presented immunodominant Human Leukocyte Antigen (HLA)-B7-restricted epitope from the tumor antigen NY-ESO-1 has been suggested to have implications for cancer vaccine design (5).

Immune responses include innate immunity and adaptive immunity. Innate immune responses are the first line and nonspecific defense against foreign agents or antigens. Adaptive immunity is the second line of defense, composed of the cellular response of T lymphocytes and the humoral response of B cells that produce antibodies.

T cell epitopes have a linear conformation (6). B cell epitopes are more difficult to predict using computational techniques because they have nonlinear conformational structures. The computational predictions of epitopes are usually made based on the identification of patterns and motifs with sequence similarities and three-dimensional (3D) structures, as well as the modeling of interactions and the binding with MHC molecules. Commonly used computational methodologies include homology modeling, decision trees, artificial neural networks (ANN), hidden Markov models (HMM), support vector machines (SVMs), and quantitative structure-activity relationship (QSAR) analysis.

Immune epitope databases and tools are very useful in vaccine or drug target design, as well as host–pathogen interaction analyses. Table 1 lists some of these resources that can be used for studies on T cells, B cells, and epitope predictions. For example, the immune epitope database and analysis resource (IEDB) is a program for analyzing epitopes (see Table 1) (7). The database has data on antibody and T cell epitopes for humans, nonhuman primates, rodents, and other animal species. It also has MHC binding data from a variety of antigenic sources.

Currently (June 2009), the database contains curated data including National Institute of Allergy and Infectious Diseases (NIAID) Category A, B, and C priority pathogens (8). It contains more than 60,000 peptidic epitopes. The program has tools for T cell and B cell epitope prediction. It also provides epitope analysis tools such as a tool for Epitope Cluster Analysis that groups epitopes into clusters based on sequence similarity.

Table 1
Immunoinformatics resources on T cells, B cells, and epitope analyses

Name	URL[a]	Explanation
Immune Epitope Database and Analysis Resource (IEDB)	http://www.immuneepitope.org	Antibody and T cell epitopes
AntiJen database	http://www.jenner.ac.uk/antijen/	Peptides binding to MHC ligands, T and B cell epitopes, TAP, protein–protein interactions
Epitome	http://cubic.bioc.columbia.edu/services/epitome/submit.php	Database of structurally inferred antigenic epitopes in proteins
CED	http://web.kuicr.kyoto-u.ac.jp/~ced/	Conformational epitope database
MIMOX	http://web.kuicr.kyoto-u.ac.jp/~hjian/mimox/	Phage display-based epitope mapping
EpiJen	http://www.jenner.ac.uk/EpiJen/	T cell epitope prediction
MHCPred	http://www.jenner.ac.uk/MHCPred/	Quantitative prediction of peptide-MHC binding
RankPep	http://bio.dfci.harvard.edu/Tools/rankpep.html	Prediction of binding peptides to MHC
EPIMHC	http://bio.dfci.harvard.edu/epimhc/	Database of MHC ligands
SYFPEITHI	http://www.syfpeithi.de/	Database of peptide sequences known to bind MHC
BIMAS	http://www-bimas.cit.nih.gov/molbio/hla_bind/	HLA peptide binding predictions
NetMHC	http://www.cbs.dtu.dk/services/NetMHC/	Predicts binding of peptides to HLA alleles
BepiPred	http://www.cbs.dtu.dk/services/BepiPred/	Predicts the location of linear B-cell epitopes
DiscoTope	http://www.cbs.dtu.dk/services/DiscoTope/	Prediction of residues in discontinuous B cell epitopes
AbCheck	http://www.bioinf.org.uk/abs/seqtest.html	Tests an antibody sequence against the Kabat database
NetCTL	http://www.cbs.dtu.dk/services/NetCTL/	Predicts CTL epitopes in protein sequences
VaxiJen	http://www.jenner.ac.uk/VaxiJen/	Prediction of protective antigens and subunit vaccines
CTLPred	http://www.imtech.res.in/raghava/ctlpred/	Predicts CTL epitopes in subunit vaccine design

(continued)

Table 1 (continued)

Name	URL[a]	Explanation
Macrophages.com	http://www.macrophages.com/content/macrophages/home/	A community website for researchers with an interest in macrophage biology
HIV Molecular Immunology Database	http://www.hiv.lanl.gov/immunology	HIV-1 cytotoxic and helper T-cell epitopes and antibody binding sites
HCV Sequence Database	http://hcv.lanl.gov/content/sequence/HCV/ToolsOutline.html	Tools including sequence and immunology analyses
Influenza Research Database	http://www.fludb.org/brc/home.do?Decorator=influenza	Database and tools for analyses of sequences and epitopes
Viral genomes	http://www.ncbi.nlm.nih.gov/genomes/GenomesHome.cgi?taxid=10239	Sequences of viral genomes
Immunoinformatics portal	http://immune.pharmtao.com	A web portal of immunoinformatics

[a]Websites were accessed in July 2009

As conservative pathogenic sequences are usually crucial for the retention of protein functions across microbial strains, they can be good targets for vaccines and antimicrobial drugs. A useful tool is Epitope Conservancy Analysis in IEDB that calculates the degree of conservancy of an epitope. Vaccines are usually designed based on the identification of the pathogen-unique epitopes that can lead to protective responses.

The identification of proteins containing unique epitopes is also useful for designing subunit vaccines. A subunit vaccine has some fragment of an organism to generate immune responses, rather than using an inactivated or attenuated organism. Tools listed in Table 1, such as VaxiJen can be used for the prediction of protective antigens and subunit vaccines (9). Another program, CTLPred provides information for the prediction of cytotoxic T lymphocyte epitopes that are crucial in subunit vaccine design (10) (see Table 1).

Many other tools are available for epitope prediction. For example, the AntiJen Database contains quantitative binding data for peptides binding to MHC Ligand, TCR-MHC Complexes, and T Cell Epitopes (see Table 1). The database also provides information on antigen peptide transporter (TAP), B cell epitope molecules, and immunological protein–protein interactions (11).

Besides resources for epitope analyses and vaccine design, Table 1 lists some resources for immunological studies in infectious diseases such as human immunodeficiency virus (HIV) and Hepatitis C Virus (HCV). For example, HCV Sequence Database contains tools for identifying epitopes within immunologically reactive peptides (called Epitope Location Finder), and for finding HLA anchor motifs.

3. Analyses of the Structure–Function Association of the Immune System

Immunoinformatics approaches in systems biology can be conducted in a systematic way, starting from the molecular level (see Fig. 1). The detailed genetic sequence analytic information includes sequence retrieval and comparison, and sequence variation information such as single nucleotide polymorphisms (SNPs). Information on sequence patterns can correlate genetic structures to functional motifs. When a time dimension is added at this level, evolutionary or phylogenetic trees can be built to compare these genetic sequences of different times.

The structure–function association is one of the key issues to investigate for the development of personalized medicine (12).

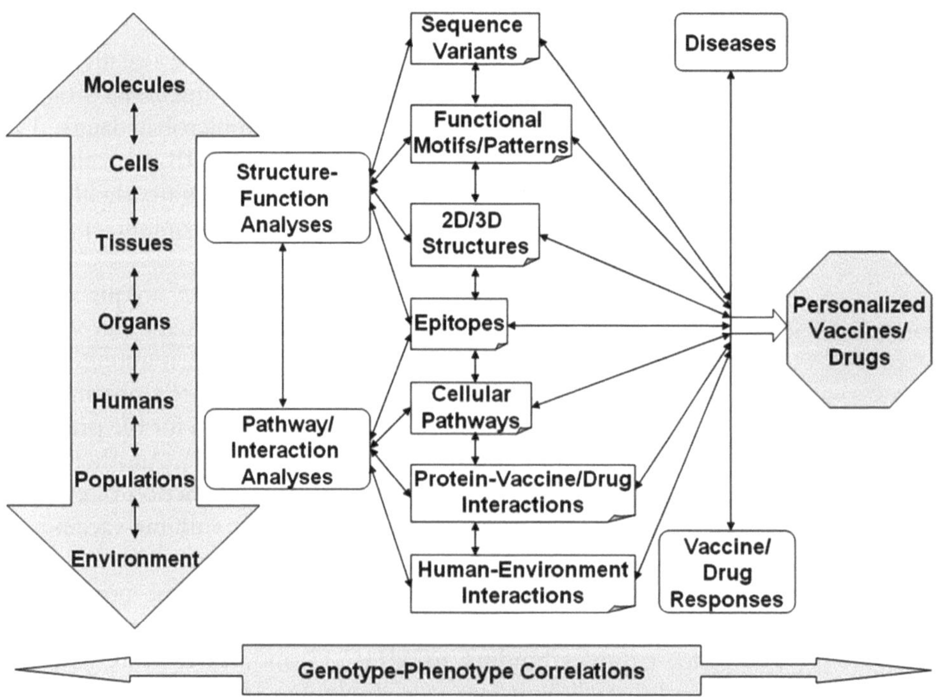

Fig. 1. The workflow and various dimensions of immunoinformatics and systems biology analyses for personalized vaccines and drugs.

Table 2 lists some tools for the structure–function analyses of the immune system. For instance, a comprehensive resource is IMGT, the international ImMunoGeneTics information system (13). IMGT is an integrated resource of the immunoglobulins (IG), TR, MHC, immunoglobulin superfamily (IgSF), MHC superfamily (MhcSF), and related proteins of the immune system (RPI) of humans and other vertebrate species (see Table 2).

IMGT consists of five databases. IMGT/LIGM-DB contains nucleotide sequences of IG and TR from more than 200 species. IMGT/MHC-DB has sequences of HLA alleles. IMGT/PRIMER-DB contains oligonucleotides (primers) of IG and TR from 11 species. IMGT/GENE-DB has more than 3,000 alleles of IG and TR genes from humans, mice, rats, and rabbits. IMGT/3Dstructure-DB is a database of 3D structures of IG, TR, MHC, and RPI with more than 1,000 entries. IMGT also has interactive online tools for sequence, genome, and 3D structure analyses. For example, IMGT/V-QUEST is a sequence alignment tool for IG, TR, and HLA.

Table 2
Tools for structure–function analyses of the immune system

Name	URL	Explanation
IMGT®, the international ImMunoGeneTics information system®	http://imgt.cines.fr/	Resource of the IG, TR, MHC, IgSF, MhcSF and RPI
ABG	http://www.ibt.unam.mx/vir/structure/structures.html	Directory of 3D structures of antibodies
Immunome	http://bioinf.uta.fi/Immunome/	Immunity-related proteins, domain structure and ontology terms
Basic Local Alignment Search Tool (BLAST)	http://www.ncbi.nlm.nih.gov/BLAST/	Comparison of novel sequences with known genes
CLUSTAL W	http://www.ebi.ac.uk/clustalw/	Sequence alignment tool for similarities and differences
Motif Scan	http://myhits.isb-sib.ch/cgi-bin/motif_scan	Finding motifs in a sequence
PredictProtein	http://www.predictprotein.org/	Protein secondary structure prediction
Protein Data Bank (PDB)	http://www.rcsb.org/pdb/	Biological macromolecular structure data
Molecular Modeling Database (MMDB)	http://www.ncbi.nlm.nih.gov/sites/entrez?db=structure	3D structures, including proteins, polynucleotides

In addition to such specific programs, general bioinformatics databases and tools can be used for immunological analyses. For example, tools such as Basic Local Alignment Search Tool (BLAST) (14) and CLUSTAL W (15) are commonly used in comparing genetic sequences, evolutionary relationships, and building phylogenetic trees for immune molecules (see Table 2). Tools for sequence pattern analysis including Motif Scan are useful for correlating sequence structure to functional motifs and epitopes (16) (see Table 2).

Three-dimensional modeling of the sequence structure provides better understanding of the structure–function relationship, such as using the database Protein Data Bank (PDB) (17) (see Table 2). The Molecular Modeling Database (MMDB) provides information on 3D macromolecular structures, including protein and polynucleotide structures of antibodies, HLA, and TCRs. The database has links to sequences, taxonomic classifications, and structure neighbors.

4. Immuno-informatics for Pharmacogenomics and Personalized Medicine

The identification of the association between a disease and genetic variations is one of the most important analyses in pharmacogenomics and the development of personalized medicine (18). Pharmacogenomics studies the genetic basis of individual variation in response to therapeutic agents (18). The information about allele frequencies of immune molecules in anthropologically defined population is especially important. This is because different patient subgroups can be identified with different vaccine or drug responses.

For example, a SNP (S427T) in the innate immune gene interferon regulatory factor 3 (IRF3) has been associated with increased risk of human papillomavirus (HPV) persistence and cervical cancer (19). The investigation of genetic diversity in the immune system would make it possible for the prediction of disease predisposition and prognosis, as well as to tailor optimal vaccine or drug prescription to the right person.

Genetic variations are also important for the understanding of immune responses at the systems level. For example, interactions have been found between the SNPs of the DNA repair and immune genes, for example, xeroderma pigmentosum group D (XPD) and Interleukin-10 (IL10), which might be associated with breast cancer predisposition (20).

Some resources on genetic variations are available specifically for the immune system (see Table 3). For example, the Immuno Polymorphism Database (IPD) has several specialist databases (21). The IPD–KIR Database contains allelic sequences of human

Table 3
Resources on genetic variations of the immune system

Name	URL	Explanation
IPD	http://www.ebi.ac.uk/ipd/	Immuno-polymorphism database
dbMHC	http://www.ncbi.nlm.nih.gov/gv/mhc/main.fcgi?cmd=init	DNA and clinical data related to MHC
dbLRC	http://www.ncbi.nlm.nih.gov/gv/lrc/main.cgi?cmd=init&user_id=0&probe_id=0&source_id=0&locus_id=0&locus_group=2&proto_id=0&kit_id=0&dummy=0	Database on human Leukocyte Receptor Complex (LRC)
MHC Haplotype Project	http://www.sanger.ac.uk/HGP/Chr6/MHC/	Association studies of all MHC-linked-diseases
SNEP	http://elchtools.de/SNEP/	Prediction of SNP-derived epitopes
SiPep (SNPBinder)	http://www.sipep.org/	Prediction of tissue-specific minor histocompatibility antigens
International HapMap Project	http://snp.cshl.org/	Databases, linkage maps, features of sequence variation
dbSNP	http://www.ncbi.nlm.nih.gov/SNP/	NCBI's database of SNPs
Allele Frequencies Database	http://www.allelefrequencies.net/	On human populations

killer-cell immunoglobulin-like receptors (KIR). IPD–MHC Database has data on MHC of various species. IPD–HPA Database contains data on human platelet antigens (HPA). IPD–ESTDAB is a database of immunologically characterized tumor cell lines. The comprehensive program of IMGT (see Table 2) described above also contains relevant information on the genetic variations of the immune system.

The National Center for Biotechnology Information (NCBI) has a database specifically on MHC, named dbMHC (see Table 3). The program contains an interactive Alignment Viewer for HLA and related genes, and an MHC microsatellite database. It also has a sequence interpretation site for Sequencing Based Typing (SBT) and a Primer/Probe database. The program has information on the HLA anthropology, including individual allele and haplotype frequencies from various populations. A parallel program is dbLRC, a database on the human Leukocyte Receptor Complex (LRC), including KIR and leukocyte Ig-like receptor (LILR).

Another program, the Sanger MHC Haplotype Project has information on association studies of MHC-linked-diseases (22) (see Table 3). The site contains data on genomic sequences of different HLA-homozygous typing haplotypes, variations or SNPs, and ancestral relationships. The program SNEP predicts SNP-derived epitopes for mHAgs (23). It predicts potential T-cell epitopes in a chosen distance around the polymorphic residue.

SiPep is a tool for the prediction of tissue-specific minor histocompatibility antigens (24) (see Table 3). Minor histocompatibility antigens (mHAgs) are T-cell epitopes containing polymorphic spots. They are additional transplantation antigens outside the MHC that may cause the rejection of tumors and skin grafts from MHC identical donors. The program predicts candidate minor histocompatibility loci through the identification of coding SNPs in an MHC-binding peptide that change the ability of the peptide to bind.

Genomic variation databases such as HapMap and dbSNP (see Table 3) provide information on individual genotype data. The Allele Frequencies Database can be used to search for polymorphic regions of various populations on histocompatibility and immunogenetics (see Table 3). The database includes polymorphism information on HLA, cytokines, and KIR. Such information provides direct connections of systemic information from the molecular level to the population level (see Fig. 1).

5. Pathway and Interaction Analyses Inside and Beyond the Immune System

With the comprehensive examination of structures, functions, and relationships between them at the molecular level, we can scale up to the higher level to gain a more complete view of how the immune system works and interacts with other systems (see Fig. 1). The host immune–pathogen interactions have crucial impacts on pathogen evolution, pathogenesis, and immunogen design. At the cellular level, the interactions and networks among those immune molecules should be examined.

For example, the innate immune response alone is a complex network of interconnected pathways and dynamic networks of molecules with multiple influences (25). The understanding of changes in molecular and cellular pathways and interactions can be useful for finding new drug targets and designing effective drugs.

Resources for epitope studies as listed in Table 1 are useful for interaction analyses. In addition, Table 4 lists some databases and tools for pathway and interaction analyses in immune responses. For example, InnateDB provides information on interactions and

Table 4
Immunoinformatics resources for pathways and interactions

Name	URL	Explanation
InnateDB	http://www.innatedb.ca/	Interactions and signaling pathways in the innate immune response
Innate Immune Database (IIDB)	http://db.systemsbiology.net/cgi-bin/GLUE/U54/IIDBHome.cgi	Genes associated with immune responses
JenPep	http://www.jenner.ac.uk/jenpep/	Immunological protein–peptide interactions
InnateImmunity-SystemsBiology.org	http://www.innateimmunity-systemsbiology.org	Innate immunity, inflammation and septic shock
Pathogen Interaction Gateway (PIG)	http://molvis.vbi.vt.edu/pig/	Host–pathogen, protein–protein interactions (PPIs)
VirusMINT	http://mint.bio.uniroma2.it/virusmint/Welcome.do	Interactions between human and viral proteins
NetPath	http://www.netpath.org/	Signal transduction pathways in humans, including immune and cancer pathways
KEGG	http://www.genome.jp/kegg/pathway.html	Pathway maps, ortholog group tables, catalogs
Reactome	www.reactome.org/	Pathways
HPRD	http://www.hprd.org/	Pathways and protein interaction networks
GenMapp	http://www.genmapp.org/	Pathway tools
BioCyc	http://www.biocyc.org/	Pathway/genome databases
Pathguide	http://www.pathguide.org/	The pathway resource list
Pathway Interaction Database	http://pid.nci.nih.gov/	Biomolecular interactions, cellular processes, human signaling pathways
Database of Interacting Proteins (DIP)	http://dip.doe-mbi.ucla.edu/	Interactions between proteins
IntAct	http://www.ebi.ac.uk/intact/main.xhtml	Protein interaction data
MINT, the Molecular INTeraction database	http://mint.bio.uniroma2.it/mint/Welcome.do	Experimentally verified protein–protein interactions

(continued)

Table 4 (continued)

Name	URL	Explanation
BioGRID	http://www.thebiogrid.org/	Repository for Interaction Datasets
Cytoscape	http://www.cytoscape.org/	Visualization software for molecular interaction networks
Cerebral	http://www.pathogenomics.ca/cerebral/	Cytoscape molecular interaction viewer
Network Analysis Tools (NeAT)	http://rsat.ulb.ac.be/rsat/index_neat.html	Programs for biological networks
PSORT	http://psort.nibb.ac.jp/	Prediction of protein localization sites in cells
Online Mendelian Inheritance in Man (OMIM)	http://www3.ncbi.nlm.nih.gov/omim/	A catalog of human genes and genetic disorders
Frequency of Inherited Disorders Database (FIDD)	http://archive.uwcm.ac.uk/uwcm/mg/fidd/introduction.html	For medical and epidemiological studies
miRBase	http://www.mirbase.org/	microRNA database
MethDB	http://www.methdb.net	DNA methylation database

signaling pathways associated with the innate immune response to microbial infections in humans and mice (26).

Pathogen Interaction Gateway (PIG) is a database of known host–pathogen interactions (27) (see Table 4). It provides functional annotations and data of the domains in the interacting proteins. JenPep contains quantitative binding data for immunological protein–peptide interactions (28). These pathways are potential targets for developing novel therapeutics. VirusMINT collects data on protein interactions between viral and human proteins (29). It has information on more than 5,000 interactions and over 490 viral proteins from more than 110 viral strains.

General protein–protein interaction databases, gene network, and pathway databases such as Kyoto Encyclopedia of Genes and Genomes (KEGG) are also useful for this level of study (30) (see Table 4). Other pathway databases containing data on immune pathways include Reactome (31) and Human Protein Reference Database (HPRD) (see Table 4).

The Database of Interacting Proteins (DIP) stores information about experimentally determined interactions between proteins (32) (see Table 4). Cytoscape is a software tool for visualizing molecular interaction networks and integration with other data (33).

For the prediction of protein localization sites in cells, the tool PSORT can be used (34).

Resources are also available for studying higher levels including the tissue and the organism level, and for making the connection between different levels (see Fig. 1). For example, some databases supply linkages between sequence variation genotypes and disease phenotypes, such as the Online Mendelian Inheritance in Man (OMIM) database (35) (see Table 4).

MicroRNAs (miRNAs) are short RNA sequences expressed from longer transcripts that may cause translational repression or transcript degradation. miRNAs may be involved in pathways such as apoptosis and metabolism, and in diseases such as cancer. miRBase is a database that contains data on nomenclature, sequences, annotation, and the target prediction of miRNAs, including information of miRNAs from immune-related tissues (36) (see Table 4).

Table 5 lists examples of some known immune pathways involved in human diseases. For instance, complement and coagulation cascades are associated with a broad range of health conditions including blood pressure, sodium homeostasis, cardioprotective effects, inflammation, and infectious diseases. ERK (extracellular signal-regulated kinase), NF-kB (nuclear factor kappa-light-chain-enhancer of activated B cells), and p38 signaling pathways are involved in cancer, microbial infections, and inflammation. Toll-like receptor signaling pathway plays an important role in microbial infections, and may be potential targets for antimicrobial therapeutics (see Table 5).

6. Epigenetics, Gene–Environment Interactions, and the Immune System

Epigenetic studies, such as those on DNA methylation and histone modifications, may help understand how environmental changes influence complex immune diseases such as allergy. Immune development has been found to be under epigenetic regulation, including the pattern of T helper Th1 and Th2 cell differentiation, and regulatory T cell differentiation (37). Epigenetic mechanisms involving chromatin may be responsible for the immune escape of cancer cells (38).

Epigenetic regulation is associated with controlling tumor antigen processing. Similar to mutations, epigenetic silencing in cancer may be a factor of gene inactivation. The reversal of epigenetic codes may have immunotherapeutic potentials for cancer therapy (39). A few epigenetic tools are available, such as the MethDB database that contains data on methylation patterns and profiles (40) (see Table 4).

Table 5
Examples of immunological pathways in humans

Pathway names	Examples of involved diseases, health conditions, and applications	Sample resources
Hematopoietic cell lineage	Blood clotting, surface markers	KEGG (see Table 4)
Complement and coagulation cascades	Blood pressure, sodium homeostasis, cardioprotective effects, inflammation, infectious diseases	KEGG, Reactome, InnateDB (see Table 4)
Toll-like receptor signaling pathway	Microbial infections, cancer	KEGG, Reactome, InnateDB
RIG-I-like receptor signaling pathway	Viral pathogens	KEGG
Natural killer cell mediated cytotoxicity	Infections with viruses, bacteria, parasites or malignant transformation	KEGG
Antigen processing and presentation	Various processes	KEGG
T cell receptor signaling pathway	Various processes	KEGG, Reactome
B cell receptor signaling pathway	Various processes	KEGG
Fc epsilon RI signaling pathway	Inflammatory responses	KEGG
Fc gamma R-mediated phagocytosis	Infectious pathogens	KEGG
Leukocyte transendothelial migration	Immune surveillance and inflammation	KEGG
Cell surface interactions at the vascular wall	Inflammatory conditions	Reactome
Chemokine signaling pathway	Inflammatory immune responses	KEGG, InnateDB
Immunoregulatory interactions between a lymphoid and a nonlymphoid cell	Cancer, viral infections	Reactome
Costimulation by the CD28 family	Autoimmunity, aging, viral infections	Reactome
ERK signaling pathway	Cancer, microbial infections	InnateDB, KEGG
NF-kB signaling pathway	Inflammation, cancer	InnateDB, KEGG
p38 signaling pathway	Cancer, longevity, microbial infections	InnateDB

7. High-Throughput Analyses

Combined with bioinformatics methodologies, high-throughput technologies enable the measurements and catalogs of genes, proteins, interactions, and behavior in various conditions. Such methods allow genome-wide association (GWA) studies for the identification of biomarkers for disease diagnosis and therapeutic outcome assessments.

Microarray has been used extensively for protein–protein interaction and pharmacogenomics studies, including the immune system. For example, using a protein microarray approach, immune responses to a broad set of antigens were identified in colon cancer (41). DNA microarray analysis helped in identifying innate immune pathways in virus-induced autoimmune diabetes (42). The DNA microarray approach found that Kilham Rat Virus (KRV)-induced innate immune pathways play a role in islet inflammation and diabetes. Another example is that statistical and gene ontology analyses of microarray data suggested modified neuro-immune signaling in nucleated blood cells among patients with Parkinson's disease (43). Such bioinformatics analyses in turn can direct further examinations of the complex network of diseases.

Various sites are available for protocols, databases, and data analysis tools on high-throughput analyses of the immune system. For example, the Innate Immune Database (IIDB) has data on gene regulatory systems underlying innate responses to pathogens (44) (see Table 4). It contains more than 2,000 genes involved in immune responses in the mouse genome. The data are from more than 150 microarray experiments. RefDIC is a database of quantitative mRNA and protein profiles constructed specifically for the immune system (45) (see Table 6).

Table 6 also lists some general sources for microarray analyses. For instance, Microarray Informatics at European Bioinformatics Institute (EBI) provides resources for microarray data management, storage, and analyses. Gene Expression Omnibus (GEO) is useful for the browsing, query, and retrieval of gene expression and array information (46).

8. Conclusion

The combination of the two emerging disciplines, immunoinformatics and systems biology, provides promising methods and novel strategies for the discovery and development of personalized medicine. Figure 1 illustrates and summarizes the overall

Table 6
Immunoinformatics resources for microarray analyses

Name	URL	Explanation
RefDIC	http://refdic.rcai.riken.jp/welcome.cgi	Database of mRNA and proteins for the immune system
Immunological Genome Project	http://www.immgen.org/index_content.html	Microarray analyses of gene expression in mouse immune system
Gene Expression Omnibus (GEO)	http://www.ncbi.nlm.nih.gov/geo/	Gene expression and array database
Microarray Informatics at EBI	http://www.ebi.ac.uk/microarray/	Microarray data management, storage, and analyses
NHGRI Microarray Project	http://research.nhgri.nih.gov/microarray/main.html	Protocols, databases and tools for arrays
Bibliography on Microarray Data Analysis	http://www.nslij-genetics.org/microarray/	Links and publications on array data analyses
Stanford Microarray Database	http://genome-www5.stanford.edu/	Microarray database and analysis tools
System for Integrative Genomic Microarray Analysis (SIGMA)	http://sigma.bccrc.ca	Visualization and analysis of data from high resolution array CGH platforms about cancer genomes

workflow and various dimensions of applying immunoinformatics methods and tools in systems biology toward personalized vaccines and drugs. Relevant resources for each analysis have been discussed in the above sections.

The core of the analyses is to understand the genotype–phenotype correlations at various systems levels, from the molecular level, cellular level, through organs, tissues, humans, populations, and the whole environment. Here phenotype is defined as visible traits, such as clinical measurements. The key aspects for immunoinformatics and systems biology investigation include structure–function analyses and pathway/interaction analyses.

The structure–function analyses include the examination of how sequence variants such as polymorphisms may have functional influences. Studies of transcription factors, functional motifs, 2D and 3D structure may help with the identification of epitopes and design of vaccines. These studies may shed light on the mechanisms of cellular pathways and protein–protein interactions. Advances in high-throughput analyses may greatly enhance such investigations.

The perception at these points may contribute to the understanding of the interaction networks among humans, vaccines, drugs, and the environment and enable new insights of disease mechanisms and therapeutic responses. The integration of all of the information at various systems levels may ultimately lead to the development of optimized vaccines and drugs tailored to individualized prevention and treatment.

References

1. Rapin N, Kesmir C, Frankild S et al (2006) Modelling the human immune system by combining bioinformatics and systems biology approaches. J Biol Phys 32: 335–353
2. Brusic V, Petrovsky N (2003) Immunoinformatics – the new kid in town. Novartis Found Symp 254:3–13, discussion 13–22, 98–101, 250–252
3. Yan Q (2005) Pharmacogenomics and systems biology of membrane transporters. Mol Biotechnol 29:75–88
4. Yan Q (2008) Bioinformatics databases and tools in virology research: an overview. In Silico Biol 8:71–85
5. Ebert LM, Liu YC, Clements CS et al (2009) A long, naturally presented immunodominant epitope from NY-ESO-1 tumor antigen: implications for cancer vaccine design. Cancer Res 69:1046–1054
6. Evans MC (2008) Recent advances in immunoinformatics: application of in silico tools to drug development. Curr Opin Drug Discov Devel 11:233–241
7. Peters B, Sidney J, Bourne P et al (2005) The design and implementation of the immune epitope database and analysis resource. Immunogenetics 57:326–336
8. NIAID Category A, B, and C Priority Pathogens. Available at http://www3.niaid.nih.gov/topics/BiodefenseRelated/Biodefense/research/CatA.html. Accessed June 2009
9. Doytchinova IA, Guan P, Flower DR (2006) EpiJen: a server for multistep T cell epitope prediction. BMC Bioinform 7:131
10. Bhasin M, Raghava GP (2004) Prediction of CTL epitopes using QM, SVM and ANN techniques. Vaccine 22:3195–3204
11. Toseland CP, Clayton DJ, McSparron H et al (2005) AntiJen: a quantitative immunology database integrating functional, thermodynamic, kinetic, biophysical, and cellular data. Immunome Res 1:4
12. Yan Q (2003) Pharmacogenomics of membrane transporters: an overview. Methods Mol Biol 227:1–20
13. Lefranc MP, Giudicelli V, Ginestoux C et al (2009) IMGT, the international ImMunoGeneTics information system. Nucleic Acids Res 37:D1006–D1012
14. Altschul SF, Gish W, Miller W et al (1990) Basic local alignment search tool. J Mol Biol 215:403–410
15. Larkin MA, Blackshields G, Brown NP et al (2007) Clustal W and Clustal X version 2.0. Bioinformatics 23:2947–2948
16. Pagni M, Ioannidis V, Cerutti L et al (2007) MyHits: improvements to an interactive resource for analyzing protein sequences. Nucleic Acids Res 35:W433–W437
17. Dutta S, Burkhardt K, Swaminathan GJ et al (2008) Data deposition and annotation at the worldwide protein data bank. Methods Mol Biol 426:81–101
18. Yan Q (2008) The integration of personalized and systems medicine: bioinformatics support for pharmacogenomics and drug discovery. Methods Mol Biol 448:1–19
19. Wang SS, Bratti MC, Rodriguez AC et al (2009) Common variants in immune and DNA repair genes and risk for human papillomavirus persistence and progression to cervical cancer. J Infect Dis 199:20–30
20. Onay VU, Briollais L, Knight JA et al (2006) SNP-SNP interactions in breast cancer susceptibility. BMC Cancer 6:114
21. Robinson J, Marsh SG (2007) IPD: the Immuno Polymorphism Database. Methods Mol Biol 409:61–74
22. Horton R, Gibson R, Coggill P et al (2008) Variation analysis and gene annotation of eight MHC haplotypes: the MHC Haplotype Project. Immunogenetics 60:1–18
23. Schuler MM, Donnes P, Nastke MD et al (2005) SNEP: SNP-derived epitope prediction program for minor H antigens. Immunogenetics 57:816–820

24. Halling-Brown M, Quartey-Papafio R, Travers PJ et al (2006) SiPep: a system for the prediction of tissue-specific minor histocompatibility antigens. Int J Immunogenet 33:289–295
25. Gardy JL, Lynn DJ, Brinkman FS et al (2009) Enabling a systems biology approach to immunology: focus on innate immunity. Trends Immunol 30:249–262
26. Lynn DJ, Winsor GL, Chan C et al (2008) InnateDB: facilitating systems-level analyses of the mammalian innate immune response. Mol Syst Biol 4:218
27. Driscoll T, Dyer MD, Murali TM et al (2009) PIG – the pathogen interaction gateway. Nucleic Acids Res 37:D647–D650
28. McSparron H, Blythe MJ, Zygouri C et al (2003) JenPep: a novel computational information resource for immunobiology and vaccinology. J Chem Inf Comput Sci 43: 1276–1287
29. Chatr-aryamontri A, Ceol A, Peluso D et al (2009) VirusMINT: a viral protein interaction database. Nucleic Acids Res 37: D669–D673
30. Okuda S, Yamada T, Hamajima M et al (2008) KEGG Atlas mapping for global analysis of metabolic pathways. Nucleic Acids Res 36:W423–W426
31. Matthews L, Gopinath G, Gillespie M et al (2009) Reactome knowledgebase of human biological pathways and processes. Nucleic Acids Res 37:D619–D622
32. Xenarios I, Salwinski L, Duan XJ et al (2002) DIP, the Database of Interacting Proteins: a research tool for studying cellular networks of protein interactions. Nucleic Acids Res 30:303–305
33. Cline MS, Smoot M, Cerami E et al (2007) Integration of biological networks and gene expression data using Cytoscape. Nat Protoc 2:2366–2382
34. Horton P, Park KJ, Obayashi T et al (2007) WoLF PSORT: protein localization predictor. Nucleic Acids Res 35:W585–W587
35. Hamosh A, Scott AF, Amberger JS et al (2005) Online Mendelian Inheritance in Man (OMIM), a knowledgebase of human genes and genetic disorders. Nucleic Acids Res 33:D514–D517
36. Griffiths-Jones S, Saini HK, van Dongen S et al (2008) miRBase: tools for microRNA genomics. Nucleic Acids Res 36: D154–D158
37. Martino DJ, Prescott SL (2009) Silent mysteries: epigenetic paradigms could hold the key to conquering the epidemic of allergy and immune disease. Allergy 65:7–15
38. Tomasi TB, Magner WJ, Khan AN (2006) Epigenetic regulation of immune escape genes in cancer. Cancer Immunol Immunother 55:1159–1184
39. Setiadi AF, David MD, Seipp RP et al (2007) Epigenetic control of the immune escape mechanisms in malignant carcinomas. Mol Cell Biol 27:7886–7894
40. Amoreira C, Hindermann W, Grunau C (2003) An improved version of the DNA Methylation database (MethDB). Nucleic Acids Res 31:75–77
41. Nam MJ, Madoz-Gurpide J, Wang H et al (2003) Molecular profiling of the immune response in colon cancer using protein microarrays: occurrence of autoantibodies to ubiquitin C-terminal hydrolase L3. Proteomics 3:2108–2115
42. Wolter TR, Wong R, Sarkar SA et al (2009) DNA microarray analysis for the identification of innate immune pathways implicated in virus-induced autoimmune diabetes. Clin Immunol 132:103–115
43. Soreq L, Israel Z, Bergman H et al (2008) Advanced microarray analysis highlights modified neuro-immune signaling in nucleated blood cells from Parkinson's disease patients. J Neuroimmunol 201–202: 227–236
44. Korb M, Rust AG, Thorsson V et al (2008) The Innate Immune Database (IIDB). BMC Immunol 9:7
45. Hijikata A, Kitamura H, Kimura Y et al (2007) Construction of an open-access database that integrates cross-reference information from the transcriptome and proteome of immune cells. Bioinformatics 23:2934–2941
46. Barrett T, Edgar R (2006) Gene expression omnibus: microarray data storage, submission, retrieval, and analysis. Methods Enzymol 411:352–369

Chapter 11

Systems Biology Approaches to the Study of Cardiovascular Drugs

Yuri Nikolsky and Robert Kleemann

Abstract

Atherogenic lipids and chronic inflammation drive the development of cardiovascular disorders such as atherosclerosis. Many cardiovascular drugs target the liver which is involved in the formation of lipid and inflammatory risk factors. With robust systems biology tools and comprehensive bioinformatical packages becoming available and affordable, the effect of novel treatment strategies can be analyzed more comprehensively and with higher sensitivity. For example, beneficial as well as adverse effects of drugs can already be detected on the gene and metabolite level, and prior to their macroscopic manifestation. This chapter describes a systems approach for a prototype CV drug with established beneficial and adverse effects. All relevant steps, for example, experimental design, tissue collection and high quality RNA preparation, bioinformatical analysis of functional processes, and pathways (targeted and untargeted) are addressed.

Key words: Cardiovascular disease, Biological networks, Drugs, Inflammation, Lipid metabolism, Pathway analysis, Signaling pathways, Side effects, Systems biology

1. Introduction

The liver is a source of many proatherogenic factors, among which lipids, lipoproteins, and inflammatory molecules (1–3). It is therefore the target organ of many established cardiovascular drugs as well as drugs in development. The liver has not only a role in the synthesis of proatherogenic molecules, it also regulates their plasma levels (2, 4) and controls cholesterol homeostasis (5). Detoxification is another important function of the liver. This involves both metabolism and the breakdown of drugs, and their structural modification for excretion. Because of these crucial functions for overall body homeostasis and for drug metabolism, hepatic gene expression and metabolite profiles can provide

Qing Yan (ed.), *Systems Biology in Drug Discovery and Development: Methods and Protocols*, Methods in Molecular Biology, vol. 662, DOI 10.1007/978-1-60761-800-3_11, © Springer Science+Business Media, LLC 2010

important insights into the effects of drugs and also allow a first estimation of their safety.

Safety aspects have become increasingly important, in particular for the development of novel combination treatment strategies that intend to combine a new pharmaceutical with an established cardiovascular drug. A considerable number of drugs have recently been withdrawn from the market or have failed in clinical trials because of delirious side effects including hepatotoxicity and cardiovascular toxicity (6, 7). There is ample evidence that individual drugs or drug combination can induce serious side effects not predictable from their presumed mode of action (8–10). Selecting the optimal drug for individual patients requires more detailed analyses of the effects of individual drugs on metabolism, transporters and signaling pathways. Liver and kidney are organs of choice for performing a comprehensive systems biology, or functional, analysis of biological effects and potential side effects.

A systems biology approach consists of applying a structured "knowledge base" of protein interactions, biological pathways, disease biomarkers, etc., to the interpretation of OMICs experimental data, gene lists, compound assay data, or compound structures. The tools of systems analysis can be divided onto ontology enrichment, networks, and interactome (11). The added value of systems biology approaches for cardiovascular research is manifold and includes (a) saving time by the early detection of the (adverse) drug effects on the level of genes and prior to their macroscopic manifestation, (b) the studying of complex processes across pathways which cannot be assessed by classical means, (c) high sensitivity because small changes in gene expression and metabolites add up when these are analyzed as groups across pathways and networks, and (d) the detection of putative side effects at an early stage in development.

In this chapter, an activator of the liver X receptor (LXR), T0901317 (12), is used as a prototype novel antiatherosclerotic drug to illustrate how systems biology tools can detect beneficial effects as well as adverse effects of a test compound at an early stage. LXRs are members of the nuclear receptor superfamily of transcription factors, and function as intracellular sensors of cholesterol excess. The ligand activation of LXR can affect lipid homeostasis as well as inflammatory gene expression (12, 13), two major determinants of atherosclerosis development.

2. Materials

2.1. Animal Model and Diets

1. Three groups of $n = 10$ female ApoE*3Leiden transgenic mice, an established humanized model of metabolic and CV disease (see Note 1).

2. Macrolon cages with saw dust in clean conventional animal rooms with relative humidity 50–60%, temperature ~21°C, light cycle 6 am to 6 pm.
3. An atherogenic Western-type diet (diet T; Hope Farms, Woerden, The Netherlands) with the following as major ingredients: (all w/w) cacao butter (15%), corn oil (1%), sucrose (40.5%), casein (20%), corn starch (10%), and cellulose (6%) as described (2).
4. Crystalline cholesterol and LXR activator T0901317 (both Sigma-Aldrich Chemie BV, Zwijndrecht, The Netherlands).

2.2. High Quality RNA Isolation and Quality Control

1. Liquid nitrogen.
2. 10 mm diameter glass beads and a Polytron homogenizer (both Merlin Diagnostics, Breda, The Netherlands).
3. Polytron 2 mL tubes (Sarstedt BV, Etten-Leur, The Netherlands).
4. Autoclaved filtertips and Eppendorf vials (Eppendorf, Cambridge, UK).
5. RNAzol (Campro Scientific, Veenendaal, The Netherlands).
6. Chloroform, isopropanol, and ethanol (Sigma-Aldrich Chemie BV, Zijndrecht, The Netherlands).
7. NucleoSpin RNA II column together with membrane desalting buffer (MDB), DNAase, buffer RA1, wash buffer RA2, elution buffer RA3, and RNAse-free water (all NucleoSpin RNA II kit, Macherey-Nagel, Düren, Germany).
8. Agilent Lab-on-a-chip bioanalyzer type 2100 in combination with RNA 6000 Nano LabChip kits (both Agilent Technologies, Amstelveen, The Netherlands).
9. One-Cycle Target Labeling and Control Reagent kit (Affymetrix #900493) and Affymetrix GeneChip® mouse full genome 430 2.0 arrays (45,037 probe sets) provided by the Leiden Genome Technology Center (LGTC; Leiden, The Netherlands).

2.3. Systems Biology Software

MetaDiscovery (GeneGo Inc., St. Joseph, MI, USA), a comprehensive OMICs data analysis tool consisting of several analytical modules including

1. MetaCore – a module for gene list ontology enrichment, network, and interactome analyses, multiexperiment comparisons, workflows for toxicity and biomarkers assessment (see Note 2).
2. MetaDrug – a "systems pharmacology" module designed for the analysis of medicinal chemistry data, for example, the molecular structures of drugs and their metabolites after

breakdown. MetaDrug predicts biological effects of novel drug-like compounds, including indications, side effects, and human toxicity (see Note 3).

2.4. Pathway and Network Analysis

The annotated content of MetaDiscovery consists of two domains: (1) binary protein interactions and gene–disease associations (over 300,000 interactions and associations) and (2) higher level, multiprotein functional units such as pathways, pathways maps, networks for normal and pathological cellular processes, sets of toxicity and disease biomarkers. Both domains are interlinked into an Oracle database with 107 tables, with entities linked by a controlled vocabulary into a semantically consistent ontology and data scheme. MetaDicovery features ten functional ontologies used for gene list enrichment analysis, by network algorithms and prioritization of experimental data as specified below.

1. *Signaling pathways* are linear multistep chains of consecutive interactions, typically consisting of a ligand–receptor interaction, an intracellular signal transduction cascade between receptor (R) and transcription factor (TF) and, finally, TF–target gene interaction. Signaling pathways are mainly used by network generation algorithms and only visualized on networks.
2. *Metabolic pathways* are multistep chains of metabolic reactions, linked into functionally linear chains and cycles. Metabolic pathways are also used for network generation and visualized on the networks and pathway maps. Both metabolites and genes/proteins from experimental data can be superimposed on metabolic pathways and networks.
3. *GeneGo canonical pathways maps* are the main level of pathway visualization in MetaDiscovery. Maps represent interactive images drawn in Java-based MapEditor and typically contain 3–6 pathways. There are over 700 maps in MetaDiscovery, comprehensively covering human signaling and metabolism, certain diseases and some drug target mechanisms. Pathway maps are primarily used as an ontology for enrichment analysis.
4. *GeneGo process network models.* This ontology represents a reconstruction of main signaling and metabolic processes in the cell, such as a “cell cycle checkpoints” or “innate immune response”. The manually built process networks typically have over 100 nodes (proteins) belonging to a certain normal cellular processes.
5. *GO processes* are a graphical user interface (GUI)-supported representation of the Gene Ontology (GO) collection of cellular processes (14) which comes with GO tree structure and access to proteins and interactions within a process.

This ontology is updated with GO standard updates. GO processes are mostly used in enrichment analysis and for the prioritization of genes on the built networks.

6. *GO molecular functions.* A GUI-supported ontology of standard protein functions from GO (14) and mostly used in enrichment analysis.
7. *Disease and toxicity biomarkers.* These are a collection of genes genetically linked to over 500 diseases and conditions, supported by the hierarchical disease tree and GUI for gene retrieval. Disease biomarkers are mostly used in enrichment analysis.
8. *GeneGo disease network models.* GeneGo reconstruction of disease mechanisms in a form of manually built networks. These are mechanistic networks linking the disease-associated genes via physical and functional protein interactions.
9. *GeneGo toxicity networks.* The GeneGo reconstruction of toxicity mechanisms in a form of manually built networks. These are mechanistic networks linking genes associated with a particular toxicity endpoint via physical and functional protein interactions.
10. *GeneGo drug metabolism enzymes and regulation.* A set of pathway maps for Phase I, II, and III drug metabolism and their regulation by nuclear hormone receptors.

3. Methods

3.1. Diet Treatment and Tissue Collection

1. Design an animal experiment following the steps indicated in the decision tree (Fig.1). Define the study design (here: treatment with LXR activator T0901317 from $t=0$, i.e., a progression study set-up). Choose an appropriate animal model and gender (here: female ApoE3Leiden transgenic mice) and define the experimental diet (here: atherogenic Western type diet). The dietary fat content determines to a large extent the bioavailability of a test compounds, in particular of lipophilic molecules. The specific research question determines whether multiple sacrifice time points are necessary (dynamical set-up for mechanistic insight) or not (here: descriptive insight by analyzing a steady state condition). Calculate the number of test animals required (here $n=30$ in total).
2. ApoE3Leiden mice are matched into three groups of $n=10$ each on basis of body weight and age (see Note 4). Atherogenic diet T is fed to a control group (group 1; Con). A second group of mice receives the same diet but supplemented with 1% w/w cholesterol (group 2; high cholesterol group; HC)

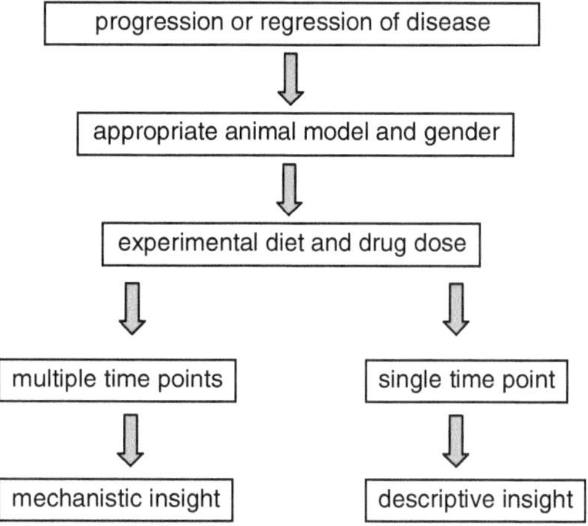

Fig. 1. Decision tree for designing cardiovascular drug intervention studies with antiatherosclerotic drugs. First, a study set-up is defined. In a progression study, the drug treatment starts from the beginning ($t=0$), mostly using healthy animals, with the aim of reducing the *development* of atherosclerotic lesions. In a regression study, drug treatment is started later, that is, when the disease has become manifest and when lesions are established. This set-up uses diseased animals and mimics the situation in humans and clinical practice. The study design often determines the animal model of choice, and the gender. Because of the significant impact of gender on atherosclerosis outcomes (16, 19), the interpretation of mixed gender systems biology studies is typically much more complicated. The study design determines which experimental diet has to be used. The fat content of the diet has a great effect on drug bioavailability, that is, the drug dose to be administered. The research question determines whether multiple or single time point analyses are required. Since gene expression changes typically precede a macroscopic effect, multiple time-point studies provide more mechanistic insight and allow making correlations between a specific change in gene expression and effects at a later stage. Single time-point studies are more descriptive and often provide no or only limited mechanistic insight.

and a third group receives diet T supplemented with 1% w/w cholesterol and 0.01% w/w T0901317 LXR agonist (LXR treatment group; LXR). The treatment period is 10 weeks with regular (monthly) plasma collection (see Note 5).

3. Animals are sacrificed without fasting by cervical dislocation or CO/CO_2. Livers are isolated, washed with phosphate-buffered saline (PBS; room temperature), dried with a laboratory tissue and weighed. The individual liver lobes are then snap-frozen immediately in liquid nitrogen (see Note 6). The tissue is then stored at –80°C in an aluminum tube.

4. For RNA extraction and subsequent microarray analysis, $n=5$ livers per group are selected randomly. Depending on the research question, a specific selection may be more appropriate (e.g., if a subgroup analysis of animals with specific characteristics is wished). It is notable that the use of the biological variation within a group often allows the performance of

correlation analysis and can provide more mechanistic insight than an analysis of a homogenous subgroup.

3.2. Preparation of High Quality RNA

1. Fill a Polytron tube with glass beads (volume of about 100 μl) so that the bottom of the tube is covered with beads and add 1 mL RNAzol.
2. Keep the tubes in which the livers are stored on dry-ice and prepare a piece of liver (3×3×3 mm) without thawing the tissue (e.g., pounding with pestle).
3. Add a piece of tissue to the RNAzol and glass beads and homogenize for 40 s in the Polytron beater.
4. Centrifuge the homogenate for 5 min at 10,000×*g* in a regular laboratory centrifuge at 4°C.
5. Following centrifugation, discard the fatty top layer (depending on the treatment, the lipid content of livers can vary considerably) and transfer the remaining suspension (RNA remains exclusively in the aqueous phase) into a clean vial.
6. Add 0.2 mL chloroform to each vial. Shake vials vigorously (by hand) for 30 s, place them on ice for 5 min, and centrifuge the homogenate for 15 min at 10,000×*g* and 4°C.
7. Transfer the upper aqueous top phase (RNA remains exclusively in the aqueous phase whereas DNA and proteins are in the interphase and organic phase) into a clean vial and add 0.5 mL isopropanol (see Note 7). Cap the tube and shake vigorously (e.g., vortex for 10 s). The samples then have to remain at room temperature for 10 min. Centrifuge the samples for 15 min at 10,000×*g* and 4°C. Discard the supernatant. Because of the presence of salt crystals, a pellet is often observed, yet a pellet consisting of pure RNA is not visible.
8. Wash the pellet with 75% ethanol (vortex or shake vigorously) and centrifuge the solution at 7,000×*g* for 5 min and at 4°C to precipitate clean (salt-free) RNA. Dry the pellet with a Pasteur pipette connected to a tap (see Note 8).
9. Solubilize the RNA pellet in 100 μl RA1 buffer plus an equal volume 70% ethanol and transfer onto NucleoSpin RNA II column. Add 350 μL MDB and centrifuge the column at 11,000×*g* for 1 min.
10. Mix 10 μl DNAse with 90 μl reaction buffer and incubate the column with 95 μl of this solution at 24°C. Wash with 200 μl RA2 buffer and centrifuge at 11,000×*g* for 30 s.
11. Incubate the column with 600 μl RA3 buffer and centrifuge at 11,000×*g* for 30 s. Incubate the column with 250 μl RA3 buffer and centrifuge at 11,000×*g* for 2 min.

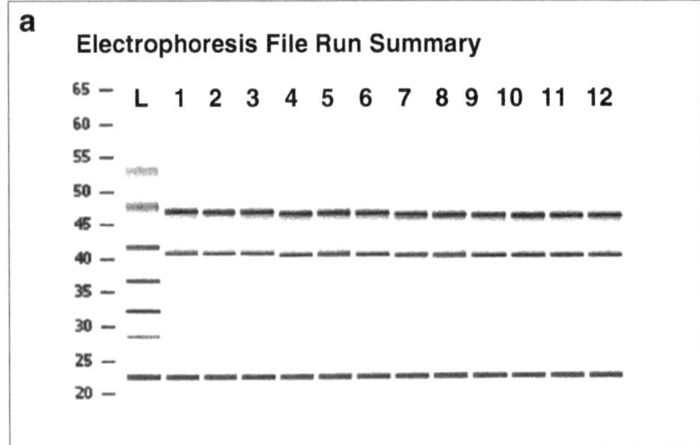

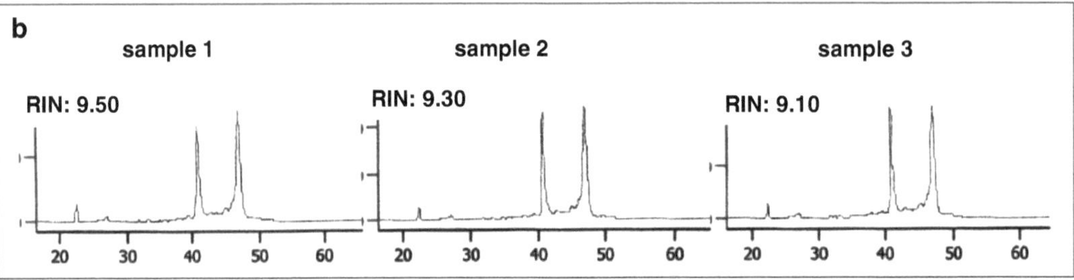

Fig. 2. Quality control analysis of RNA for microarray analysis. (**a**) The electrophoresis file run summary shows that all RNA samples (samples 1–12) have a comparable high quality. There are no breakdown products or impurities (e.g., DNA). L represents the RNA ladder. (**b**) The corresponding RIN value (15) is calculated automatically and provided in the corresponding histograms as indicated for sample 1, 2, and 3.

12. Incubate the column with 60 μl RNAse-free water, centrifuge at 11,000×*g* for 1 min, and collect the eluting RNA in a clean tube.
13. Load RNA samples onto RNA 6000 Nano LabChip and perform a quality control analysis prior to RNA labeling and hybridization to microarray (see Note 9) by determining RNA integrity number (RIN) values (15) with a Agilent Lab-on-a-chip bioanalyzer type 2100 (see Note 10) as shown in Fig. 2. Store the high quality RNA at –80°C.

3.3. Data Upload in MetaDiscovery

MetaDiscovery is designed for the analysis of a large variety of gene/protein/compound lists, small molecule structures and "OMICs data". The data types include the microarray and serial genome-wide analysis of gene expression (SAGE), single nucleotide polymorphisms (SNP) arrays genotyping data, DNA sequencing data (methylation, gene copy number, somatic mutations and SNPs), proteomics and metabolomics data. On the chemistry side, MetaDiscovery handles Simplified Molecular Input Line Entry Specification (SMILES) strings, brutto formulas, molecular weights and structures (see Note 11).

1. Upload gene expression data with the data parsers. Most of experimental data, such as gene expression, are uploaded by a universal parser which recognizes most common systems of gene/protein/compound identifiers (IDs). The majority of commercial microarrays including Illumina, Affymetrix, ABI, and GE Health care for human, mouse, rat, dog, bovine, and chimpanzee are recognized directly (see Note 12).
2. Upload metabolite data (e.g., lipid and clinical chemistry data) obtained at the time point of sacrifice and corresponding with the gene expression data together with the chemical structure of the test compound (Fig. 3a for the LXR activator T0901317) and see Note 13.

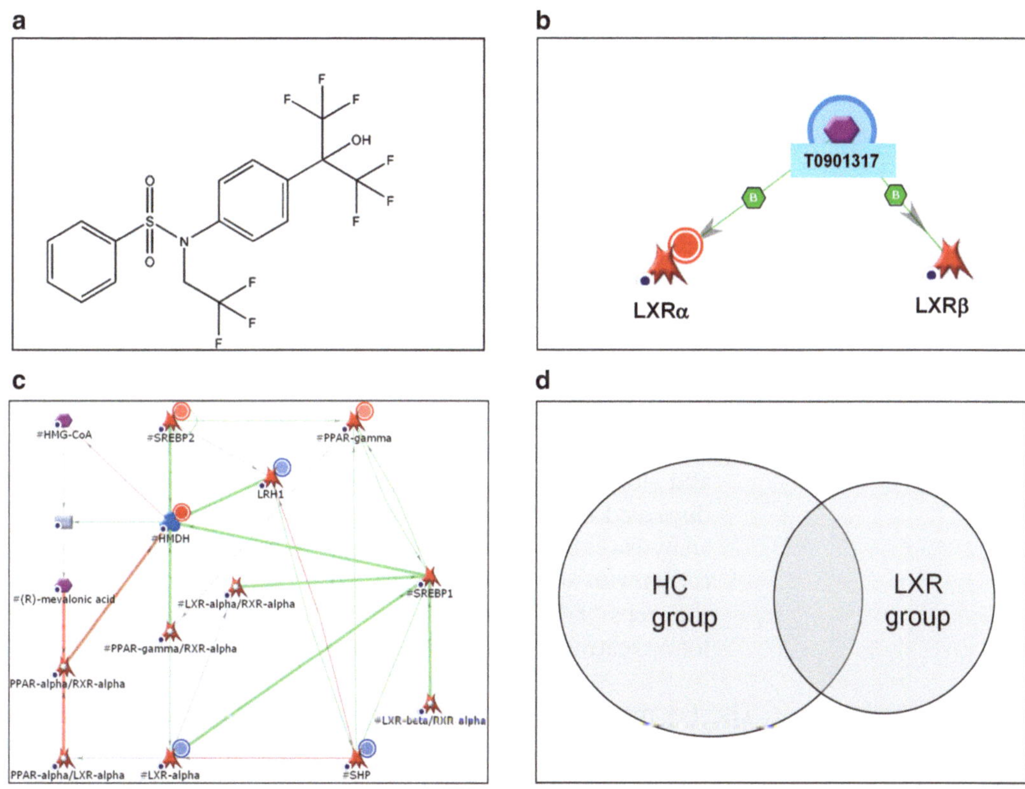

Fig. 3. (**a**) Chemical structure of the test compound, LXR activator T0901317. (**b**) Mapping numerical experimental data on pathway maps and networks. Interaction of T0901317 with its molecular targets, the nuclear hormone receptors LXRα and LXRβ. The symbol B on the arrows indicates physical binding as the interaction mechanism. Solid *red circle* indicates than LXRα is overexpressed in response to T0901317 treatment. (**c**) Visualization of gene expression data, protein abundance data, and metabolite concentrations on a network as exemplified for the modulation of a key enzyme of cholesterol biosynthesis, HMG-CoA reductase, via sterol regulatory element-binding proteins (SREBPs). *Solid red circle* indicates overexpression and high protein abundance, solid blue circle indicate under-expression and low abundance. (**d**) Venn diagram representing the significantly affected genes by atherogenic diet treatment in the HC group (*left circle*) and by atherogenic diet plus LXR activator T0901317 (*right circle*) obtained by comparing the respective gene expression profiles to the gene expression profile of the untreated Con group.

3.4. Automated Data Analysis in the Compare Experiments Workflow

Uploaded gene expression, protein, metabolite datasets, and compound lists all are analyzed in a similar way, namely by matching gene/protein/metabolite/compound IDs from the datasets with the internal MetaBase IDs. These IDs are then used as seed nodes for network and interactome analysis, and for enrichment analysis. Experimental numerical data such as level of gene expression on a microarray, protein abundance or a metabolite concentration in body liquids are visualized as histograms or as a solid circle of gradient intensity on pathway maps and networks (Fig. 3b, c).

It is useful to manually prepare a Venn-diagram (Fig. 3d) to get a first impression of the number of genes affected by the treatments. The number of genes significantly affected by atherogenic diet treatment is obtained from a comparison of the HC group and the Con group (left circle) and that of the genes affected in the LXR group is obtained by analyzing the LXR group versus the Con group (right circle). The intersection is formed by genes that are affected by HC treatment and modulated in the LXR group. The intersection thus represents genes that can putatively contribute to disease development and that are significantly modulated by LXR activator T0901317.

To perform a Compare Experiments Workflow

1. Activate the datasets in the Data Manager, and choose an automated "compare experiments" workflow in the Tools menu. After choosing a desired threshold for experimental values, the intersection between the datasets is calculated based on matching internal IDs.
2. The common subset of IDs (intersection), unique subsets, and similar IDs (i.e., present in all but one experiment) are displayed as a histogram (Fig. 3c), and a standard three-step analysis (enrichment analysis (EA) – interactome – networks) is then run automatically (see Note 14). Output examples for the steps "EA" and "networks" are provided in the following for a treatment with LXR agonist T0901317.

3.5. Dataset Analysis by Gene List Enrichment Analysis (EA)

The EA module calculates the probability of a random intersection between the uploaded dataset and an ontology's subfolder (say "cell adhesion") based on a hypergeometric distribution (Fig. 4). The *p*-value essentially represents the probability of a particular mapping arising by chance, given the numbers of genes in the set of all genes on maps/networks/processes, genes on a particular map/network/process and genes in your experiment. The negative natural logarithm of the *p*-value is displayed so that a larger bar represents a higher significance. The False Discovery Rate (FDR) correction procedure is standard. FDR threshold can be custom changed or switched off. There are two important issues in the EA calculation (see Note 15).

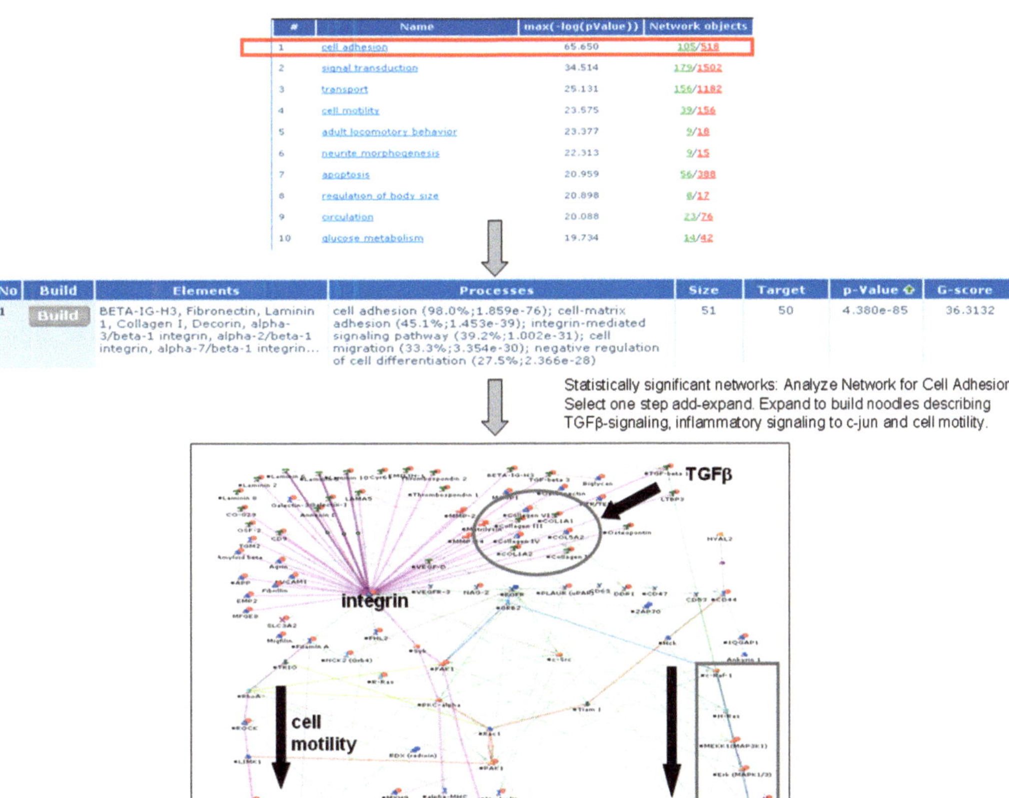

Fig. 4. A workflow for generating statistically significant networks. Based on a gene enrichment analysis (EA), a list of the top ten processes which are mainly affected by LXR activator T0901317 is produced. The highest ranked process "cell adhesion" is selected and its gene content further analyzed by network analysis. The *grey arrows* trace the steps that were followed in generating the signature network for "cell adhesion" (threshold 1.5) using the Analyze Network (AN) algorithm applying one step add-expand to expand the network. The upper black top indicates a possible activation of the TGFβ pathway and the positive association with collagen I, III, IV, VI gene expression (*grey rectangle*) suggesting that the T0901317 treatment may promote liver fibroses. The lower *black arrow* (*right*) marks the activated inflammatory route to c-Jun (via integrin and MAP3K1 and ERK1/3) indicating that T0901317 has a proinflammatory effect. The *lower arrow* (*left*) indicates possible pathway to cell motility activation.

3.6. Network Generation

Network algorithms use the uploaded network objects (converted from gene and protein IDs) as seed nodes, link them together by pulling interactions from the database and display the built networks on the screen.

1. Generate significant networks starting with the top ten affected biological processes (as shown in Fig. 4 for the most significantly T0901317-sensitive process, "cell adhesion").
2. Apply the Analyze Network (AN) algorithm. This algorithm starts with building a super network. This network, which is never visualized, connects all objects from the input list with all other objects. In a next step, this large network is "divided" into smaller fragments of chosen size, from 2 to 100 nodes.

This is done in a cyclical manner, that is, fragments are created sequentially one by one. Edges used in a fragment are never reused in subsequent fragments. Nodes may be reused, but with different edges leading to them in different fragments. The end result of the AN algorithm is a list of overlapping multiple networks (usually ~30), which can be prioritized based on five parameters: the number of nodes from the input list among all nodes on the network, the number of canonical pathways on the network, and three statistical parameters: *p*-value, z-score, and g-score (for other network algorithms see Note 16).

3. Apply the auto-expand (AE) algorithm: AE algorithm creates subnetworks around *every* object from the uploaded list. The expansion halts when the subnetworks intersect. Objects that do not contribute to connecting subnetworks are automatically truncated.
4. Identify the pathways activated or deactivated and downstream targets. For LXR activator T0901317, a positive association between TGFβ and profibrotic genes is observed as well as an induction of proinflammatory genes including c-jun proinflammatory transcription factor (Fig. 4).

3.7. Analysis of Specific Genes

In addition to the above untargeted approaches of biological process and pathways analysis using networks and gene enrichment tools, a targeted approach (i.e., the analysis of specific genes) often has an added value, in particular for experts in the field. For the development of cardiovascular disorders, the handling of cholesterol and fatty acid by the liver is crucial, and so is the inflammatory status of the liver. Table 1 illustrates a targeted approach specifically analyzing transporter genes. Similar analyses can be performed for genes involved in cholesterol metabolism, fatty acid metabolism, peroxisome proliferator activated receptors target genes (see Note 17), or inflammation (see Table 1).

3.7.1. Transporters

Significant upregulation or downregulation by LXR activator (T0901317) treatment on the gene expression level is indicated in the right column (up/down). Results are obtained by comparing the gene expression data of group 2 (HC group) with group 3 (LXR group). See Note 18.

3.7.2. Inflammation

Treatment with a 1% cholesterol-containing atherogenic diet (group 2) induced the expression of 170 genes belonging to the category “immune and inflammatory response to stress” when compared to group 1 (control group). 35% of these genes were suppressed in the LXR group indicating that the LXR activator T0901317 has an inflammation quenching effect in the liver (Fig. 5a).

Table 1
Targeted gene expression analysis for a series of transporters relevant in cardiovascular disease

Transporter	Abbreviation	Drug effect
Niemann Pick C1-like 1 (NPC1-like 1)	Npc1l1	–
Niemann Pick C2-like 1 (NPC1-like 1)	Npc2l1	Up
Na^+-taurocholate cotransporting polypeptide (reabsorption of bile acids at basolateral membrane of hepatocytes)	Slc10 a1	Down
Na^+-taurocholate cotransporting polypeptide (reabsorption of bile acids from the intestinal lumen, the bile duct and kidney)	Slc10 a2	–
Organic anion-transporting polypeptides OATP-C and -B (statin transporter)	Oatp-b	Down
Creatinin transporter (high urinary *creatine* to *creatinine* ratio is linked to creatinine transporter deficiency)	Slc6a8	Up
Glycerol-3-phosphate transporter multipass membrane protein (transmembrane sugar transport)	Slc 37a2	Up
Glycerol-3-phosphate transporter multipass membrane protein (transmembrane sugar transport)	Slc37a3	Up
UDP-glucaronic acid transporter (movement of UDP-glucuronic acid into, out of, within or between cells)	Slc35d1	Up
Fatty acid transport protein-1 (FFA blood→liver)	Slc27a1	–
Carnitine-acylcarnitine translocases (fatty acid transport protein a20 transporting both carnitine and carnitine-fatty acid complexes into and out of the mitochondria)	Slc25a20	–
H^+/peptide transporter (Proton-coupled intake of oligopeptides of two to four amino acids)	Slc15a2	Up
Proton-coupled Me^{2+} transporter (iron metabolism; mutations associated with susceptibility for infectious diseases such as Crohn's disease and rheumatic arthritis)	Slc11a1	Up
DTM; divalent metal transporter (iron transport)	Slc11a2	Up
MRP4 (prostaglandin transport)	Abc C4	Up

Treatment with 1% cholesterol induced 350 genes that encode for extracellularly acting factors (i.e., factors that can be secreted by the liver) having proatherosclerotic effects. T0901317 treatment suppressed the induction of 30% of these genes as determined from a comparison of group 2 (HC group) with group 3 (LXR group) and shown in Fig. 5b. A confirmation of microarray results obtained from Affymetrix or comparable advanced platforms with RT-PCR is in general not very useful since RT-PCR is a very limited approach and less powerful (see Note 19).

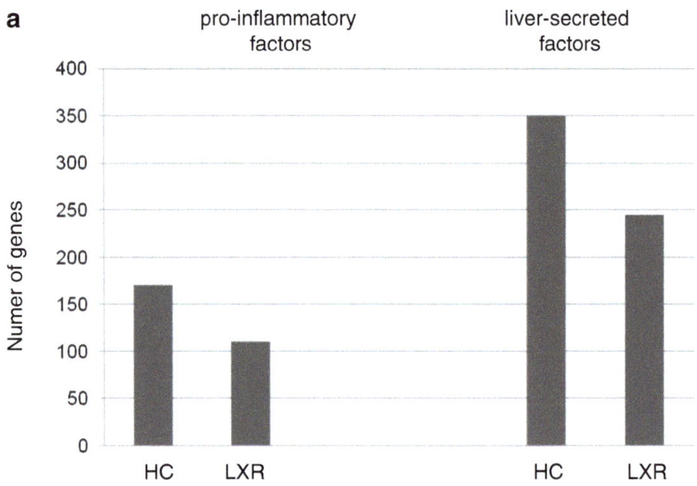

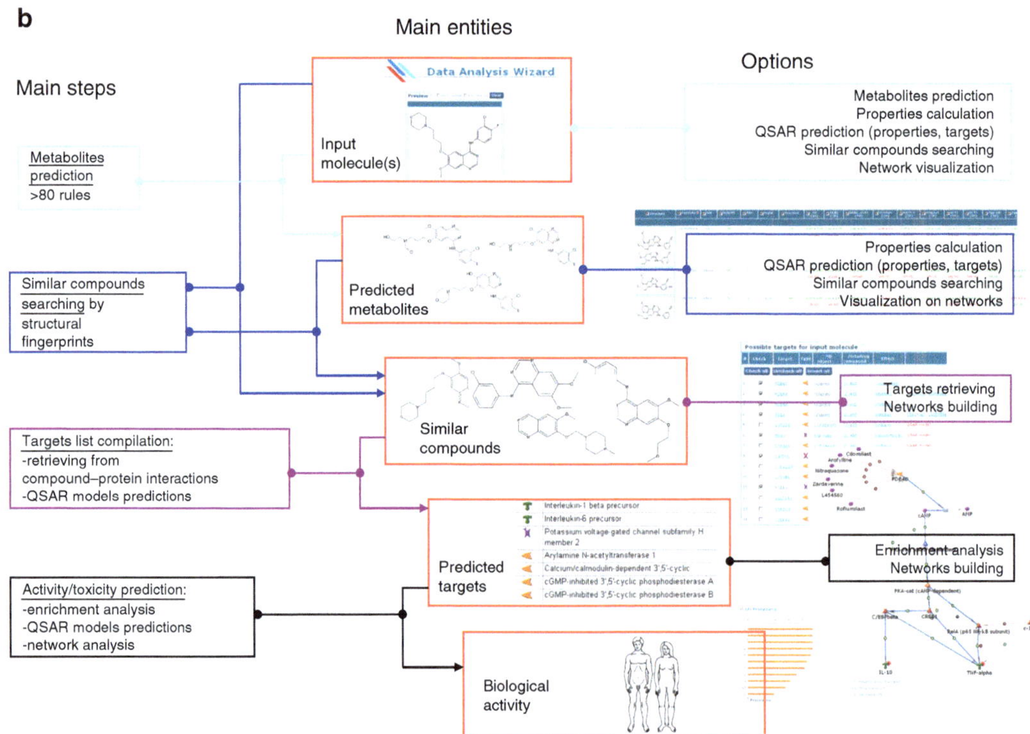

Fig. 5. (**a**) Targeted analysis of the effects of LXR activator T0901317 on genes of the category "inflammation and immune response" and "extracellularly acting factors". (**b**) General workflow of functional analysis of small molecules compounds in MetaDrug module. The structured are processed into potential human metabolites by empirical rules, followed by similarity search against 700,000 molecules in the "knowledge base". Protein targets are retrieved by protein–compound interactions and processed as a gene list by ontology enrichment, interactome, and network analysis tools.

3.8. Metabolites Prediction Tool with MetaDrug

The MetaDrug module of the platform is designed for the prediction of biological effects of small molecules (heterocyclic) compounds of arbitrary structure. Essentially, it uses chemoinformatics tools for the conversion of a compound structure to a list of proteins – possible targets and metabolizing enzymes which are then processed via ontology enrichment, networks, and interactome analysis as any other gene/protein list. The first step in the conversion of a chemical structure into a protein list is to split the molecule into a series of predicted human metabolites. It is well established that in many cases, the active (and often toxic) ingredients in many drug molecules are their metabolites which human (mostly liver) enzymes break down the original molecules. MetaDrug uses a set of 90 empirical rules based on the manual curation of xenobiotic metabolism literature and metabolism prioritization algorithms in order to deduce Phase I and Phase II metabolites. The capabilities of the MetaDrug module also include QSAR models which can be used for the prediction of toxicity, activity and physio-chemical properties of novel compounds, and that of human metabolites for heterocyclic compounds of differing structure (see Note 20).

4. Notes

1. The ApoE*3Leiden mouse is sensitive to established CV drugs (e.g., statins and fibrates) at doses that which these drugs are typically used in humans (16). Animals respond to statins and fibrates with cholesterol-lowering and triglyceride-lowering, which distinguishes this model from other atherosclerosis models such as $ApoE^{-/-}$ and $Ldlr^{-/-}$ mice.
2. MetaDiscovery contains a set of gateways within the MetaCore interface which provide access to mouse and rat interactions content, pathway maps, and networks. MetaRodent emphasizes on the differences between human and animal models at the level of protein complexes, signaling and metabolic pathways, which is – since most animal models for cardiovascular disorders are rodent models – an important advantage of the MetaDisovery software.
3. MetaDiscovery also contains a MapEditor which is a standalone Java editing application for the generation of custom pathway maps from scratch, editing of standard maps from the MetaDiscovery collection, and conversion of networks into pathway view.
4. For studying the effects of CV drugs, test animals should be between 10 to 15 weeks of age at the start of the experiment. For optimal microarray analysis, an *age-matched* control group is required (here HC group which develops the disease maximally). Addition of a second age-matched control group (here Con which does not develop disease) improves the

subsequent interpretation of treatment effects because it defines the maximal disease-specific response of each gene.

5. An important aspect that has to be considered is the treatment time which mainly depends on the specific research questions to be addressed. For most research questions, steady state conditions are required. In the first 2–3 weeks of almost any dietary treatment, many (metabolic) adaptations will take place in the liver. Therefore, longer treatment periods are preferable. (As a rule of thumb, the achievement of a new steady state of a messenger RNA requires about three times the messenger RNA half-life of a particular gene of interest). Another aspect which is often neglected is the activation of hepatic cytochrome P450 enzymes which affect the half-life of drugs. In rodents, the plasma half-life of statins varies considerably between the first days and after several weeks of intervention (17).
6. For optimal systems biology analyses, the tissues have to be collected under consistent and very well-defined conditions. Because of the zonation of livers, the same liver lobe should be used for microarray analysis as well as possible histological examinations performed in parallel. Preparation of slices of fresh liver lobes or cutting the fresh tissue in pieces results in excessive RNA breakdown and should be avoided. Also, if fasting is applied, the fasting period has to be exactly the same for all animals. Because of the diurnal variations in gene expression, the tissue collection of large studies often requires several days for sacrifice and tissue collection. In such cases, the individual animals of a group should be equally divided over the various sacrifice time points. In case anesthesia is applied, it is important to verify whether the anesthetics interfere with the key readout parameters or principle genes of interest since many anesthetics have an effect on (intrahepatic) lipids and gene expression. Instead of cervical dislocation, treatment with CO/CO_2 is an often applied sacrifice procedure; yet a stress response has then to be considered.
7. Traces of DNA are often observed as contaminant in RNA preparations. Investigators who fail to remove DNA are recommended to use colons on which a DNAse treatment can be executed, for example, using a NucleoSpin® *RNA* II kit (Macherey-Nagel, Germany). Of note, although of great importance, this step is inadequate with many commercial kits.
8. It is important to carefully remove all ethanol and water from the RNA pellet. However, the pellet should not dry out completely because it hardly resolves when the vial is, for example, left opened overnight. Heating a Pasteur pipette in a flame

allows the extension of its tip to a small capillary. When connected to a tap, the suction of running water is sufficient to remove ethanol and water leaving the pellet intact. After another 20 min at 24°C, the RNA pellet can be further processed.

9. The reproducible preparation of high quality RNA from specific parts of an organ (e.g., specific liver lobes) is a prerequisite for a meaningful systems biology analysis and is therefore described in detail. The subsequent steps carried out with high quality RNA, namely labeling and hybridization to microarrays, are often performed by specialized laboratories since they require an infrastructure typically not present at a standard laboratory. Because of the rapid technological changes of the various transcriptomics platforms (e.g., Affymetrix, Illumina), we here emphasize on the generic aspects of high quality RNA preparation for studying drug effects on the transcriptome level and proceed with the data analysis of transcriptomics data obtained from Affymetrix gene chips (.cel files).

10. A determination of the E260/E280 ratio, although often applied, is not sufficient for determining the quality of an RNA preparation. The described RNA quality control procedure is standard for transcriptomics analysis and should also be applied for other quantitative mRNA measurements (e.g., RT-PCR analysis).

11. MetaDiscovery is used in three main modes: Browser, Combinatorial search, and for Analysis and Editing. The content of functional ontologies, gene, protein, and compound annotations can be accessed from multiple pages in MetaDiscovery's different applications. The main menu includes GeneGo maps as a separate entry, as well as GeneGo process networks and disease networks. GeneGo process networks and disease networks can be opened up from enrichment analysis distribution or called from the main menu. Annotations for genes and proteins are available by either clicking on an object on maps or networks, or found by search genes/proteins. The gene/protein pages, which contains links to outside databases such as Swissprot, EntezGene, etc., has information on protein isoforms, gene variants, as well as information on SNPs and mutations, etc. The Compound page is common for all exobiotics in the database, and includes pharmacological information such as prime and secondary indication, toxicity, drug–drug interactions, drug–target interactions, etc.

12. The metabolic parser is designed for uploading endogenous small molecule compounds and recognizes AC numbers,

SMILES strings, molecular weights, and KEGG IDs. Xenobiotic compounds are uploaded with the help of the integrated Accord module (Accelrys) in the form of SDF and MOL files.

13. A special metabolic parser is designed for uploading endogenous small molecule compounds and recognizes AC numbers, SMILES strings, molecular weights, and KEGG IDs. Chemical drug structures can also be drawn using the ChemDraw plug-in. Importantly, all compounds in MetaDiscovery are included in ISIS index, a system of choice for drug screening assays. The assays can be parsed into MetaDiscovery via ISIS identifiers.

14. Standard data analysis overview. The uploaded experimental data are subjected to four levels of analysis:
    (a) *The experimental set(s) are custom filtered* according to the user's needs. Filters include gene expression in human tissues and cellular organelles, matching with orthologs in ten organisms, specific cellular processes, etc. In addition, the uploaded gene lists can be normalized against microarray content or a custom dataset.
    (b) *Enrichment analysis (EA)* in multiple functional ontologies. EA is a "classical" tool which shows relative prevalence of genes from certain cellular processes, pathways, diseases, etc., in the uploaded dataset(s).
    (c) The *interactome analysis* feature calculates relative connectivity (number of interactions) of individual proteins/genes within the set compared to the whole database. Proteins are divided by protein classes such as transcription factors, receptors, ligands (secreted proteins), kinases, phosphatases, proteases, and endogenous metabolic enzymes.
    (d) *Network analysis.* Genes/proteins in the dataset(s) can be connected to each other via protein interactions, forming signaling, and metabolic networks.

15. *Functional ontologies.* EA analysis is only as informative as the ontology behind it. Using only one ontology (for instance, GO molecular functions) provides a rather insufficient overview of large datasets. For instance, GO processes help little in the evaluation of a toxicogenomics expression dataset, for which a specialized ontology of toxic categories and pathological processes is needed.

    Standard datasets and normalization. EA calculates the relative enrichment of a dataset on a background of a larger database of IDs the set of interest is part of. For instance, a subset of genes differentially expressed in atherosclerotic vasculature has to be "normalized" to the gene ID content of the microarray it was generated on.

16. *Analyze network (Transcription Factors – TFs) and Analyze network (Receptors) algorithms*: Both algorithms start with creating two lists of objects expanded from the initial list: the list of transcription factors and the list of receptors. Next, the algorithm calculates the shortest paths from the receptors to TFs. Then, the shortest paths are prioritized in a similar way. The first algorithm, AN(TFs) connects every TF with the closest receptor by all shortest paths and delivers one specific network per TF in the list. Similarly, the second algorithm AN(Receptors) delivers a network consisting of all the shortest paths from a receptor in the list to the closest TF; one network per receptor. Since all the edges, and therefore, paths are directional, the resulted networks are not reciprocal.
17. Other examples for a targeted analysis of specific genes relevant for drug effects in the field of cardiovascular disorders are "intrahepatic cholesterol handling" and "fatty acid handling and PPAR target genes" (see Table 2).
18. In general, SLC transporters are uptake or exchange transporters while ABC transporters are efflux transporters. Multidrug resistance protein (MRP) transporters have a low specificity and transport a broad spectrum of lipophilic molecules. Lipophilic compounds are typically taken up in the intestine and reach the bloodstream via the lymph, finally reaching the liver. Hydrophilic compounds are typically taken up by enterocytes and enter the bloodstream via a network of small capillaries surrounding the intestine, and reach the portal vein and liver.
19. Messenger quantification with the current microarrays (e.g., Affymetrix, Illumina etc.) is much more sophisticated and sensitive than qRT-PCR. The Affymetrix technology applied herein analyzes 11 gene regions (segments) per gene while qRT-PCR primer/probe sets measure only one segment, namely the segment that they are amplifying. Affymetrix also checks the specificity of a hybridization signal by comparing the 11 hybridization signals to 11 mismatch sequences (i.e., same sequence but containing one mismatch base). To pass the Affymetrix criteria for specific hybridization, a signal has to be positive on the 11 correct segments and must be absent in the corresponding mismatch sequences. In previous studies (2, 18), we have compared the microarray technology with qRT-PCR and have found a higher sensitivity for the array technology. In our experience, a control of the results obtained with qRT-PCR does not have added value. However, when less sophisticated arrays are used, qRT-PCR controls can be of importance.

Table 2
Factors involved in hepatic cholesterol handling and lipid metabolism

Intrahepatic cholesterol handling	
3-Hydroxy-3-methylglutaryl-Coenzyme A reductase	Hmgcr
3-Hydroxy-3-methylglutaryl-Coenzyme A synthase 1	Hmgcs1
3-Hydroxy-3-methylglutaryl-Coenzyme A synthase 2	Hmgcs2
Farnesyl diphosphate farnesyl transferase 1; squalene synthase	Fdft1
Sterol regulatory element binding transcription factor 2	Srebf2
Acetyl-Coenzyme A acetyltransferase 1	Acat1
Low density lipoprotein receptor	Ldlr
Proprotein convertase subtilisin/kexin type 9	Pcsk9
Cytochrome P450, family 7, subfamily a, polypeptide 1	Cyp7a1
ATP-binding cassette, subfamily G (WHITE), member 5	Abcg5
ATP-binding cassette, subfamily G (WHITE), member 8	Abcg8
ATP-binding cassette, subfamily B (MDR/TAP), member 4	Abcb4
Niemann Pick C1-like 1 (NPC1-like 1)	Npc1l1
Scavenger receptor class B, member 1	Scarb1
ATP-binding cassette, subfamily A (ABC1), member 1	Abca1
Apolipoprotein A-I	Apoa1
Phospholipid transfer protein	Pltp
Fatty acid handling and PPAR target genes	
Lipoprotein lipase	Lpl
Hepatic lipase	Lipc
Apolipoprotein C-III	Apoc3
Apolipoprotein C-II	Apoc2
Apolipoprotein C-I	Apoc1
Apolipoprotein A-V	Apoa5
Sterol regulatory element binding transcription factor 1	Srebf1
Acetyl-Coenzyme A carboxylase alpha	Acaca
Acetyl-Coenzyme A carboxylase beta	Acacb
Fatty acid synthase	Fasn
Diacylglycerol O-acyltransferase 1	Dgat1
Diacylglycerol O-acyltransferase 2	Dgat2

(continued)

Table 2 (continued)

Carnitine palmitoyltransferase 1a, liver	Cpt1a
Carnitine palmitoyltransferase 1c	Cpt1c
Acyl-Coenzyme A oxidase 1, palmitoyl	Acox1
Acyl-Coenzyme A oxidase 2, branched chain	Acox2
Acyl-Coenzyme A oxidase 3, pristanoyl	Acox3
Enoyl-Coenzyme A, hydratase/3-hydroxyacyl Coenzyme A dehydrogenase	Ehhadh
Acetyl-Coenzyme A acyltransferase 1A , thiolase A	Acaa1a
Acetyl-Coenzyme A acyltransferase 1B, thiolase B	Acaa1b
Acetyl-Coenzyme A acyltransferase 2 (mitochondrial 3-oxoacyl-Coenzyme A thiolase)	Acaa2
Acyl-CoA synthetase long-chain family member 1, 3, 4, 5, 6	Acsl1, 3, 4, 5, 6
Acyl-CoA synthetase long-chain family member 3, 4, 5, 6	Acsl3, 4, 5, 6
Acyl-CoA synthetase short-chain family member 1–3	Acss1–3
Acyl-CoA synthetase medium-chain family member 1–5	Acsm1–5
Apolipoprotein B	Apob
Apolipoprotein B mRNA editing enzyme, catalytic polypeptide 1–4	Apobec1–4
Microsomal triglyceride transfer protein	Mttp
Nuclear receptor subfamily 1, group H, member 3; LXRalpha	Nr1h3
Nuclear receptor subfamily 1, group H, member 2; LXRbeta	Nr1h2
Peroxisome proliferator activated receptor alpha	Ppara
Peroxisome proliferator activator receptor delta	Ppard
Peroxisome proliferator activated receptor gamma	Pparg
Fatty acid binding protein 1, liver	Fabp1
Fatty acid binding protein 2, intestinal	Fabp2
Fatty acid binding protein 4, adipocyte	Fabp4
Fatty acid binding protein 6, ileal (gastrotropin)	Fabp6
Solute carrier family 27 (fatty acid transporter), member 1–5	Slc27a1 to a5
CD36 antigen	Cd36

20. *QSAR models.* The compound structure and the predicted metabolites are tested for bioactivity by the calculation of quantitative structure-activity relationship (QSAR) models. MetaDiscovery uses the ChemTree modeling module

developed by Golden Helix Inc. (http://www.goldenhelix.com) for model generation. There are over 100 models in MetaDiscovery for the evaluation of a compound's physiochemical properties, reactivity, metabolic hepatotoxicity (phase I and II drug metabolism), general toxicity (Herg, transporters, etc.), as well as activity on potential drug-able targets (Fig. 5). Some models are built around specific proteins, Phase II drug metabolism enzymes, transporters, membrane and nuclear receptors, kinases, etc. These proteins can be selected by a user for the follow-up functional analysis.

Acknowledgments

The authors gratefully acknowledge support from the TNO research program "Personalized Health VP9" and from the National Cancer Institute (USA) SBIR grant "Systems Biology Platform to Study of Nutrient Compounds". The authors thank Mrs Karin Toet for developing and optimizing an excellent RNA isolation protocol.

References

1. Kleemann R, Kooistra T (2005) HMG-CoA reductase inhibitors: effects on chronic subacute inflammation and onset of atherosclerosis induced by dietary cholesterol. Curr Drug Targets Cardiovasc Haematol Disord 5: 441–453
2. Kleemann R, Verschuren L, van Erk MJ et al (2007) Atherosclerosis and liver inflammation induced by increased dietary cholesterol intake: a combined transcriptomics and metabolomics analysis. Genome Biol 8:R200
3. Vergnes L, Phan J, Strauss M, Tafuri S, Reue K (2003) Cholesterol and cholate components of an atherogenic diet induce distinct stages of hepatic inflammatory gene expression. J Biol Chem 278:42774–42784
4. Kooistra T, Verschuren L, de Vries-van der Weij J et al (2006) Fenofibrate reduces atherogenesis in ApoE*3Leiden mice: evidence for multiple antiatherogenic effects besides lowering plasma cholesterol. Arterioscler Thromb Vasc Biol 26:2322–2330
5. Tannock LR, O'Brien KD, Knopp RH et al (2005) Cholesterol feeding increases C-reactive protein and serum amyloid A levels in lean insulin-sensitive subjects. Circulation 111:3058–3062
6. Nissen SE, Wolski K, Topol EJ (2005) Effect of muraglitazar on death and major adverse cardiovascular events in patients with type 2 diabetes mellitus. JAMA 294:2581–2586
7. Doggrell SA (2008) The failure of torcetrapib: is there a case for independent preclinical and clinical testing? Expert Opin Pharmacother 9:875–878
8. Armitage J (2007) The safety of statins in clinical practice. Lancet 370:1781–1790
9. Rubenstrunk A, Hanf R, Hum DW, Fruchart JC, Staels B (2007) Safety issues and prospects for future generations of PPAR modulators. Biochim Biophys Acta 1771: 1065–1081
10. Choe SS, Choi AH, Lee JW et al (2007) Chronic activation of liver X receptor induces beta-cell apoptosis through hyperactivation of lipogenesis: liver X receptor-mediated lipotoxicity in pancreatic beta-cells. Diabetes 56:1534–1543

11. Nikolsky Y, Kirillov E, Zuev R, Rakhmatulin E, Nikolskaya T (2009) Functional analysis of OMICs data and small molecule compounds in an integrated "knowledge-based" platform. From the "Protein networks and pathway analysis" book. Humana, UK
12. Verschuren L, de Vries-van der Weij J, Zadelaar S, Kleemann R, Kooistra T (2009) LXR agonist suppresses atherosclerotic lesion growth and promotes lesion regression in ApoE*3Leiden mice: time course and mechanisms. J Lipid Res 50:301–311
13. Tontonoz P, Mangelsdorf DJ (2003) Liver X receptor signaling pathways in cardiovascular disease. Mol Endocrinol 17:985–993
14. The Gene Ontology project in 2008 (2008) Nucleic Acids Res 36:D440–D444
15. Schroeder A, Mueller O, Stocker S et al (2006) The RIN: an RNA integrity number for assigning integrity values to RNA measurements. BMC Mol Biol 7:3
16. Zadelaar S, Kleemann R, Verschuren L et al (2007) Mouse models for atherosclerosis and pharmaceutical modifiers. Arterioscler Thromb Vasc Biol 27:1706–1721
17. Black AE, Sinz MW, Hayes RN, Woolf TF (1998) Metabolism and excretion studies in mouse after single and multiple oral doses of the 3-hydroxy-3-methylglutaryl-CoA reductase inhibitor atorvastatin. Drug Metab Dispos 26:755–763
18. Verschuren L, Kooistra T, Bernhagen J et al (2009) MIF-deficiency reduces chronic inflammation in white adipose tissue and impairs the development of insulin resistance, glucose intolerance and associated atherosclerotic disease. Circ Res 105: 99–107
19. Kleemann R, Zadelaar S, Kooistra T (2008) Cytokines and atherosclerosis: a comprehensive review of studies in mice. Cardiovasc Res 79:360–376

Chapter 12

Cancer Systems Biology

Dana Faratian, James L. Bown, V. Anne Smith, Simon P. Langdon, and David J. Harrison

Abstract

Cancer is a complex and heterogeneous disease, not only at a genetic and biochemical level, but also at a tissue, organism, and population level. Multiple data streams, from reductionist biochemistry in vitro to high-throughput "-omics" from clinical material, have been generated with the hope that they encode useful information about phenotype and, ultimately, tumour behaviour in response to drugs. While these data stand alone in terms of the biology they represent, there is the enticing prospect that if incorporated into systems biology models, they can help understand complex systems behaviour and provide a predictive framework as an additional tool in understanding how tumours change and respond to treatment over time. Since these biological data are heterogeneous and frequently qualitative rather than quantitative, at the present time a single systems biology approach is unlikely to be effective; instead, different computational and mathematical approaches should be tailored to different types of data, and to each other, in order to test and re-test hypotheses. In time, these models might converge and result in usable tractable models which accurately represent human cancer. Likewise, biologists and clinicians need to understand what the requirements of systems biology are so that compatible data are produced for computational modelling. In this review, we describe some theoretical approaches (data-driven and process-driven) and experimental methodologies which are being used in cancer research and the clinical context where they might be applied.

Key words: Systems biology, Cancer, S-systems, Bayesian networks, Targeted therapeutics, Oncology

1. Introduction

1.1. What Is Cancer, and Why Do We Need Cancer Systems Biology?

Cancer is the term applied to over 200 different diseases in which cells acquire a set of characteristic biological properties, namely autonomous growth, evasion of death, and the ability to invade and spread to distant sites (metastasise) (1). The underlying cause of cancer is genetic, with either inherited or acquired abnormalities of genes or the control of genes giving rise to the cancerous phenotype (2).

Qing Yan (ed.), *Systems Biology in Drug Discovery and Development: Methods and Protocols*, Methods in Molecular Biology, vol. 662, DOI 10.1007/978-1-60761-800-3_12, © Springer Science+Business Media, LLC 2010

While cancer can arise from any cell in the body, the commonest cancers in man (such as breast and lung, which together account for over 83,000 new cases in the UK per annum) arise from the epithelial cells, which line the cavities, ducts and surfaces of the body, and are called carcinomas (to distinguish them from sarcomas, which are rarer cancers arising from mesenchymal cells such as muscle cells or vascular endothelial cells). The ready isolation and growth properties of some human carcinoma epithelial cells in vitro have made them excellent experimental models to study cancer for several decades, and in some cases represent patterns of genomic aberration in human disease (3), but it is also recognised that cell line models are imperfect representations of complex phenotypes in vivo. This is partly because cancer cell lines may carry more complex genetic abnormalities than those seen in vivo as part of their acquired ability to survive in vitro. Also, simple 2D cultures lack the cells which normally support cancerous epithelium, such as stroma and blood vessels, which are intrinsic to the tumour in vivo, and these cellular models therefore fail to represent the complex interplay between epithelium and stroma which can influence both how cancers form and how they respond to treatment (1). Although this additional complexity leaves theoreticians and experimentalists with the uncomfortable prospect of not only trying to understand epithelial biology but also complex tissue biology (and indeed interactions with the organism as a whole), cancer cells do not exist in isolation and the growth, death and invasive phenotypes seem exquisitely sensitive to the spatial context. Therefore, tumours do not grow without new blood vessels, and invasion does not occur without the degradation of the extracellular matrix. Furthermore, metastases in distant sites (such as lymph nodes or bone marrow) may have different sensitivity to therapy than primary tumour. A further problem is that the cancer and its environment continually evolve and change over time, particularly in the face of therapeutic insult. These spatial and temporal challenges need to be considered in advancing our approach to understanding cancer and responses to therapy.

In clinical practice, the diagnosis of cancer is rarely a problem, with histopathology still the gold standard by the microscopic examination of stained sections of tissue. Traditionally, the objective of histopathology has been to categorise and classify disease by grade (how closely the tumour cells resemble their normal counterparts), or stage (how far the tumour has spread), because this has prognostic value (4). More recently, however, there has been a move to stratify patients for optimal therapy on the basis of molecular biomarkers (5), so that the appropriate drug can be given to the patient. Thus, those who are not likely to respond should be spared ineffective therapy. The commonest clinical setting where this applies is in breast cancer, where estrogen receptor (ER) and HER2 protein or DNA copy number are measured by immunohistochemistry or Fluorescence In Situ Hybridisation (FISH), to identify patients who should be given endocrine therapy (tamoxifen or aromatase inhibitors) and trastuzumab (Herceptin), respectively (6). Nevertheless, these markers, while very good at excluding patients who will not respond to therapy (high negative predictive value), are poor at identifying patients who will respond to therapy (low positive predictive value), either because those tumours are intrinsically resistant to that therapy or they develop resistance over time. The empirical approach to overcome this problem has been to measure more biological variables (usually using gene expression microarrays) and calculate statistical associations with disease outcome, but to date the clinical usefulness has been disappointing. We have already discussed the reasons for this situation (5, 7), but essentially the current approach to translational research by the analysis of candidate biomarkers, even within large trials, requires re-evaluation. The emerging evidence indicates that a failure to recognise the dynamic properties of signalling can result in costly mistakes. For example, a loss of feedback inhibition in tumours treated with the mammalian target of rapamycin (mTOR) inhibitors results in the induction of AKT signalling, and may be responsible for the disappointing efficacy of mTOR antagonists in the clinic (8). Negative feedback signalling mechanisms are likely to contribute to the poor efficacy of agents when studied in phase II and phase III cancer trials and to the high rate of attrition of drugs (approximately 30% due to efficacy), which is both time-consuming and expensive (9). Empirical testing of every possible agent or combination of agents in the preclinical or clinical setting becomes prohibitively expensive and impractical.

No single experimental or theoretical methodology can be used in isolation in order to de-convolute complex biological behaviour. On the contrary, the current state is that different methodologies have different strengths and weaknesses. For instance, while experimentally it is at present relatively difficult to measure variables over time (either in real-time or at sufficiently high density to be meaningful), kinetic modelling permits the simulation of biological behaviour in real time. On the other hand, it is relatively easy to quantify complex spatial data experimentally but significantly harder to do so mathematically. Likewise, given the heterogeneity of experimental data available to theoreticians, such as high density and high volume but static data gene expression microarray data from clinical trials or very low density, highly quantitative data from reductionist biology, a "one size fits all approach" to mathematical modelling is unrealistic.

The ultimate goal of these investigations is to translate knowledge from both diagnostic and biomarker data into an individualised treatment protocol, informed by a predicted outcome. This requires a predictive framework that is able to absorb experimental results, reflect the dynamical states of signaling pathways in aberrant cells and represent the impact of treatment agents on those pathways. Increasingly, it is becoming clear that multi-target treatment regimens are necessary to overcome the inherent robustness of the cell cycle, and it is likely that combinatorial approaches to therapy will similarly be required to overcome the robustness intrinsic in most druggable signaling pathways (such as PI3K or MAPK pathways) (10). Importantly, to support the identification of such regimens in a predictive framework requires a representation of the dynamical state of the essential pathway network and their interconnections (10). However, the construction of such a systems-level model attracts several challenges. First, there is wide variation in the level of detail known for different parts of this network and it is necessary to best utilise the descriptions available. Second, the scale of a dynamical system-level model means that it is not possible to fully describe all parts and it is difficult to interpret any predictions. Third, experimentation is not able to provide a complete description of the system over space and time. To address these challenges, we turn to the interoperability of experimentation and modelling.

For the first challenge, we consider *process-driven models* that exploit areas of the network about which at least some of the molecular species interactions are known and show that these models are able to make mechanistic predictions, test assumptions about unknowns, and determine target areas to measure experimentally. We also consider the role of *data-driven schemes* to derive hitherto unknown associations among measurables, even when these comprise multiple data types. To address the second challenge, we exploit network robustness analysis for reducing model complexity, the need for a standardised representation of network subcomponents to support integration, and optimisation approaches to fill gaps in knowledge by mixing data-driven and process-driven approaches. For the third challenge, we relate state-of-the-art *experimental methods* for profiling the spatio-temporal dynamics of cancer. In the following sections, we will therefore discuss a number of modelling approaches, which types of data generated by cancer biologists and clinicians they are best suited to, and then describe advances in cancer experimental biology which may aid systems biology approaches.

2. Approaches to Modelling Cancer

2.1. Linking Experiment and Theory

Process-based models afford a mechanistic representation of the underlying cell dynamics and may be parameterised directly by experimental data. These models are formulated in terms of ordinary differential equations that describe the kinetics of the concentrations of molecular species within the network over time. Here, we present two illustrative examples of the approach and the opportunity to inform subsequent experiments.

In Faratian et al. (11), we employed a process-driven approach in order to study resistance factors to receptor tyrosine kinase (RTK)-inhibitors such as trastuzumab and pertuzumab. Trastuzumab (Herceptin) is widely used as breast cancer therapy in patients who overexpress the HER2 oncogene. Unfortunately, HER2 protein expression or gene amplification status is a poor predictor of response with a very low positive predictive value (12, 13). The documented actual benefit of adjuvant trastuzumab combined with chemotherapy vs chemotherapy alone in terms of overall survival is only modest (96% vs. 95% respectively at 1 year) (12) and 91% vs. 87% respectively at 4 years (13). A large proportion of patients therefore unnecessarily receive ineffective and expensive treatments with toxic side-effects, and there is a need to identify markers which predict therapeutic response. Since the reported resistance mechanisms to trastuzumab seem to relate to aberrant MAPK/PI3K signalling (*PIK3CA* mutations and inactivation of the tumour suppressor gene *PTEN* (14, 15), we reasoned that a systems analysis of these pathways, which are the best studied process-driven models to date, would be a useful application of systems biology to a clinical problem in oncology. These canonical pathways have only been modelled in order to explain and predict physiological phenomena, such as the binding of ligand to growth factor receptors (e.g. EGF to EGFR) (16–22), but have not been so helpful for understanding therapeutic interventions, since they frequently fail to include important oncogene and tumour suppressor nodes, which are known to be fundamental to carcinogenesis and proven resistance proteins (such as HER2, PTEN, and SRC in PI3K and MAPK signalling models). A new model of MAPK/PI3K was developed to describe HER2-inhibitor antibody/receptor binding, HER2/HER3 dimerisation and inhibition, AKT/MAPK crosstalk, and the kinetic and regulatory properties of PTEN, and was based on modelling studies of the HER signalling network (19, 23–25). The inclusion of the tumour suppressor protein PTEN was deemed particularly important since it is a key negative regulator of the PI3K signalling pathway. We successfully demonstrated that resistance to RTK-inhibitors was governed by the PTEN:activated PI3K ratio (integrated resistance factor γ), and that PTEN, appropriately measured in the clinical setting, could stratify patients for HER2-inhibitor or combinatorial therapy, particularly an RTK-inhibitor and PI3K-inhibitor in cancers with low γ. This is one of few "success stories" of how a systems biology approach can generate hypotheses that can be tested experimentally in preclinical models and which can then be applied to clinical evaluation.

Further examples of applied systems biology are required so that it might gain credibility and be accepted within the clinical community.

Clyde et al. (26), used a process-based model parameterised by experimental data to generate a hypothesis for a new and important mechanism in the ATM intracellular pathway. ATM contributes to the co-ordination of the DNA damage response pathways that protect cells from potentially harmful mutations. In doing this, ATM also has the capacity to initiate the repair of treatment-driven damage, for example, radio-therapy, and so limits the impact of some treatments. A better understanding of the mechanisms of ATM regulation is therefore important both in the prevention and treatment of disease. Clyde et al. (26) investigated the behaviour of the damage response signaling pathway by treating cells with a DNA damaging agent and measuring ATM expression levels, and demonstrated that ATM gene expression is unaffected by the damaging agent. However, following the application of a specific ATM-inhibitor, a significant increase in ATM and ATR transcription was observed. Importantly, these results cannot be explained in terms of known cellular processes. Using a process-based model of the interaction network for all of the protein species considered in the experimental data together with the impact of the inhibitor and damaging agents used in the experiment, a novel feedback process was identified which was able to explain the anomalies in the data. Model predictions are consistent both with these in vitro experiments and with in vivo studies by another group (27). The model predictions point to a possible new target for ATM inhibition that overcomes the restorative potential of the proposed feedback.

Such process-based approaches are highly dependent on the assumptions made in model formulation. Crucial assumptions relate to the architecture of the network and the strength of the interdependencies among the measurables in the system. Where existing biological knowledge is limited, statistical data-driven approaches can make a valuable contribution to determining those associations. Here we describe two such data-driven approaches, Bayesian networks and S-systems modelling.

A Bayesian network (BN) is a graphical representation of statistical dependencies among a number of variables (28–30). Variables are drawn as nodes linked by arrows, forming a network. In BN parlance, the variable at the root of an arrow is known as the *parent* and the variable at the point is the *child*: the arrow indicates a direct statistical dependence of the child on the parent. While BNs were initially developed as "expert systems" (29) – a process-driven model where domain experts were consulted to form a network that could make predictions – BNs are now commonly used as a data-driven method, where data is used to learn the structure of a BN (30, 31). A feature of learned BNs

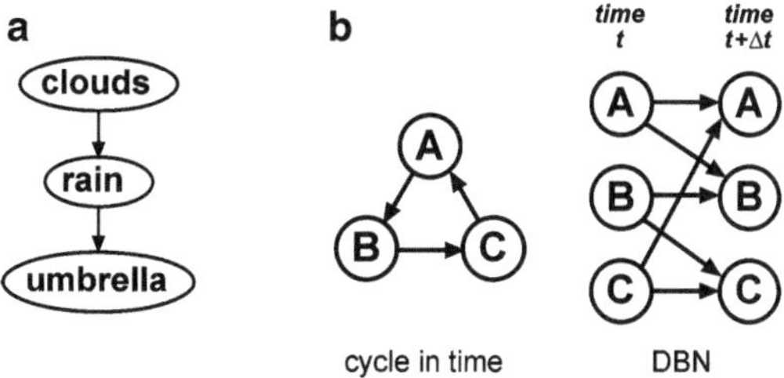

Fig. 1. Bayesian networks. (**a**) An example Bayesian network. The BN is able to distinguish that the presence of rain is a more direct predictor of people carrying umbrellas than are the presence of clouds. This prediction is in an informative, not a causal, sense, however; the *arrows* could conceivably all be reversed, for example, using the presence of umbrellas seen from a high office window to conclude that it must be raining. (**b**) A cyclic causal interaction in time represented by a dynamic Bayesian network. When time series are available, DBNs can infer causal interactions by making informative predictions about the future: a set of causal relationships (*left*) can be represented by a DBN (*right*) that represents all variables at two points in time, where Δt represents the time between data samples. Every variable is predictive of itself in the future, and thus identical variables are linked across time slices; the informative predictions of one variable (e.g., A) of another in the future (e.g., B) can be interpreted to be causal (if A, then, later, B).

is that they find a minimal set of direct dependencies necessary to explain statistical structure in the data; thus, they are well-suited to distinguish direct from indirect relationships among (potentially correlated) measured variables (Fig. 1a) (28). As such, BNs can be useful for identifying direct relationships to be used in further fine-grained modelling with process-driven methods. Caution must be taken, however, with relationships identified in a BN, as they are not necessarily causal (28, 29). The statistical dependence indicated by an arrow in the BN is most usefully conceptualised as “is useful for predicting”. For example, rain does not cause clouds, but the presence of rain is a useful predictor for the presence of clouds; similarly, the value of parent in a BN would be a useful predictor for the value of its child (Fig. 1a). Causal relationships can be discovered when time-series data are available: here, a dynamic Bayesian network (DBN) represents variables across time, and arrows from a parent to a child indicate that the past value of the parent is useful for predicting the future value of the child (Fig. 1b). As time series data becomes more common in biology, more biological data is likely to be modelled with DBNs (31); however, the continued need for analysis of non-time series data, such as clinical samples, means that static BNs will continue to play an important role.

Bayesian networks can be applied to a variety of different data types. Commonly, continuous data, such as from gene or protein expression, is discretised, and relationships found among variables with states such as low/medium/high. Such discrete data enables the discovery of multiple types of relationships, including linear, non-linear, and non-monotonic (e.g., U-shaped or combinatoric (28)). Continuous data can also be analysed directly with BNs, although such models are often limited to additive (i.e., non-combinatoric) relationships (32). The most common biological application of BNs has been to microarray gene expression data, to discover networks that represent transcriptional regulation interactions (30, 31). Metabolic flux, protein expression, and phosphoprotein expression have also been used to discover networks representing signalling pathways (33–35). BNs can also incorporate multiple data types into a single network. For example, clinical data can be included as variables in a network alongside gene expression (36). This flexibility in the data types BNs can model means it is possible to determine statistical relationships between very different variables: for example, identifying which genes are most directly dependent on an experimental condition (37). However, this same flexibility means that care must be taken in interpreting networks. A discovered relationship between two gene expression values may be interpreted as translational regulation; a relationship between a metabolite and a protein may be interpreted as enzymatic activity. But the networks represent only statistical dependence without any suggestion of mechanism; interpretations are based on the user's biological knowledge and thus are only as good as our understanding of the system. Bayesian networks are thus most useful for (1) identifying direct relationships among a number of associated variables, and (2) identifying biological species that are most relevant to broader variables such as experimental condition or clinical outcome. Identified relationships can then be followed up with more detailed mechanistic modelling methods.

An alternative data-driven approach to Bayesian networks is one based on general power-law formalisms and Biochemical Systems Theory, (38) and is particularly suited to interaction networks within the cell. The Biochemical Systems Theory is based on an underlying S-system, which is a mathematical representation of non-linear systems, based on power-laws, an approach that explicitly represents the dynamics of the network in terms of differential equations that describe the rate of change of variables such as protein concentrations or gene-expression levels (38). The equations characterise the rate of change of the variables in terms of the interaction between components of the system as products of power-laws of the concentrations (or expression levels) of these components. This allows component interactions to be described in terms of rates of production and degradation of concentrations, and it can be shown that any kind of interaction can be approximated by this form, at least locally (39).

An S-system approach allows the construction of an interaction network for a given set of measured variables by fitting the

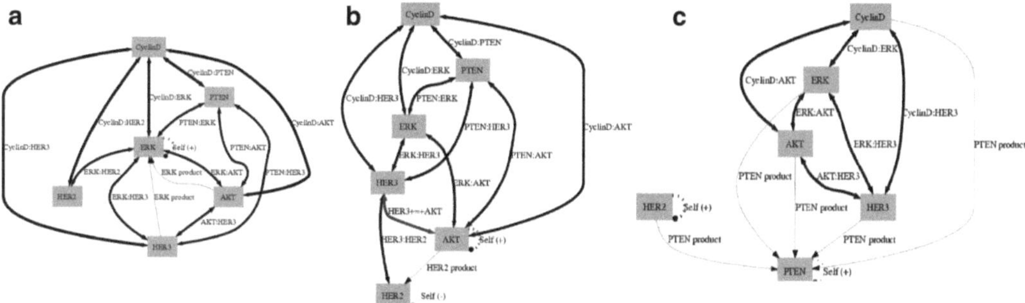

Fig. 2. Figures are visualisations of results of S-systems analyses. The protein expression data relates to a breast cancer cell line (BT474), treated with heregulin in the presence of the HER2 dimerisation inhibitor, pertuzumab. In the diagrams, *lines* represent complexes (either formation or degradation), positive feedback interactions, or production influence; *line thickness* indicates the strength of interaction. Figures (**a**)–(**c**) represent analyses on timecourse data and show the evolution of the network in response to treatment. The approach requires three at least time points to derive a network. Network (**a**) represents the first three time points (0, 1 and 2 min); (**b**) represents the next three (*overlapping*) time points, that is, 1, 2 and 5 min; (**c**) represents 2, 5 and 10 min. Among other results, the diagrams show the dissociation of HER2 from the interaction network in response to treatment with a receptor dimerisation inhibitor, which would be expected from the known biology.

equations to time series data in terms of a set of power-law exponents. S-systems are flexible in terms of type of data, so long as the kinetics are described by making a series of measurements over time. Where this fit returns an exponent value of zero, one can infer that the variable in question is independent of that component. Positive associations indicate activating influences; negative associations indicate inhibition. A network of interactions can be defined from this fitting by considering only the non-zero exponents for each variable in the system. This network can be visualised to indicate those links that most sensitively affect the rate of change of connected components (Fig. 2). In this sense, the derived network can be used to identify components as potential targets for changing the dynamical state, such as a new drug target or tumour-suppressor nodes. The prediction is valid provided the system is not changed too much from the dynamical state in which the original measurements were made or where combination therapies are being considered. The S-systems approach is beginning to be used in relevant areas of biology such as the analysis of experimentally-derived time courses of cDNA gene expression array data, albeit in yeast (40). This is promising since it is proof of principle that the technique may be used in gene expression data generated from cancer specimens, such as those taken at different time points in a neoadjuvant trial (see below).

2.2. Systems-Level Modelling

The cell is a complex interaction network that must maintain its functioning while being subjected to continuous ecological and evolutionary pressures. This inherent robustness in behaviour is a systems-level property and is pervasive across a wide range of biological systems (41). Effective treatment of cancer cells is dependent on changing existing cell functioning, and so any treatment regime must overcome this robustness. However, it is likely that interventions aimed at affecting the cell behaviour must target multiple pathways (42, 43), as the dynamics may be extremely robust to uncoordinated changes in individual pathways (41). It is clear that any model that can contribute to our understanding of how to impact on cell behaviour by overcoming a systems-level property must be constructed at a systems-level, and so represents a sufficient level of complexity and detail. However, the formulation of such a model presents challenges in terms of constructing a model that integrates knowledge across the signalling pathway network and is able to operate in spite of the requirement of a large number of parameters.

The integration of knowledge arising from the research output of multiple, disparate groups require a common currency of exchange. Often, diagrams are used to describe the current understanding of pathway interactions but these diagrams, while heavily annotated, are informal and ambiguous (44). In recognition of both the utility of annotated diagrams for explaining pathway interactions and the benefits of formalising those diagrams, effort has been invested in developing a standardised language for reporting results. One increasingly pervasive scheme for this is SBGN, for example, as used in Calzone et al. (45), the Systems Biology Graphical Notation (46). SBGN supports the syntactic representation of biological entities and their interactions, where these interactions are semantically and visually unambiguous. This provides to different research communities a homogeneous reporting platform that enables knowledge synthesis. Because the entities in the diagrams are defined in terms of their syntax and semantics, it becomes possible to interconnect, for example, a pair of networks where the same, unambiguously defined node exists in each. In principle, this is all that is required to support knowledge integration. However, SBGN offers further support to bridge the gap between experiment and theory. SBGN offers tooling support for creating and verifying the diagrams used to describe pathway architecture and interactions. SBGN also offers tooling to translate these diagrams into formal models that aid computational model construction. Given the large number of groups working on different regions of the cell and the need for systems-level modelling, the value of such an integrative framework is clear.

Systems-level modelling can lead to very large-scale models with many parameters describing the kinetics of the system. Experimental data may only support the identification of some of these parameters and so, in principle, the sensitivity of the model to uncertainty in parameter values may be explored using sensitivity analysis (47). Sensitivity analysis offers a highly structured approach to identify the subset of parameters that the model is most sensitive to changes in value. In its simplest form, this is achieved on a parameter-by-parameter basis: for each parameter, fix all other parameter values and systematically vary that selected parameter, measuring the extent to which the system-scale behaviour changes in response to value changes. This knowledge may be used to inform experiment and theory. It may direct experimentation, as it provides the set of parameters that are most important to system-level dynamics. It may inform model formulation as it offers the potential for model simplification. This may be in the form of determining regions of the network where the model is robust to large parameter changes, and these regions can then be simplified into an abstracted form or even removed from the network thus reducing the complexity and need for experimental parameterisation. In extreme cases, it may be used to completely reformulate the model in terms of only those sensitive parameters, as in Pachepsky et al. (48).

This form of sensitivity analysis ignores interactions among parameters, and so a more holistic approach to this form of analysis is required here to account for the interconnectivity among pathways. Saltelli et al. (49) describe global sensitivity analysis, where multiple parameters may be varied simultaneously and the impact of this measured again in terms of the degree to which parameter value modifications drive changes in system-level behaviour. The approach has been used in the optimisation of (synthetic) genetic circuits to inform experimental designs (50) and, importantly, the authors highlight the applicability of the approach to other biological networks.

In addition to sensitivity analysis, the fact that the model describes the whole system means that knowledge of that system-level behaviour, that is, model output, can be used to constrain the parameters, that is, model input. Techniques from artificial intelligence, such as Genetic Algorithms (51) can be used to reverse engineer missing parameter values that are consistent with known system-scale behaviour when combined with other, known input parameters. This means that not all variables in the system require explicit measurement; a subset of variables can be derived from simulation. The reverse engineering approach operates by comparing system-scale model predictions with observed system-scale behaviour, and refining through iteration the unknown input parameters to reduce the difference between predicted and observed behaviour. Clearly, the success of the scheme is dependent on defining an appropriate measure of difference, termed a fitness function. The definition of this fitness function is challenging but also the key to combining the strengths of process-based and data-driven approaches to address gaps in knowledge.

When considering the output from the process-based model, the fitness function can be defined to account for, for example, particular concentration values and/or particular rates of change in those values. However, there are many possible parameter configurations that could match the predicted behaviour to that which is observed. To reduce the parameter space further and determine more accurate values for unknown parameters, it is possible to integrate the results of data-driven models into the evaluation of specific process-based parameter configurations. This integration may be effected through the fitness function. The data-driven models described above provide a mechanism to infer the network structure from measurables without specifying causal links in the network. This provides a meta-level description of the experimental system that is not dependent on any assumptions relating to cell structure or architecture. By deriving the same data-driven meta-level description of the predictions of the process-based model, based on the equivalent output of those same (simulated) measurables, it is possible to further constrain the set of parameters that may be reverse-engineered to be consistent with the observed system-scale behaviour. Thus, data-driven models offer the possibility of substantially reducing the potentially large parameter search space in a rigorous way and without making a priori assumptions about parameter ranges and/or interaction network architecture.

3. Experimental Methods for Data Generation in Systems Biology

3.1. Spatial Resolution

We have already highlighted that cancer is more than just an epithelial disease. Therefore, in tissues, at least two spatial levels must be resolved; tissue compartments (i.e., epithelial, stromal, inflammatory component, vascular and interstitial) and cellular compartments (e.g., immediate extracellular environment, membrane, cytoplasmic, nuclear and organelles). Since the modelling approaches described above, particularly DBNs, require sufficient density of data, high-throughput experimental approaches are required. We use multiple strategies to combine compartment-specific analysis with high-throughput molecular analysis; including tissue microdissection, in situ protein quantification and reverse-phase protein arrays (RPPA; see below). Microdissection is becoming a standard strategy in gene-expression microarray (52) and genomic hybridization protocols (53) in order to enrich for epithelial-cell populations, either to overcome the inherent limitations of sensitivity of the assay (as in array CGH) or to infer compartment-specific biology (as in the case of gene expression) (52, 53). Microdissection often means a relatively crude dissection of the epithelial area of a tumour by hand under a dissecting microscope, or alternatively the use of laser capture/catapult microdissection techniques in order to obtain populations of pure cells (54). The latter approach may be more suited for systems biology because the methodology can also be extended to separate any tissue compartment (e.g., blood vessels, stroma) and also morphologically heterogeneous elements within the epithelial areas of tumours, which have differential gene and protein expression signatures and therefore potentially different responses to therapy. A second approach to compartment-specific analysis is in situ protein quantification on automated image analysis platforms for total and phosphorylated (usually active) states of proteins within signalling pathways of interest (Fig. 3a and (55)). These methods multiplex antibodies against particular compartments (usually epithelial, although any compartment may be discriminated) with one or more targets of interest, so that compartment-specific protein expression can be quantified. The advantage of this technique over microdissection methods is the ability to discriminate protein expression levels at the compartmental and subcellular levels. However, disadvantages include the limited number of targets that can be measured from a single section (which is governed by the number of filters (usually up to five) on the fluorescence microscope), availability of high-quality specific antibodies, tissue

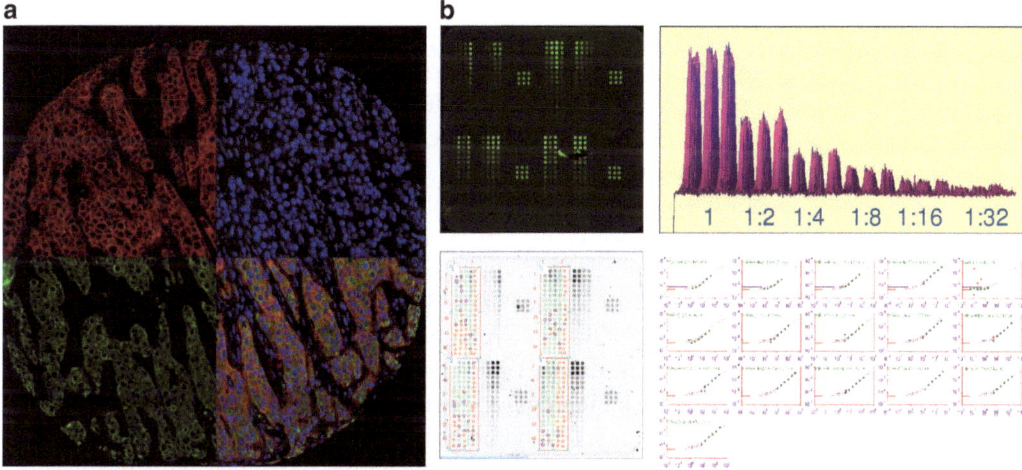

Fig. 3. Examples of quantitative methodologies for systems biology research in vivo and in vitro. (**a**) HistoRx AQUA fluorescence image analysis. The expression of protein per unit area of a molecularly defined compartment (tissue and subcellular) is measured using this technique. In this case, membrane expression of the HER2 oncoprotein (*top left quadrant*) in invasive cancer (*bottom left quadrant*; cytokeratin mask) has been quantified. Expression of proteins in the nuclear compartment may be measured by quantifying signal intensity in the DAPI compartment (*top right quadrant*). (**b**) Reverse phase protein arrays (RPPA). Cell or tissue lysates are spotted onto nitrocellulose-coated glass slides in replicates and as a dilution series, and probed for a protein of interest using specific antibodies and a fluorescent secondary (*top*). Each spot intensity is measured and plotted as a dilution curve, and the expression of protein derived from a regression analysis of all the data generated for a single sample (*bottom*). In this way, the expression of protein is guaranteed to be derived from the linear range of detection (i.e., not saturated) and the technical variance of the data is measured.

auto-fluorescence and difficulties with absolute quantification. While most in situ methods are limited to protein techniques, all macromolecules may be extracted from microdissected tissues, including RNA, microRNA, DNA and protein. Each macromolecule is amenable to high-throughput array-based technologies and, in the case of protein, the tantalizing prospect of reliable clinical mass-spectrometry proteomics (56) can offer quantitative advantages or assessment of functional post-translational modifications (e.g., phosphorylation or acetylation).

3.2. Temporal Resolution

Although increasing temporal resolution seems straightforward for in vitro-based experiments, it is more difficult within the clinical setting. There are a range of biological models available for analyzing complex biological systems. Nonetheless, even the simplest in vitro models require reconsideration of design in order to generate data of sufficient quality to populate mathematical models, particularly data-hungry methodologies such as DBNs. While it is relatively trivial to increase data points in an experiment examining pathway responses to targeted therapy in vitro, it is only recently that downstream assays are sufficiently "high-throughput" to generate robust, quantitative data of sufficient quality and quantity to be used to populate mathematical models. Although robotics can help meet the demands of high-performance throughput, new techniques need to be considered to meet the data-rich demands of systems approaches.

Since many of the components assessed within mathematical models are proteins and their activated forms, newer protein assays may be used in order to address the problems outlined above. Reverse-phase protein arrays (RPPA) are high-throughput, high-density protein arrays in which protein lysates from in vitro or in vivo biological samples are immobilised as spots on nitrocellulose-coated glass slides and protein targets are detected with specific antibodies, similar to immunohistochemistry or immunofluorescence (Fig. 3b and (57)). In this way, hundreds to thousands of lysates (including technical and biological replicates) can be assayed under the same conditions on a single slide. As only picoliter quantities are spotted on each spot, tens to hundreds of identical slides may be produced for multiple target analysis from a single cellular or tissue lysate, particularly if robotic spotting is used. In addition, since each lysate is spotted as a dilution series, the signal detected can always be guaranteed to be in the linear range of detection (that is while signal intensity is still unsaturated), which is rarely the case for western blotting, and if recombinant peptide or protein controls are spotted on the same slide, accurate quantification is possible. As well as facilitating high throughput protein analysis in vitro, the small amount of lysate required and the high number of assays that can be performed make this a useful technique for assessing clinical material. This means that large models, requiring tens to hundreds of biological measurements (e.g., the ordinary differential equation-based approach) may be populated with ease.

A second challenge to achieving sufficient temporal resolution for systems approaches is the selection of appropriate in vitro or in vivo models. While transgenic mice offer the attraction of stable genetic perturbations that can be applied to computational models, the long generation times and high numbers of time points required do not lend themselves to the iterative nature of systems biology, that is the need to refine the model on the basis of re-experimentation in order to improve it. Nevertheless, if coupled with live imaging techniques, such as the relatively new technique of intra-vital microscopy, which has been used to image tumour cell invasion in real-time (58), these animal models may become attractive models for detailed pharmacodynamic studies. We have used a combination of 3-dimensional in vitro primary and human adenocarcinoma cell line cultures on contracted collagen matrices (59) and cell line xenografts to model the pharmacodynamics of targeted therapies. Three-dimensional culture simplifies the tumour context but offers a flexible analysis of epithelial and stromal compartments, where both compartments may be genetically manipulated and subjected to both destructive and non-destructive temporal analysis, such as by reverse-phase protein arrays and immunofluorescence. Primary and cell-line xenografts capture the complexity of whole tissues but what makes these models ideal for high-throughput, spatially resolved analyses is their ability to assess tumours from multiple time points, and the availability of abundant tissue for fresh frozen and formalin-fixed paraffin-embedded analysis. In addition, cell lines are readily manipulated in vitro for the knockdown or overexpression of specific targets with small hairpin RNA and stable transfection of genetic constructs, respectively, and are then ready for re-implantation and re-testing of the system with specific perturbations. These models are an important intermediate step in validating computational models before they are sufficiently reliable to be used in clinical decision making.

4. Clinical Considerations for Data Generation

The real challenge lies in achieving sufficient temporal resolution using real human disease as the model. Nevertheless, there is now extensive experience in gathering tissue and biological samples from three time points in the neoadjuvant setting (that is, patients treated with drugs or radiotherapy before surgery), in individual patients with breast cancer, such that limited pharmacodynamic studies may be performed. In this model, patients are given endocrine or chemotherapy for 3 months prior to definitive resection and samples taken at diagnosis, 2 weeks and 3 months at the time of resection (60). If basic pathological endpoints such as proliferation (immunohistochemistry analysis of Ki67 expression levels) are measured, then the proliferation index at 2 weeks (but not at diagnosis) is predictive of long-term survival in response to aromatase inhibitor therapy (61, 62). Breast cancer is amenable to this type of temporal intervention, since there is the added benefit that tumour shrinkage in the neoadjuvant setting can result in the use of breast-conserving surgery rather than mastectomy. Nevertheless, other cancers may also be amenable to multiple sampling, such as ovarian tumours treated with intraperitoneal chemotherapy (63), or colorectal tumours treated with pre-operative radiotherapy (64), which may be achieved with minimal discomfort or inconvenience to the patient. By exploiting carefully selected human models, we can begin to determine the true nature of the dynamics of tumour responses and move away from inferred biology based on static biomarker analysis. In the short term, gathering high quality, temporal data from real clinical material is essential to populate and validate computational models. In time, such mathematical models are likely to become applicable and may avoid the need for multiple biological measurements.

5. Conclusions

In this review, we have discussed a number of possible general approaches to applying systems biology to understanding therapeutic responses in cancer. It should be apparent that no one computational, mathematical or experimental methodology can be used in isolation to de-convolute the complexity of cancer. We have discussed some approaches which are being used in this infant field. On their own, some of these approaches are starting to produce results, but by using different approaches, such as process- and data-driven models in tandem in order to refine and validate models, the hope is that clinically relevant and useful models may become a reality sooner. In order for this to occur, data from clinical trials will also have to be incorporated, which will also require coordinated multidisciplinary efforts from the clinical and basic science communities.

References

1. Hanahan D, Weinberg RA (2000) The hallmarks of cancer. Cell 100:57–70
2. Wood LD, Parsons DW, Jones S, Lin J, Sjoblom T, Leary RJ, Shen D, Boca SM, Barber T, Ptak J, Silliman N, Szabo S, Dezso Z, Ustyanksky V, Nikolskaya T, Nikolsky Y, Karchin R, Wilson PA, Kaminker JS, Zhang Z, Croshaw R, Willis J, Dawson D, Shipitsin M, Willson JK, Sukumar S, Polyak K, Park BH, Pethiyagoda CL, Pant PV, Ballinger DG, Sparks AB, Hartigan J, Smith DR, Suh E, Papadopoulos N, Buckhaults P, Markowitz SD, Parmigiani G, Kinzler KW, Velculescu VE, Vogelstein B (2007) The genomic landscapes of human breast and colorectal cancers. Science 318:1108–1113
3. Neve RM, Chin K, Fridlyand J, Yeh J, Baehner FL, Fevr T, Clark L, Bayani N, Coppe JP, Tong F, Speed T, Spellman PT, DeVries S, Lapuk A, Wang NJ, Kuo WL, Stilwell JL, Pinkel D, Albertson DG, Waldman FM, McCormick F, Dickson RB, Johnson MD, Lippman M, Ethier S, Gazdar A, Gray JW (2006) A collection of breast cancer cell lines for the study of functionally distinct cancer subtypes. Cancer Cell 10:515–527
4. Sobin LH, Wittekind CH (2002) UICC: TNM classification of malignant tumors. Wiley-Liss, New York
5. Faratian D, Bartlett J (2008) Predictive markers in breast cancer – the future. Histopathology 52:91–98
6. Payne SJ, Bowen RL, Jones JL, Wells CA (2008) Predictive markers in breast cancer – the present. Histopathology 52:82–90
7. Faratian D, Moodie SL, Harrison DJ, Goryanin I (2007) Dynamic computational modeling in the search for better breast cancer drug therapy. Pharmacogenomics 8:1757–1761
8. O'Reilly KE, Rojo F, She QB, Solit D, Mills GB, Smith D, Lane H, Hofmann F, Hicklin DJ, Ludwig DL, Baselga J, Rosen N (2006) mTOR inhibition induces upstream receptor tyrosine kinase signaling and activates Akt. Cancer Res 66:1500–1508
9. Kola I, Landis J (2004) Can the pharmaceutical industry reduce attrition rates? Nat Rev Drug Discov 3:711–715
10. Kholodenko BN (2006) Cell-signalling dynamics in time and space. Nat Rev Mol Cell Biol 7:165–176
11. Faratian D, Goltsov A, Lebedeva G, Sorokin A, Mullen P, Kay C, Um I, Langdon SP, Goryanin I, Harrison DJ (2009) Systems biology reveals new strategies for personalising cancer medicine and confirms PTEN"s role in resistance to trastuzumab. Cancer Res 69:6713–6720
12. Piccart-Gebhart MJ, Procter M, Leyland-Jones B, Goldhirsch A, Untch M, Smith I, Gianni L, Baselga J, Bell R, Jackisch C, Cameron D, Dowsett M, Barrios CH, Steger G, Huang CS, Andersson M, Inbar M, Lichinitser M, Lang I, Nitz U, Iwata H, Thomssen C, Lohrisch C, Suter TM, Ruschoff J, Suto T, Greatorex V, Ward C, Straehle C, McFadden E, Dolci MS, Gelber RD (2005) Trastuzumab after adjuvant chemotherapy in HER2-positive breast cancer. N Engl J Med 353:1659–1672
13. Romond EH, Perez EA, Bryant J, Suman VJ, Geyer CE Jr, Davidson NE, Tan-Chiu E, Martino S, Paik S, Kaufman PA, Swain SM, Pisansky TM, Fehrenbacher L, Kutteh LA, Vogel VG, Visscher DW, Yothers G, Jenkins RB, Brown AM, Dakhil SR, Mamounas EP, Lingle WL, Klein PM, Ingle JN, Wolmark N (2005) Trastuzumab plus adjuvant chemotherapy for operable HER2-positive breast cancer. N Engl J Med 353:1673–1684
14. Berns K, Horlings HM, Hennessy BT, Madiredjo M, Hijmans EM, Beelen K, Linn SC, Gonzalez-Angulo AM, Stemke-Hale K, Hauptmann M, Beijersbergen RL, Mills GB, van de Vijver MJ, Bernards R (2007) A functional genetic approach identifies the PI3K pathway as a major determinant of trastuzumab resistance in breast cancer. Cancer Cell 12:395–402
15. Nagata Y, Lan KH, Zhou X, Tan M, Esteva FJ, Sahin AA, Klos KS, Li P, Monia BP, Nguyen NT, Hortobagyi GN, Hung MC, Yu D (2004) PTEN activation contributes to tumor inhibition by trastuzumab, and loss of PTEN predicts trastuzumab resistance in patients. Cancer Cell 6:117–127
16. Fuss H, Dubitzky W, Downes CS, Kurth MJ (2005) Mathematical models of cell cycle regulation. Brief Bioinform 6:163–177
17. Hatakeyama M, Kimura S, Naka T, Kawasaki T, Yumoto N, Ichikawa M, Kim JH, Saito K, Saeki M, Shirouzu M, Yokoyama S, Konagaya A (2003) A computational model on the modulation of mitogen-activated protein kinase (MAPK) and Akt pathways in heregulin-induced ErbB signalling. Biochem J 373:451–463
18. Hendriks BS, Cook J, Burke JM, Beusmans JM, Lauffenburger DA, de Graaf D (2006) Computational modelling of ErbB family phosphorylation dynamics in response to

transforming growth factor alpha and heregulin indicates spatial compartmentation of phosphatase activity. Syst Biol (Stevenage) 153: 22–33

19. Kholodenko BN, Demin OV, Moehren G, Hoek JB (1999) Quantification of short term signaling by the epidermal growth factor receptor. J Biol Chem 274:30169–30181
20. Markevich NI, Hoek JB, Kholodenko BN (2004) Signaling switches and bistability arising from multisite phosphorylation in protein kinase cascades. J Cell Biol 164: 353–359
21. Shankaran H, Wiley HS, Resat H (2006) Modeling the effects of HER/ErbB1-3 coexpression on receptor dimerization and biological response. Biophys J 90:3993–4009
22. Steuer R (2007) Computational approaches to the topology, stability and dynamics of metabolic networks. Phytochemistry 68: 2139–2151
23. Birtwistle MR, Hatakeyama M, Yumoto N, Ogunnaike BA, Hoek JB, Kholodenko BN (2007) Ligand-dependent responses of the ErbB signaling network: experimental and modeling analyses. Mol Syst Biol 3:144
24. Moehren G, Markevich N, Demin O, Kiyatkin A, Goryanin I, Hoek JB, Kholodenko BN (2002) Temperature dependence of the epidermal growth factor receptor signaling network can be accounted for by a kinetic model. Biochemistry 41:306–320
25. Schoeberl B, Eichler-Jonsson C, Gilles ED, Muller G (2002) Computational modeling of the dynamics of the MAP kinase cascade activated by surface and internalized EGF receptors. Nat Biotechnol 20:370–375
26. Clyde RG, Craig AL, de Breed L, Bown JL, Forrester L, Vojtesek B, Smith G, Hupp T, Crawford J (2009) A novel ataxia-telangiectasia mutated autoregulatory feedback mechanism in murine embryonic stem cells. J R Soc Interface 6:1167–1177
27. Gueven N, Fukao T, Luff J, Paterson C, Kay G, Kondo N, Lavin MF (2006) Regulation of the Atm promoter in vivo. Genes Chromosomes Cancer 45:61–71
28. Heckerman D, Geiger D, Chickering D (1995) Learning Bayesian networks: the combination of knowledge and statistical data. Mach Learn 20:197–243
29. Pearl J (1988) Probabilistic reasoning in intelligent systems. Morgan Kaufmann, San Francisco
30. Friedman N (2004) Inferring cellular networks using probabilistic graphical models. Science 303:799–805
31. Kim SY, Imoto S, Miyano S (2003) Inferring gene networks from time series microarray data using dynamic Bayesian networks. Brief Bioinform 4:228–235
32. Imoto S, Kim S, Goto T, Miyano S, Aburatani S, Tashiro K, Kuhara S (2003) Bayesian network and nonparametric heteroscedastic regression for nonlinear modeling of genetic network. J Bioinform Comput Biol 1:231–252
33. Sachs K, Perez O, Pe'er D, Lauffenburger DA, Nolan GP (2005) Causal protein-signaling networks derived from multiparameter single-cell data. Science 308:523–529
34. Guha U, Chaerkady R, Marimuthu A, Patterson AS, Kashyap MK, Harsha HC, Sato M, Bader JS, Lash AE, Minna JD, Pandey A, Varmus HE (2008) Comparisons of tyrosine phosphorylated proteins in cells expressing lung cancer-specific alleles of EGFR and KRAS. Proc Natl Acad Sci USA 105:14112–14117
35. Li Z, Chan C (2004) Inferring pathways and networks with a Bayesian framework. FASEB J 18:746–748
36. Gevaert O, De Smet F, Timmerman D, Moreau Y, De Moor B (2006) Predicting the prognosis of breast cancer by integrating clinical and microarray data with Bayesian networks. Bioinformatics 22:e184–e190
37. Matthäus F, Smith VA, Fogtman A, Sommer WH, Leonardi-Essmann F, Lourdusamy A, Reimers MA, Spanagel R, Gebicke-Haerter PJ (2009) Interactive molecular networks obtained by computer-aided conversion of microarray data from brains of alcohol-drinking rats. Pharmacopsychiatry 42:S118–S128
38. Sorribas A, Savageau MA (1989) A comparison of variant theories of intact biochemical systems. I. Enzyme-enzyme interactions and biochemical systems theory. Math Biosci 94:161–193
39. Savageau MA, Voit EO (2008) Power-law approach to modeling biological systems. 1. Theory, 60th edn. pp 519–544
40. Voit EO (2002) Models-of-data and models-of-processes in the post-genomic era. Math Biosci 180:263–274
41. Kitano H (2004) Biological robustness. Nat Rev Genet 5:826–837
42. Bild AH, Yao G, Chang JT, Wang Q, Potti A, Chasse D, Joshi MB, Harpole D, Lancaster JM, Berchuck A, Olson JA Jr, Marks JR, Dressman HK, West M, Nevins JR (2006) Oncogenic pathway signatures in human cancers as a guide to targeted therapies. Nature 439:353–357

43. Wulfkuhle J, Espina V, Liotta L, Petricoin E (2004) Genomic and proteomic technologies for individualisation and improvement of cancer treatment. Eur J Cancer 40: 2623–2632
44. Moodie SL, Sorokin A, Goryanin I, Ghazal P (2009) Graphical notation to describe the logical interactions of biological pathways. J Integr Bioinform 3:36
45. Calzone L, Gelay A, Zinovyev A, Radvanyi F, Barillot E (2008) A comprehensive modular map of molecular interactions in RB/E2F pathway. Mol Syst Biol 4:173
46. Le Novere N, Moodie SL, Sorokin A, Hucka M, Schreiber F, Demir E, Mi H, Matsuoka Y, Wegner K, Kitano H (2008) Systems biology graphical notation: process diagram level 1. Nature Precedings
47. Saltelli A, Ratto M, Andres T, Campolongo F, Cariboni J, Gatelli D, Saisana M, Tarantola S (2008) Global sensitivity analysis: the primer. Wiley, Chichester
48. Pachepsky E, Crawford JW, Bown JL, Squire G (2001) Towards a general theory of biodiversity. Nature 410:923–926
49. Saltelli A, Tarantola S, Chan K (1999) Quantatative model-independent method for sensitivity analysis of model output. Technometrics 41:39–56
50. Feng XJ, Hooshangi S, Chen D, Li G, Weiss R, Rabitz H (2004) Optimizing genetic circuits by global sensitivity analysis. Biophys J 87:2195–2202
51. Goldberg D (1989) Genetic algorithms in search, optimization, and machine learning. Addison-Wesley, Reading, MA
52. Kirby J, Heath PR, Shaw PJ, Hamdy FC (2007) Gene expression assays. Adv Clin Chem 44:247–292
53. Kennett JY, Watson SK, Saprunoff H, Heryet C, Lam WL (2008) Technical demonstration of whole genome array comparative genomic hybridization. J Vis Exp, 870
54. Edwards RA (2007) Laser capture microdissection of mammalian tissue. J Vis Exp, 309
55. Camp RL, Chung GG, Rimm DL (2002) Automated subcellular localization and quantification of protein expression in tissue microarrays. Nat Med 8:1323–1327
56. Pan S, Aebersold R, Chen R, Rush J, Goodlett DR, McIntosh MW, Zhang J, Brentnall TA (2008) Mass spectrometry based targeted protein quantification: methods and applications. J Proteome Res 8(2):787–797
57. Tibes R, Qiu Y, Lu Y, Hennessy B, Andreeff M, Mills GB, Kornblau SM (2006) Reverse phase protein array: validation of a novel proteomic technology and utility for analysis of primary leukemia specimens and hematopoietic stem cells. Mol Cancer Ther 5:2512–2521
58. Kedrin D, Gligorijevic B, Wyckoff J, Verkhusha VV, Condeelis J, Segall JE, van Rheenen J (2008) Intravital imaging of metastatic behavior through a mammary imaging window. Nat Methods 5:1019–1021
59. Edward M (2001) Melanoma cell-derived factors stimulate glycosaminoglycan synthesis by fibroblasts cultured as monolayers and within contracted collagen lattices. Br J Dermatol 144:465–470
60. Dixon JM (2004) The scientific value of preoperative studies and how they can be used. Breast Cancer Res Treat 87(Suppl 1):S19–S26
61. Dowsett M, Smith IE, Ebbs SR, Dixon JM, Skene A, Griffith C, Boeddinghaus I, Salter J, Detre S, Hills M, Ashley S, Francis S, Walsh G, A'Hern R (2006) Proliferation and apoptosis as markers of benefit in neoadjuvant endocrine therapy of breast cancer. Clin Cancer Res 12:1024s–1030s
62. Dowsett M, Smith IE, Ebbs SR, Dixon JM, Skene A, A'Hern R, Salter J, Detre S, Hills M, Walsh G (2007) Prognostic value of Ki67 expression after short-term presurgical endocrine therapy for primary breast cancer. J Natl Cancer Inst 99:167–170
63. Alberts DS, Markman M, Armstrong D, Rothenberg ML, Muggia F, Howell SB (2002) Intraperitoneal therapy for stage III ovarian cancer: a therapy whose time has come! J Clin Oncol 20:3944–3946
64. Nagtegaal ID, Gaspar CG, Peltenburg LT, Marijnen CA, Kapiteijn E, van de Velde CJ, Fodde R, van Krieken JH (2005) Radiation induces different changes in expression profiles of normal rectal tissue compared with rectal carcinoma. Virchows Arch 446:127–135

Chapter 13

Systemic Lupus Erythematosus: From Genes to Organ Damage

Vasileios C. Kyttaris

Abstract

Systemic lupus erythematosus (SLE) is a disease characterized by inappropriate response to self-antigens. Genetic, environmental and hormonal factors are believed to contribute to the development of the disease. We think of SLE pathogenesis as occurring in three phases of variable duration. A series of regulatory failures during the ontogeny of the immune system lead to the emergence of auto-reactive clones and the production of auto-antibodies (phase I). As the immune response to self-antigens broadens, the auto-antibody repertoire is enriched (phase II) and clinical manifestations eventually ensue (phase III). The final result is tissue damage that if not treated will lead to the functional failure of such important organs as the kidney and brain.

Key words: Systemic Lupus Erythematosus, Auto-antibodies, Complement, Cell signaling, CD3ζ chain, Fcγ receptor, NFAT, Nephritis

1. Introduction

A prototypic autoimmune disease, systemic lupus erythematosus (SLE) is a syndrome characterized by inappropriate immune response to self-antigens. In SLE, the immune system fails to control auto-reactive T, B, and antigen-presenting cells that produce an array of auto-antibodies and cytokines leading to the cellular infiltration of various organs and the local activation of complement (1). Eventually organ damage occurs that, if left untreated, leads to serious and even fatal complications such as renal failure.

Although the clinical and laboratory findings in SLE are well described, the exact events that lead to the development of the disease are unclear. Schematically, we can hypothesize

Qing Yan (ed.), *Systems Biology in Drug Discovery and Development: Methods and Protocols*, Methods in Molecular Biology, vol. 662, DOI 10.1007/978-1-60761-800-3_13, © Springer Science+Business Media, LLC 2010

based on the current evidence that the development of SLE occurs in three phases:

1. The initial break in immunologic tolerance toward certain self-antigens induced by a poorly characterized interplay between environmental, hormonal and genetic factors.
2. The propagation of the abnormal immune response and the appearance of laboratory evidence of immunological dysfunction, such as antinuclear antibodies.
3. The clinical manifestations, with one or more organs such as the joints, skin or kidneys displaying inflammation-induced damage.

In this chapter, we will discuss initially the epidemiology and clinical manifestations of SLE, before addressing the pathogenetic mechanisms that lead to the expression of the disease.

2. Epidemiology: Heredity, Gender and the Environment

SLE is a relatively rare disease affecting approximately 40–50 individuals per 100,000 people in the United States (2), with an incidence of approximately two to three new cases per year per 100,000 persons (3). Worldwide studies, although showing variable incidence and prevalence among different populations (4–6), agree in that the majority of patients with SLE are women of childbearing age; the female: male ratio is estimated between 6 and 14:1 (7–10). SLE is less common at the extremes of age but can affect both children (11) and elderly individuals (12, 13); interestingly, the difference in incidence between men and women is not as striking in these age groups (13). These observations led to the hypothesis that hormonal factors, either the excess of estrogens or lack of androgens, may be involved in the development of the disease by influencing the development and/or reactivity of the immune system.

In addition to the gender differences in prevalence, SLE prevalence and severity differ among different populations. For example, individuals of African and Asian descent in the United States are more commonly and more severely affected than European-Americans (3, 14, 15). This observation underscores the fact that genetic factors predispose individuals to the development of SLE. Adding further credence to this argument, twin studies have shown a higher concordance rate for SLE among monozygotic versus dizygotic twins (24% vs. 2%) (16).

Nevertheless, genetic factors can not account for all the cases of SLE and therefore both natural factors and infectious agents have been implicated in the development of the disease. One of the first environmental factors that was found to be associated with SLE is sunlight: indeed, a significant proportion of patients display skin sensitivity to light and in particular ultraviolet light (17). No definite association with a particular infectious agent has been made; one exception is the Epstein-Barr virus (EBV), a virus associated with B cell hyperactivity. Patients with SLE are almost universally seropositive for EBV as compared to lower rates in healthy controls. Given this as well as similarity of EBV antigens and auto-antigens targeted by auto-antibodies found in SLE patients' sera, EBV has been suggested as a possible instigator of SLE (18).

The above described epidemiological characteristics of SLE suggest that a host of environmental, infectious and hormonal factors when applied to a genetically predisposed individual may lead to the development of SLE. The exact factors and the processes that trigger the disease are unclear and are the focus of many studies in humans with SLE and animal models of lupus.

3. Clinical Manifestations: Evaluating and Treating Systemic Inflammation

The initial clinical presentation, course and outcome of SLE are highly variable. The diagnosis is based on the presence of certain clinical and laboratory findings that are listed in Table 1 (19). Typically the disease course in most patients is characterized by

Table 1
Criteria for the diagnosis of systemic lupus erythematosus

1.	Malar rash
2.	Discoid rash
3.	Photosensitivity
4.	Oral ulcers
5.	Arthritis
6.	Serositis
7.	Renal disorder
8.	Neurologic disorder
9.	Hematologic disorder
10.	Immunologic disorder
11.	Antinuclear antibody (ANA)

periods of high activity (flare) and remission. The duration and frequency of these flares, their severity and precise clinical picture differ significantly among patients. This makes SLE a challenging disease to diagnose and treat.

The symptoms and signs of SLE can be broadly categorized as constitutional (a result of systemic inflammation), and organ-specific (20–23). Constitutional symptoms that patients with SLE may experience include high fevers, fatigue, and weight loss. More importantly the overwhelming majority of patients with SLE will present with specific organ involvement. Approximately 60–90% of the patients will develop some form of inflammatory skin rash which is oftentimes caused or exacerbated by sunlight (ultraviolet radiation) (17). The joints are affected in a significant percentage of patients in the form of inflammatory arthritis with pain and swelling. Inflammation involving the serous membranes (pleuritis, pericarditis) manifesting as chest or abdominal pain, is also common (10–30% of the patients) (20, 21). SLE affects the hematologic system with a decrease in the levels of white cells, platelets, and red cells; life threatening thrombocytopenia and severe anemia although uncommon can be seen in patients with SLE. In the most severe forms of SLE, the kidney and the central nervous systems are affected (23). The patients with renal lupus will present with abnormalities in the urine (blood and/or protein in the urine) and oftentimes edema. If left untreated, renal lupus may lead to severe complications including renal failure and vascular disease. Patients with central nervous involvement present with neurological complications (e.g., strokes, pain due to nerve damage) and/or psychiatric manifestations (mania, depression). SLE may also affect other organs such as the muscles (myositis), the lung (pneumonitis), and the heart (myocarditis). Overall, SLE symptoms and signs are caused by local inflammation in various organs that if left untreated may result in permanent organ damage.

The treatments to date are based on immunosuppressive regiments that nonspecifically inhibit the immune system. When intervening early on, before permanent damage occurs, these medications can be effective albeit with significant side effects. For skin and joint manifestations, the antimalarial hydroxychloroquine, nonsteroidal anti-inflammatory medications, and low to moderate doses of corticosteroids are often times sufficient. For moderately severe disease such as persistent skin rashes, pleuritis, severe arthritis, higher doses of corticosteroids are used with or without the introduction of an immunosuppressive medication such as methotrexate or azathioprine (24, 25). For the most severe life or organ-threatening manifestations (kidney, nervous system lupus), cytotoxic medications (cyclophosphamide) are used (22, 23). Clinical trials are ongoing for the use of less toxic medications such as mycophenolate mofetil in renal lupus (26, 27).

4. Clinical Laboratory Findings: A Window in the Pathogenesis of SLE

Studies of patients with SLE have shown that the above described clinical manifestations are caused by an exuberant immunological activation in the absence of a readily recognizable infectious agent. Both the humoral and cellular components of the immune system are activated. The serum of patients with SLE contains an array of antibodies that recognize self antigens and in particular nuclear antigens (reviewed in (28)). Some of these antibodies have been associated with specific manifestations of the disease; for example anti-dsDNA and anti-Sm antibodies are associated with nephritis (29), while anti-Ro antibody is associated with dry mouth (sicca) and neonatal lupus (20, 28). Some, but not all of these autoantibodies have been shown to cause damage. Examples include the antiplatelet antibodies that can reduce the number of platelets leading to bleeding (immune thrombocytopenia); the antiphospholipid antibodies directed against the components of the cell membrane that can mitigate platelet aggregation and thrombosis. It has to be noted that auto-antibodies can be found in the sera of patients with SLE long before the clinical manifestations of the disease, as shown in a study evaluating sera of military recruits who developed SLE (30). In this landmark study it was shown that auto-antibodies can be found in the serum of SLE patients long before the diagnosis is made. Importantly, there was a temporal progression of the autoantibody repertoire whereas antinuclear (ANA) antibodies appeared first, followed by anti-dsDNA and antiribonucleoprotein antibodies.

Another characteristic of the serum of patients with SLE is the low level of complement proteins C3 and C4; this is thought to be due to complement activation by immune complexes in the tissues and circulation. Importantly, in a significant number of patients (especially patients with nephritis), falling C3 and rising anti-dsDNA levels may predate clinical deterioration (31). These findings in the peripheral blood of patients with SLE show that there is a hyperactive B cell compartment of the immune system that produces autoantibodies against cellular components.

Several studies have addressed the nature of organ and tissue damage in SLE. Biopsies of the skin and the kidneys, two organs that are affected in a significant proportion of patients with SLE, have clearly shown that the components of the immune system, both cellular and humoral, are found in these target-organs. Skin biopsies in cutaneous lupus show a lymphocytic infiltration of the dermis and deposition of immunoglobulin and complement along the dermal–epidermal junction (32). Depending on the level of inflammation and the location of the infiltrating cells, various clinical manifestations ensue: acute superficial rashes, chronic discoid lesions, bullae formation or deep layer inflammation (panniculitis); oftentimes ulcers and scars form. Similarly, kidney studies have shown that lymphocytes infiltrate the interstitium while immunoglobulin (IgG, IgM, and IgA) and complement deposit in the glomeruli. The result of this inflammatory process is scarring of the glomerular tuft and tubular damage. Clinically these processes are manifested by leakage of protein and cells in the urine, and in the extreme cases by uremia (22, 23).

5. Etio-Pathogenesis: Integrating Genes, Environment and the Immune System

The epidemiological features, clinical manifestations and clinical laboratory findings have been invaluable in creating a model for the development of SLE and the propagation of the autoimmune response. The immunological abnormities that lead to SLE seem to start long before the clinical manifestations become apparent; genes, infections and environmental factors as well as hormones influence the development of the immune system in the early years of life, facilitating the emergence and activation of autoimmune lymphocytic clones.

Given the difficulties studying the preclinical phases of SLE at least in humans, most research efforts to understand the etio-pathogenesis of SLE have focused on two fronts:

(a) The characterization of the immune deregulation in SLE and the identification of the key pathways involved in it.

(b) The unraveling of the genetic background of patients with SLE and the potential role of these genes to disease pathology.

In the next sections, we will try to provide the links between SLE risk conferring genes and the immunological abnormalities found in patients with active disease. In addressing this issue, both intra- and intercellular signaling aberrations as well as changes in soluble mediators will be examined. It has to be noted that no single abnormality has been recognized to date as dominant in SLE but rather an array of signaling aberrations lead to the expression of the syndrome.

6. Cell Signaling: Auto-Reactivity, Deficient Regulation, and Inappropriate Activation

One of the first and the most significant susceptibility locus that has been found to be associated with SLE resides within the highly polymorphic major histo-compatibility complex (MHC) class II locus. In particular, Caucasians that have the DR2

(*HLA-DRB1*1501*) or DR3 (*HLA-DRB1*0301*) alleles have a two- to threefold increase in their relative risk to develop SLE; the risk conferred by these alleles though, is not as prominent in the other populations (33, 34). The MHC class II locus encodes proteins that are involved in antigen presentation to CD4+ T cells. Beyond being simply associated with SLE risk, certain *MHC class II* genes have been found to be associated with the production of specific autoantibodies. For example, MHC class II allele *HLA-DRB1*03* was associated with the expression of anti-Ro and anti-La antibodies in patients with SLE (35). Similarly modest associations were found between *HLA-DQA1*0601*, and *DQB*0201* with anti-Ro and *HLA-DQA1*0501*, and *DQB*0201* with anti-La antibodies. Although certain clinical manifestations tend to be more prevalent in patients with these alleles, the associations between clinical picture and genotype are modest at best. Given these findings, it has been hypothesized that certain MHC class II molecules expressed on antigen-presenting cells (APC) of SLE patients may preferentially or aberrantly present certain (auto)-antigens to helper CD4+ T cells leading to abnormal T cell responses to antigenic stimuli that should have been ignored under normal conditions.

Indeed, multiple studies have shown that T cells that recognize and react against auto-antigens are present in the peripheral blood of patients with SLE. Among the self-antigens, the T cells from SLE patients have been shown to recognize histones, native DNA, and small nuclear ribonucleoproteins. These T cells are able to provide cognate help to B cells and lead to the production of potentially pathogenic anti-dsDNA auto-antibodies (36, 37).

In addition to auto-reactivity, T cells in SLE show a variety of signaling defects and abnormalities. Once the T cell receptor (TCR) is engaged, SLE T cells show a very robust and early influx of calcium and phosphorylation of Tyrosine residues on early signaling molecules (38). Two main reasons have been recognized as underlying the abnormal early signaling events of SLE T cells: Preaggregated lipid rafts (39) and the substitution of the CD3ζ chain by the FcεRIγ chain (40). The lipid rafts are platforms on the membrane of lymphocytes that help bring together all the important molecules that participate in the activation of the cells once their receptor has been engaged by its cognate ligand. In SLE T cells, as opposed to control T cells from healthy individuals and patients with other autoimmune diseases, the lipid rafts are already aggregated thus facilitating the early signaling events that lead to the influx of calcium and activation of kinases and phosphatases. In addition to the ready-made signaling platform, SLE T cells employ a different (rewired) (41) TCR/CD3 molecular complex that is more efficient in transducing the activating signal than the TCR/CD3 complex found in control nonactivated cells. Under normal conditions the main signaling molecule responsible for the transduction of the signal is the CD3ζ chain. SLE T cells have decreased CD3ζ chain levels and instead express in its place the FcεRIγ (40), a molecule initially recognized to be associated with the Fcε receptor on mast cells. This change results in the recruitment of Syk kinase instead of the Zap70 to the CD3 complex and leads to a more robust downstream signaling (41, 42). The expression of CD3ζ and FcεRIγ are at least in part dictated by the activity of their respective genes, both of which are controlled by the transcriptional factor Elf-1: Elf-1 binding to its *cis* element on the *CD3ζ* promoter leads to gene transcription while it acts as a transcription repressor on the *FcεRIγ* gene. Elf-1 production is defective (43) in SLE T cells thus tilting the balance toward the production of FcεRIγ. Other mechanisms such as alternative splicing of the *CD3ζ* gene also contribute to the decreased CD3ζ chain levels in SLE T cells (44, 45).

In many respects, SLE T cells appear to be hyperactive, but nevertheless their ability to produce interleukin-2 (IL-2) upon activation is limited (46). IL-2 is a cytokine that is important for T cell activation, activation-induced cell death (AICD) and the survival of T regulatory cells (Treg). Its deficient production may therefore be linked to the prolonged survival of T cells (including auto-reactive cells) and insufficient inhibition of autoreactive processes by Treg (47, 48). Multiple transcription factors have to cooperate in order for the *IL-2* gene to be transcribed. In SLE T cells the transcription activators NF-kB (49), AP-1 (a c-fos/c-jun dimer) (50) and p-CREB (51) are deficient while the transcription repressor CREM (52, 53) binds strongly to the promoter of *IL-2*. This combination of increased repressors and deficient activators is responsible for the inappropriately low production of IL-2 by activated SLE T cells.

The robust calcium flux once the T cell receptor is engaged does translate though into high NFAT (a calcium dependent transcription activator) translocation into the nucleus of T cells (54). NFAT because of deficiency of AP-1 does not stimulate the production of IL-2 but is able to bind to the promoter and stimulate the production of CD154 (also called CD40 ligand). CD154 is a costimulatory molecule that helps the T cells provide help to B cells therefore leading to the production of pathogenic autoantibodies (55). In addition SLE T cells evade activation induced death (AICD) by yet another mechanism. These cells over-express cyclo-oxygenase 2 (Cox-2) (56), a molecule that facilitates their survival after activation.

Recently, it has been identified that SLE T cells upregulate CD44, a molecule that facilitates their migration in tissues such as the kidneys (57). There they produce proinflammatory cytokines such as IL-17 (58) that enables the recruitment of other immune cells and the propagation of the inflammatory process resulting in tissue damage.

Genetic studies have tried to shed light into the underlying causes for the aberrant function of SLE T cells. Of particular interest has been the observed association of SLE with genes encoding molecules that prevent or dampen T cell activation. One such molecule is the cytotoxic T cell antigen-4 (CTLA-4). CTLA-4 is similar in structure to CD28, which by binding to the CD80/86 molecules on APC augments T cell activation. CTLA-4 blocks the CD28:CD80/86 interaction by potently binding to the CD80/86 and at the same time delivers an inhibitory signal into the cell (59). Therefore CTLA-4 brings T cell activation to an end and induces a state of anergy. This function is an important check-point that prevents over-activation of the immune system and is thought to prevent autoimmune diseases by promoting long-lived anergy (60). Preliminary studies looking at several polymorphisms of the *CTLA-4* gene showed that a T/C substitution at the −1,722 site is associated with SLE (61). Of note though, soluble CTLA-4 was shown to be increased in SLE patients, especially ones with active disease (62). More studies are therefore needed to address the potential role of CTLA-4 in SLE.

Another gene that encodes a negative regulator of T cell activation and has been associated with SLE is *PTPN22*. This gene encodes the lymphoid tyrosine phosphatase (LYP) that prevents T cell activation (63, 64). A missense polymorphism (R620W) (65) in the *PTPN22* gene may impair the negative signaling transduced by LYP thus lowering the threshold for T cell activation.

The serum of patients with SLE contains various auto-antibodies, some of which have clear pathogenic capacity. These are produced by activated auto-reactive B cells, which are clearly abnormal. Although auto-reactive B cells are part of the immunological repertoire of normal individuals, SLE B cells show clear evidence of aberrant function possibly resulting during their maturation process. It has been shown that the peripheral B cell population in patients with active SLE has an increased number of CD127 high plasma cells and decreased number of naïve B cells (66). Looking even further at the naïve B cell population, a study of a limited number of young patients with SLE showed that a large proportion of B cells from these patients (up to 50%) are auto-reactive even before their first encounter with antigens (67). These studies suggest failure of important checkpoints in the maturation of B cells with the resulting survival of a higher number of auto-reactive B cells that may produce auto-antibodies as well as secrete cytokines and act as (auto)antigen-presenting cells. Given these findings, a recent report identifying the *BLK/C8orf13* as a risk conferring area for SLE (68) is very interesting. *BLK* encodes for a src family tyrosine kinase that signals downstream to the B cell receptor. A polymorphism upstream of the *BLK* gene that in B cell lines led to the decreased production of BLK mRNA was associated with SLE. BLK is mainly found in immature B cells and therefore its decreased expression may affect B cell ontogeny (69). Under normal conditions, immature B cells that encounter and react to auto-antigens during their maturation process will undergo receptor editing or apoptosis (negative selection); these mechanisms prevent the emergence of auto-reactive B cells in the periphery. Immature B cells that recognize self-antigens but due to impaired signaling (as in the case of decreased BLK activity) do not undergo negative selection, will escape to the periphery.

Abnormal maturation and increased help from T cells cause multiple signaling aberrancies of SLE B cells. On the one hand, similar to T cells, SLE B cells upon the engagement of their receptor have increased tyrosine phosphorylation and intracellular calcium flux (70). On the other hand, inhibitory signaling such as via the Fc receptor FcγRIIb (a negative regulator of B cell receptor (BCR) signaling) are depressed (71). In addition, memory SLE B cells do not upregulate FcγRIIb as readily as controls (72). Genetic studies have shown an association of a polymorphism (FcγRIIb-232 I/T substitution) with SLE (33). The final result of this imbalance between inhibitory and activation signals in SLE B cells is augmented antibody production.

SLE B cells can also influence the function of T cells as they have been shown to produce antibodies that react with the CD3 molecule on T cells and activate the calcium and calmodulin dependent kinase IV (CaMKIV) (73) in these cells. In turn, CaMKIV causes binding of the repressor c-AMP response element modulator (CREM) to the *IL-2* promoter, blocking the transcription of the gene.

In summary, the hyperresponsive SLE B and T cells (see Fig. 1) contribute to the production of autoantibodies, cytokine imbalance and cell tissue infiltration that lead to the clinical manifestations of SLE.

7. Humoral Factors: Apoptosis, the Complement System and Cytokines

Two of the most important findings in the serum of patients with SLE are the presence of various auto-antibodies and the decrease in the concentration of complement. Under normal conditions, natural auto-antibodies and complement are important as they play a major role in the clearance of auto-antigens contained in apoptotic material (74). It has been hypothesized that the appearance of auto-antibodies and the activation of complement seen in patients with SLE are a result of inappropriate cellular debris (waste) disposal (75). According to this theory, debris from cells

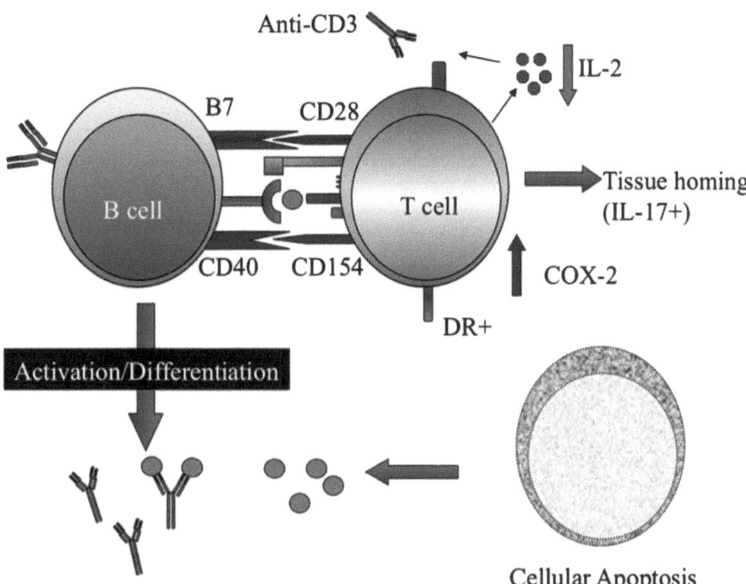

Fig. 1. Aberrant T:B cell cooperation in SLE. SLE T cells display an aberrant phenotype upon activation. They express persistently CD154 (CD40 ligand) that provides help to B cells and do not undergo readily activation-induced cell death due to the upregulation of cyclo-oxygenase 2 and deficient production of interleukin-2. At the same time, IL-17+ T cells infiltrate tissues and contribute to local inflammation. Impaired toleragenic mechanisms lead to the escape to the periphery of auto-reactive B cells, which are prone to produce autoantibodies with help by T cells. Apoptosis of cells (such as keratinocytes after UV exposure) leads to ample auto-antigen availability that is not handled appropriately by the reticulo-endothelial system. In turn, immune complexes containing autoantigens and autoantibodies deposit in tissues causing complement-mediated injury.

that have undergone cellular death (by apoptosis or necrosis) is not handled correctly by the body with a resultant stimulation of autoantibody production, creation and tissue deposition of immune complexes, and complement activation leading to tissue damage.

This theory has been further strengthened by the fact that several complement components encoding genes have been associated with susceptibility to SLE. Individuals with deficiency in C1q, C2 or C4 have a much higher risk to develop SLE than control individuals. In the case of C1q deficiency, a very rare genetic trait, the rate of development of SLE is very high, approximately 90%. Similarly, 10% of patients who have a genetic deficiency in C2 develop SLE as well as up to 75% of patients with a complete lack of C4 (75–77). These observations point to the fact that complement is essential not only as a first line of defense against pathogens, but also is instrumental in averting the emergence of autoimmunity. Complement coats (opsonizes) auto-antigen:antibody complexes facilitating their removal from the circulation. This prevents the inappropriate presentation of auto-antigens to immune cells and emergence of autoimmunity (74) as is probably the case in SLE patients with complement deficiency.

Besides complement components of the classical pathway, SLE has also been associated with components of the lectin pathway of complement activation. More precisely a proportion of SLE patients have been shown to have deficiency of the mannose binding lectin (MBL) (78), a molecule that is similar to C1q in structure and function. Another study has associated polymorphisms in the promoter or coding region of *MBL* gene with SLE (79). The MBL pathway is crucial for the opsonization of bacteria and its deficiency in SLE may be important for both decreased antimicrobial vigilance (80) and the poor handling of auto-antigens.

Complement fragments C3b and C4b coating apoptotic material bind through the complement receptor 1 (CR1) (CD35) onto erythrocytes. An association between a structural variant of CR1 (CR1 S) and SLE in Caucasians has been suggested in a meta-analysis (75) of genetic studies in SLE; this observation suggests that even in patients with normal complement activity in the serum, slow removal of the immune complexes from the circulation may result in further immune activation of cells by the auto-antigens as well as the increased deposition of immune complexes in the tissues.

Cellular waste is also cleared by the cells of the reticulo-endothelial system (RES) mainly via the interaction of IgG with its receptor. IgG-coated auto-antigens (debris) bind to one or more of the various Fcγ receptors (FcγR) on the surface of the cells of the RES. The affinity of the FcγR for the different IgG subclasses can be influenced by the substitution of a single aminoacid in their extracellular domain. It is therefore not surprising that variants of the *FcγR* gene that result in changes of the aminoacid sequence of these chains are associated with altered binding to IgG. These subtle changes can in turn influence such important immune functions as phagocytosis, antibody-dependent cell cytotoxicity (ADCC), and the clearance of immune complexes. Several studies have recognized the 1q23 region that contains the genes *FCGR2A*, *FCGR3A*, *FCGR3B* and *FCGR2B* which encode for low affinity IgG receptors as a susceptibility locus (33) for SLE.

FCGR2A has two codominant alleles that encode the different forms of FcγRIIa (CD32), a receptor found primarily on polymorphonuclear cells, mononuclear phagocytes and platelets. FcγRIIa binds IgG2 and C-reactive protein (CRP) (81). The two forms of FcγRIIa differ from each other by one single aminoacid at position 131. IgG2 binds stronger to the FcγRIIa that bears histidine at position 131 (H131) than to the molecule that has arginine at the same position (R131) (82). IgG2 is a poor activator of classical complement pathway and therefore its binding to the FcγRIIa is important for the clearance of immune complexes that contain IgG2. Multiple studies in different populations concluded that patients with the FcγRIIa-R131 polymorphism are at higher risk for SLE (33) but not nephritis (83, 84).

Altered CRP expression may independently or in conjunction with altered expression of FcγRIIa contribute to the development of SLE. Under normal conditions CRP is important for the disposal of cellular debris. In SLE, the defective expression of CRP may lead to deficient disposal of products of apoptosis making them available for presentation to T cells. SLE has been associated with a single nucleotide polymorphism (CRP-4) in the 3′ region of the *CRP* gene (85).

Similar to *FCGR2A*, *FCGR3A* has two codominant alleles that encode for two different forms of FcγRIIIa (CD16), a receptor expressed on natural killer (NK) and mononuclear cells. FcγRIIIa binds IgG1 and IgG3. The two forms of FcγRIIIa differ at position 176 with the one form having valine (V176) and the other phenylalanine (F176). The FcγRIIIa from individuals that are homozygous for the V176 form bind IgG1 and IgG3 more efficiently than FcγRIIIa from individuals homozygous for the F176 (82). A meta-analysis of data derived from 11 independent studies of the weaker FcγRIIIa-F176 form found a modest association with SLE (84).

Along the same lines SLE patients lacking the enzyme DNase I were reported (86). DNase I deficiency may lead to a decreased breakdown of DNA-protein complexes, eventually giving rise to immunological targeting of native DNA and its associated proteins.

These data suggest that proper waste handling fails at multiple levels in SLE with defective coating of the auto-antigens by complement and CRP and decreased binding to the Fc receptors.

Besides auto-antibodies and complement, the immune function in SLE is influenced by an array of cytokines that are aberrantly produced or missing (87). Genes encoding cytokines such as tumor necrosis factor (TNFα) (88) and interleukin-10 (IL-10) (89) as well as the TNF receptor (90) have been associated with SLE. It has been shown that both IL-10 and TNF are aberrantly produced in SLE. More importantly though it has been found that peripheral blood mononuclear cells (PBMC) taken from SLE patients, especially ones with active disease, bear an interferon signature (91); in essence this means that genes that depend on type I interferons are activated in SLE patients. Genetic studies have shown that genetic polymorphisms of two genes that encode transcription factors related to interferon,

STAT4 (T allele; rs7574865) (92) and *IRF5* (T allele; rs2004640) (93) are associated with higher risk for development of SLE. Multiple inflammatory cytokines such as tumor necrosis factor and interleukin-6 are upregulated by these transcription factors (94). It is therefore plausible that the cells of SLE patients over-interpret interferon-mediated signals, such as those elicited by viral responses leading to the inappropriate activation of the immune system.

8. Conclusion

Genetic and functional studies have shown that SLE pathogenesis is complex. On the one hand, the failure of toleragenic mechanisms results in the generation of autoreactive immune cell clones. On the other hand, the deficient removal of apoptotic debris due to complement and/or Fc receptor abnormalities leads to the increased total burden of self-antigens. With these mechanisms in place, external stimuli are over- or mis-interpreted by the immune system, and result in aberrant antigen presentation, lymphocyte activation, and the production of an array of inflammatory cytokines. These in turn lead to tissue deposition of immune complexes, complement activation, and target-organ cell infiltration. Current and future trials aim at understanding the complex mechanisms in every step of SLE pathogenesis so that we may design more effective and less toxic therapeutic interventions.

Acknowledgments

This work was supported by the National Institute of Arthritis, Musculoskeletal and skin diseases grant No 1K23 AR055672-01A1

References

1. Krishnan S, Chowdhury B, Juang Y-T, Tsokos GC (2007) Overview of the pathogenesis of systemic lupus erythematosus. In: Tsokos GC, Gordon C, Smolen JS (eds) Systemic lupus erythematosus: a companion to Rheumatology, 1st edn. Mosby Inc, Philadelphia, pp 55–63
2. Lawrence RC, Helmick CG, Arnett FC, Deyo RA, Felson DT, Giannini EH, Heyse SP, Hirsch R, Hochberg MC, Hunder GG, Liang MH, Pillemer SR, Steen VD, Wolfe F (1998) Estimates of the prevalence of arthritis and selected musculoskeletal disorders in the United States. Arthritis Rheum 41:778–799
3. McCarty DJ, Manzi S, Medsger TA Jr, Ramsey-Goldman R, LaPorte RE, Kwoh CK (1995) Incidence of systemic lupus erythematosus. Race and gender differences. Arthritis Rheum 38:1260–1270

4. Siegel M, Lee SL (1973) The epidemiology of systemic lupus erythematosus. Semin Arthritis Rheum 3:1–54
5. Nossent JC (1992) Systemic lupus erythematosus on the Caribbean island of Curacao: an epidemiological investigation. Ann Rheum Dis 51:1197–1201
6. Vilar MJ, Sato EI (2002) Estimating the incidence of systemic lupus erythematosus in a tropical region (Natal, Brazil). Lupus 11:528–532
7. McCarty DJ, Manzi S, Medsger TA Jr et al (1995) Incidence of systemic lupus erythematosus. Race and gender differences. Arthritis Rheum 38:1260–1270
8. Lahita RG (1999) The role of sex hormones in systemic lupus erythematosus. Curr Opin Rheumatol 11:352–356
9. Voulgari PV, Katsimbri P, Alamanos Y, Drosos AA (2002) Gender and age differences in systemic lupus erythematosus. A study of 489 Greek patients with a review of the literature. Lupus 11:722–729
10. Soto ME, Vallejo M, Guillen F, Simon JA, Arena E, Reyes PA (2004) Gender impact in systemic lupus erythematosus. Clin Exp Rheumatol 22:713–721
11. Tucker LB, Menon S, Isenberg DA (1995) Systemic lupus in children: daughter of the Hydra? Lupus 4:83–85
12. Mak SK, Lam EK, Wong AK (1998) Clinical profile of patients with late-onset SLE: not a benign subgroup. Lupus 7:23–28
13. Baker SB, Rovira JR, Campion EW, Mills JA (1979) Late onset systemic lupus erythematosus. Am J Med 66:727–732
14. Hochberg MC (1985) The incidence of systemic lupus erythematosus in Baltimore, Maryland, 1970–1977. Arthritis Rheum 28:80–86
15. Ward MM, Studenski S (1990) Clinical manifestations of systemic lupus erythematosus. Identification of racial and socioeconomic influences. Arch Intern Med 150:849–853
16. Deapen D, Escalante A, Weinrib L, Horwitz D, Bachman B, Roy-Burman P, Walker A, Mack TM (1992) A revised estimate of twin concordance in systemic lupus erythematosus. Arthritis Rheum 35:311–318
17. Werth VP (2005) Clinical manifestations of cutaneous lupus erythematosus. Autoimmun Rev 4:296–302
18. James JA, Neas BR, Moser KL, Hall T, Bruner GR, Sestak AL, Harley JB (2001) Systemic lupus erythematosus in adults is associated with previous Epstein-Barr virus exposure. Arthritis Rheum 44:1122–1126
19. Hochberg M (1997) Updating the American College of Rheumatology Revised criteria for the classification of systemic lupus erythematosus. Arthritis Rheum 40:1725
20. Cervera R, Khamashta MA, Font J, Sebastiani GD, Gil A, Lavilla P, Domenech I, Aydintug AO, Jedryka-Goral A, de Ramon E et al (1993) Systemic lupus erythematosus: clinical and immunologic patterns of disease expression in a cohort of 1, 000 patients. The European Working Party on Systemic Lupus Erythematosus. Medicine (Baltimore) 72:113–124
21. Pons-Estel BA, Catoggio LJ, Cardiel MH, Soriano ER, Gentiletti S, Villa AR, Abadi I, Caeiro F, Alvarellos A, Alarcon-Segovia D (2004) The GLADEL multinational Latin American prospective inception cohort of 1, 214 patients with systemic lupus erythematosus: ethnic and disease heterogeneity among "Hispanics". Medicine (Baltimore) 83:1–17
22. Boumpas DT, Fessler BJ, Austin HA III, Balow JE, Klippel JH, Lockshin MD (1995) Systemic lupus erythematosus: emerging concepts. Part 2: dermatologic and joint disease, the antiphospholipid antibody syndrome, pregnancy and hormonal therapy, morbidity and mortality, and pathogenesis. Ann Intern Med 123:42–53
23. Boumpas DT, Austin HA III, Fessler BJ, Balow JE, Klippel JH, Lockshin MD (1995) Systemic lupus erythematosus: emerging concepts. Part 1: renal, neuropsychiatric, cardiovascular, pulmonary, and hematologic disease. Ann Intern Med 122:940–950
24. Al-Maini M, Urowitz M (2007) Systemic steroids. In: Tsokos GC, Gordon C, Smolen JS (eds) Systemic lupus erythematosus: a companion to Rheumatology, 1st edn. Mosby Inc, Philadelphia, pp 487–498
25. Papadimitraki E, Boumpas DD (2007) Cytotoxic drug treatment. In: Tsokos GC, Gordon C, Smolen JS (eds) Systemic lupus erythematosus: a companion to Rheumatology, 1st edn. Mosby Inc, Philadelphia, pp 498–510
26. Ginzler EM, Dooley MA, Aranow C, Kim MY, Buyon J, Merrill JT, Petri M, Gilkeson GS, Wallace DJ, Weisman MH, Appel GB (2005) Mycophenolate mofetil or intravenous cyclophosphamide for lupus nephritis. N Engl J Med 353:2219–2228
27. Houssiau FA, Ginzler EM (2008) Current treatment of lupus nephritis. Lupus 17:426–430
28. Sherer Y, Gorstein A, Fritzler MJ, Shoenfeld Y (2004) Autoantibody explosion in systemic lupus erythematosus: more than 100 different antibodies found in SLE patients. Semin Arthritis Rheum 34:501–537

29. Rahman A, Hiepe F (2002) Anti-DNA antibodies – overview of assays and clinical correlations. Lupus 11:770–773
30. Arbuckle MR, McClain MT, Rubertone MV, Scofield RH, Dennis GJ, James JA, Harley JB (2003) Development of autoantibodies before the clinical onset of systemic lupus erythematosus. N Engl J Med 349:1526–1533
31. Illei GG, Lipsky PE (2004) Biomarkers in systemic lupus erythematosus. Curr Rheumatol Rep 6:382–390
32. Jerdan MS, Hood AF, Moore GW, Callen JP (1990) Histopathologic comparison of the subsets of lupus erythematosus. Arch Dermatol 126:52–55
33. Tsao BP (2004) Update on human systemic lupus erythematosus genetics. Curr Opin Rheumatol 16:513–521
34. Graham RR, Ortmann WA, Langefeld CD, Jawaheer D, Selby SA, Rodine PR, Baechler EC, Rohlf KE, Shark KB, Espe KJ, Green LE, Nair RP, Stuart PE, Elder JT, King RA, Moser KL, Gaffney PM, Bugawan TL, Erlich HA, Rich SS, Gregersen PK, Behrens TW (2002) Visualizing human leukocyte antigen class II risk haplotypes in human systemic lupus erythematosus. Am J Hum Genet 71:543–553
35. Galeazzi M, Sebastiani GD, Morozzi G, Carcassi C, Ferrara GB, Scorza R, Cervera R, de Ramon Garrido E, Fernandez-Nebro A, Houssiau F, Jedryka-Goral A, Passiu G, Papasteriades C, Piette JC, Smolen J, Porciello G, Marcolongo R (2002) HLA class II DNA typing in a large series of European patients with systemic lupus erythematosus: correlations with clinical and autoantibody subsets. Medicine (Baltimore) 81:169–178
36. Rajagopalan S, Zordan T, Tsokos GC, Datta SK (1990) Pathogenic anti-DNA autoantibody-inducing T helper cell lines from patients with active lupus nephritis: isolation of CD4-8- T helper cell lines that express the gamma delta T-cell antigen receptor. Proc Natl Acad Sci USA 87:7020–7024
37. Shivakumar S, Tsokos GC, Datta SK (1989) T cell receptor alpha/beta expressing double-negative (CD4-/CD8-) and CD4+ T helper cells in humans augment the production of pathogenic anti-DNA autoantibodies associated with lupus nephritis. J Immunol 143:103–112
38. Liossis SN, Ding XZ, Dennis GJ, Tsokos GC (1998) Altered pattern of TCR/CD3-mediated protein-tyrosyl phosphorylation in T cells from patients with systemic lupus erythematosus. Deficient expression of the T cell receptor zeta chain. J Clin Invest 101:1448–1457
39. Krishnan S, Nambiar MP, Warke VG, Fisher CU, Mitchell J, Delaney N, Tsokos GC (2004) Alterations in lipid raft composition and dynamics contribute to abnormal T cell responses in systemic lupus erythematosus. J Immunol 172:7821–7831
40. Enyedy EJ, Nambiar MP, Liossis SN, Dennis G, Kammer GM, Tsokos GC (2001) Fc epsilon receptor type I gamma chain replaces the deficient T cell receptor zeta chain in T cells of patients with systemic lupus erythematosus. Arthritis Rheum 44:1114–1121
41. Tsokos GC, Nambiar MP, Tenbrock K, Juang YT (2003) Rewiring the T-cell: signaling defects and novel prospects for the treatment of SLE. Trends Immunol 24:259–263
42. Krishnan S, Warke VG, Nambiar MP, Tsokos GC, Farber DL (2003) The FcR gamma subunit and Syk kinase replace the CD3 zeta-chain and ZAP-70 kinase in the TCR signaling complex of human effector CD4 T cells. J Immunol 170:4189–4195
43. Juang YT, Tenbrock K, Nambiar MP, Gourley MF, Tsokos GC (2002) Defective production of functional 98-kDa form of Elf-1 is responsible for the decreased expression of TCR zeta-chain in patients with systemic lupus erythematosus. J Immunol 169:6048–6055
44. Chowdhury B, Tsokos CG, Krishnan S, Robertson J, Fisher CU, Warke RG, Warke VG, Nambiar MP, Tsokos GC (2005) Decreased stability and translation of T cell receptor zeta mRNA with an alternatively spliced 3′-untranslated region contribute to zeta chain down-regulation in patients with systemic lupus erythematosus. J Biol Chem 280:18959–18966
45. Tsuzaka K, Setoyama Y, Yoshimoto K, Shiraishi K, Suzuki K, Abe T, Takeuchi T (2005) A splice variant of the TCR zeta mRNA lacking exon 7 leads to the down-regulation of TCR zeta, the TCR/CD3 complex, and IL-2 production in systemic lupus erythematosus T cells. J Immunol 174:3518–3525
46. Linker-Israeli M, Bakke AC, Kitridou RC, Gendler S, Gillis S, Horwitz DA (1983) Defective production of interleukin 1 and interleukin 2 in patients with systemic lupus erythematosus (SLE). J Immunol 130: 2651–2655
47. Valencia X, Yarboro C, Illei G, Lipsky PE (2007) Deficient CD4+CD25high T regulatory cell function in patients with active systemic lupus erythematosus. J Immunol 178:2579–2588
48. Miyara M, Amoura Z, Parizot C, Badoual C, Dorgham K, Trad S, Nochy D, Debre P,

Piette JC, Gorochov G (2005) Global natural regulatory T cell depletion in active systemic lupus erythematosus. J Immunol 175: 8392–8400

49. Herndon TM, Juang YT, Solomou EE, Rothwell SW, Gourley MF, Tsokos GC (2002) Direct transfer of p65 into T lymphocytes from systemic lupus erythematosus patients leads to increased levels of interleukin-2 promoter activity. Clin Immunol 103:145–153
50. Kyttaris VC, Juang YT, Tenbrock K, Weinstein A, Tsokos GC (2004) Cyclic adenosine 5′-monophosphate response element modulator is responsible for the decreased expression of c-fos and activator protein-1 binding in T cells from patients with systemic lupus erythematosus. J Immunol 173:3557–3563
51. Katsiari CG, Kyttaris VC, Juang YT, Tsokos GC (2005) Protein phosphatase 2A is a negative regulator of IL-2 production in patients with systemic lupus erythematosus. J Clin Invest 115:3193–3204
52. Solomou EE, Juang YT, Gourley MF, Kammer GM, Tsokos GC (2001) Molecular basis of deficient IL-2 production in T cells from patients with systemic lupus erythematosus. J Immunol 166:4216–4222
53. Tenbrock K, Juang YT, Gourley MF, Nambiar MP, Tsokos GC (2002) Antisense cyclic adenosine 5′-monophosphate response element modulator up-regulates IL-2 in T cells from patients with systemic lupus erythematosus. J Immunol 169:4147–4152
54. Kyttaris VC, Wang Y, Juang YT, Weinstein A, Tsokos GC (2007) Increased levels of NF-ATc2 differentially regulate CD154 and IL-2 genes in t cells from patients with systemic lupus erythematosus. J Immunol 178:1960–1966
55. Desai-Mehta A, Lu L, Ramsey-Goldman R, Datta SK (1996) Hyperexpression of CD40 ligand by B and T cells in human lupus and its role in pathogenic autoantibody production. J Clin Invest 97:2063–2073
56. Xu L, Zhang L, Yi Y, Kang HK, Datta SK (2004) Human lupus T cells resist inactivation and escape death by upregulating COX-2. Nat Med 10:411–415
57. Li Y, Harada T, Juang YT, Kyttaris VC, Wang Y, Zidanic M, Tung K, Tsokos GC (2007) Phosphorylated ERM is responsible for increased T cell polarization, adhesion, and migration in patients with systemic lupus erythematosus. J Immunol 178:1938–1947
58. Crispin JC, Oukka M, Bayliss G, Cohen RA, Van Beek CA, Stillman IE, Kyttaris VC, Juang YT, Tsokos GC (2008) Expanded double negative T cells in patients with systemic lupus erythematosus produce IL-17 and infiltrate the kidneys. J Immunol 181:8761–8766
59. Baroja ML, Vijayakrishnan L, Bettelli E, Darlington PJ, Chau TA, Ling V, Collins M, Carreno BM, Madrenas J, Kuchroo VK (2002) Inhibition of CTLA-4 function by the regulatory subunit of serine/threonine phosphatase 2A. J Immunol 168:5070–5078
60. Salomon B, Bluestone JA (2001) Complexities of CD28/B7: CTLA-4 costimulatory pathways in autoimmunity and transplantation. Annu Rev Immunol 19:225–252
61. Hudson LL, Rocca K, Song YW, Pandey JP (2002) CTLA-4 gene polymorphisms in systemic lupus erythematosus: a highly significant association with a determinant in the promoter region. Hum Genet 111: 452–455
62. Wong CK, Lit LCW, Tam LS, Li EK, Lam CWK (2005) Aberrant production of soluble costimulatory molecules CTLA-4, CD28, CD80 and CD86 in patients with systemic lupus erythematosus. Rheumatology (Oxford) 44:989–994
63. Cloutier JF, Veillette A (1996) Association of inhibitory tyrosine protein kinase p50csk with protein tyrosine phosphatase PEP in T cells and other hemopoietic cells. EMBO J 15:4909–4918
64. Cloutier JF, Veillette A (1999) Cooperative inhibition of T-cell antigen receptor signaling by a complex between a kinase and a phosphatase. J Exp Med 189:111–121
65. Kyogoku C, Langefeld CD, Ortmann WA, Lee A, Selby S, Carlton VE, Chang M, Ramos P, Baechler EC, Batliwalla FM, Novitzke J, Williams AH, Gillett C, Rodine P, Graham RR, Ardlie KG, Gaffney PM, Moser KL, Petri M, Begovich AB, Gregersen PK, Behrens TW (2004) Genetic association of the R620W polymorphism of protein tyrosine phosphatase PTPN22 with human SLE. Am J Hum Genet 75:504–507
66. Sims GP, Ettinger R, Shirota Y, Yarboro CH, Illei GG, Lipsky PE (2005) Identification and characterization of circulating human transitional B cells. Blood 105:4390–4398
67. Yurasov S, Wardemann H, Hammersen J, Tsuiji M, Meffre E, Pascual V, Nussenzweig MC (2005) Defective B cell tolerance checkpoints in systemic lupus erythematosus. J Exp Med 201:703–711
68. Hom G, Graham RR, Modrek B, Taylor KE, Ortmann W, Garnier S, Lee AT, Chung SA, Ferreira RC, Pant PV, Ballinger DG, Kosoy R, Demirci FY, Kamboh MI, Kao AH, Tian C,

Gunnarsson I, Bengtsson AA, Rantapaa-Dahlqvist S, Petri M, Manzi S, Seldin MF, Ronnblom L, Syvanen AC, Criswell LA, Gregersen PK, Behrens TW (2008) Association of Systemic Lupus Erythematosus with C8orf13-BLK and ITGAM-ITGAX. N Engl J Med 358:900–909
69. Tretter T, Ross AE, Dordai DI, Desiderio S (2003) Mimicry of pre-B cell receptor signaling by activation of the tyrosine kinase Blk. J Exp Med 198:1863–1873
70. Liossis SN, Kovacs B, Dennis G, Kammer GM, Tsokos GC (1996) B cells from patients with systemic lupus erythematosus display abnormal antigen receptor-mediated early signal transduction events. J Clin Invest 98:2549–2557
71. Enyedy EJ, Mitchell JP, Nambiar MP, Tsokos GC (2001) Defective FcgammaRIIb1 signaling contributes to enhanced calcium response in B cells from patients with systemic lupus erythematosus. Clin Immunol 101:130–135
72. Mackay M, Stanevsky A, Wang T, Aranow C, Li M, Koenig S, Ravetch JV, Diamond B (2006) Selective dysregulation of the Fc{gamma}IIB receptor on memory B cells in SLE. J Exp Med 203:2157–2164
73. Juang YT, Wang Y, Solomou EE, Li Y, Mawrin C, Tenbrock K, Kyttaris VC, Tsokos GC (2005) Systemic lupus erythematosus serum IgG increases CREM binding to the IL-2 promoter and suppresses IL-2 production through CaMKIV. J Clin Invest 115:996–1005
74. Mevorach D, Mascarenhas JO, Gershov D, Elkon KB (1998) Complement-dependent clearance of apoptotic cells by human macrophages. J Exp Med 188:2313–2320
75. Manderson AP, Botto M, Walport MJ (2004) The role of complement in the development of systemic lupus erythematosus. Annu Rev Immunol 22:431–456
76. Ghebrehiwet B, Peerschke EI (2004) Role of C1q and C1q receptors in the pathogenesis of systemic lupus erythematosus. Curr Dir Autoimmun 7:87–97
77. Slingsby JH, Norsworthy P, Pearce G, Vaishnaw AK, Issler H, Morley BJ, Walport MJ (1996) Homozygous hereditary C1q deficiency and systemic lupus erythematosus. A new family and the molecular basis of C1q deficiency in three families. Arthritis Rheum 39:663–670
78. Senaldi G, Davies ET, Peakman M, Vergani D, Lu J, Reid KB (1995) Frequency of mannose-binding protein deficiency in patients with systemic lupus erythematosus. Arthritis Rheum 38:1713–1714
79. Sullivan KE, Wooten C, Goldman D, Petri M (1996) Mannose-binding protein genetic polymorphisms in black patients with systemic lupus erythematosus. Arthritis Rheum 39:2046–2051
80. Iliopoulos AG, Tsokos GC (1996) Immunopathogenesis and spectrum of infections in systemic lupus erythematosus. Semin Arthritis Rheum 25:318–336
81. Bharadwaj D, Stein MP, Volzer M, Mold C, Du Clos TW (1999) The major receptor for C-reactive protein on leukocytes is fcgamma receptor II. J Exp Med 190:585–590
82. Salmon JE, Pricop L (2001) Human receptors for immunoglobulin G: key elements in the pathogenesis of rheumatic disease. Arthritis Rheum 44:739–750
83. Karassa FB, Trikalinos TA, Ioannidis JP (2002) Role of the Fcgamma receptor IIa polymorphism in susceptibility to systemic lupus erythematosus and lupus nephritis: a meta-analysis. Arthritis Rheum 46: 1563–1571
84. Karassa FB, Trikalinos TA, Ioannidis JP (2003) The Fc gamma RIIIA-F158 allele is a risk factor for the development of lupus nephritis: a meta-analysis. Kidney Int 63:1475–1482
85. Russell AI, Cunninghame Graham DS, Shepherd C, Roberton CA, Whittaker J, Meeks J, Powell RJ, Isenberg DA, Walport MJ, Vyse TJ (2004) Polymorphism at the C-reactive protein locus influences gene expression and predisposes to systemic lupus erythematosus. Hum Mol Genet 13: 137–147
86. Yasutomo K, Horiuchi T, Kagami S, Tsukamoto H, Hashimura C, Urushihara M, Kuroda Y (2001) Mutation of DNASE1 in people with systemic lupus erythematosus. Nat Genet 28:313–314
87. Kyttaris VC, Juang YT, Tsokos GC (2005) Immune cells and cytokines in systemic lupus erythematosus: an update. Curr Opin Rheumatol 17:518–522
88. Wilson AG, Gordon C, di Giovine FS, de Vries N, van de Putte LB, Emery P, Duff GW (1994) A genetic association between systemic lupus erythematosus and tumor necrosis factor alpha. Eur J Immunol 24:191–195
89. D'Alfonso S, Rampi M, Bocchio D, Colombo G, Scorza-Smeraldi R, Momigliano-Richardi P (2000) Systemic lupus erythematosus candidate genes in the Italian population: evidence for a significant association with interleukin-10. Arthritis Rheum 43:120–128

90. Morita C, Horiuchi T, Tsukamoto H, Hatta N, Kikuchi Y, Arinobu Y, Otsuka T, Sawabe T, Harashima S, Nagasawa K, Niho Y (2001) Association of tumor necrosis factor receptor type II polymorphism 196R with Systemic lupus erythematosus in the Japanese: molecular and functional analysis. Arthritis Rheum 44:2819–2827
91. Kirou KA, Lee C, George S, Louca K, Peterson MG, Crow MK (2005) Activation of the interferon-alpha pathway identifies a subgroup of systemic lupus erythematosus patients with distinct serologic features and active disease. Arthritis Rheum 52: 1491–1503
92. Remmers EF, Plenge RM, Lee AT, Graham RR, Hom G, Behrens TW, de Bakker PIW, Le JM, Lee H-S, Batliwalla F, Li W, Masters SL, Booty MG, Carulli JP, Padyukov L, Alfredsson L, Klareskog L, Chen WV, Amos CI, Criswell LA, Seldin MF, Kastner DL, Gregersen PK (2007) STAT4 and the risk of rheumatoid arthritis and systemic lupus erythematosus. N Engl J Med 357:977–986
93. Graham RR, Kozyrev SV, Baechler EC, Reddy MV, Plenge RM, Bauer JW, Ortmann WA, Koeuth T, Gonzalez Escribano MF, Pons-Estel B, Petri M, Daly M, Gregersen PK, Martin J, Altshuler D, Behrens TW, Alarcon-Riquelme ME (2006) A common haplotype of interferon regulatory factor 5 (IRF5) regulates splicing and expression and is associated with increased risk of systemic lupus erythematosus. Nat Genet 38:550–555
94. Takaoka A, Yanai H, Kondo S, Duncan G, Negishi H, Mizutani T, Kano S, Honda K, Ohba Y, Mak TW, Taniguchi T (2005) Integral role of IRF-5 in the gene induction programme activated by Toll-like receptors. Nature 434:243–249

Chapter 14

Systems Biology of Influenza: Understanding Multidimensional Interactions for Personalized Prevention and Treatment

Qing Yan

Abstract

Influenza virus infection is a public health threat worldwide. It is urgent to develop effective methods and tools for the prevention and treatment of influenza. Influenza vaccines have significant immune response variability across the population. Most of the current circulating strains of influenza A virus are resistant to anti-influenza drugs. It is necessary to understand how genetic variants affect immune responses, especially responses to the HA and NA transmembrane glycoproteins. The elucidation of the underlying mechanisms can help identify patient subgroups for effective prevention and treatment. New personalized vaccines, adjuvants, and drugs may result from the understanding of interactions of host genetic, environmental, and other factors. The systems biology approach is to simulate and model large networks of the interacting components, which can be excellent targets for antiviral therapies. The elucidation of host–influenza interactions may provide an integrative view of virus infection and host responses. Understanding the host–influenza–drug interactions may contribute to optimal drug combination therapies. Insight of the host–influenza–vaccine interactions, especially the immunogenetics of vaccine response, may lead to the development of better vaccines. Systemic studies of host–virus–vaccine–drug–environment interactions will enable predictive models for therapeutic responses and the development of individualized therapeutic strategies. A database containing such information on personalized and systems medicine for influenza is available at http://flu.pharmtao.com.

Key words: Influenza, Viruses, Systems biology, Personalized medicine, Pathways, Interactions, Host–pathogen, Vaccines, Drugs, Immune, Infection, Antiviral, Prevention, Treatment, Systems medicine

1. Introduction

Influenza virus infection is a public health threat worldwide. Each year in the U.S., an average of about 36,000 people die from influenza, and more than 200,000 people are hospitalized as a

Qing Yan (ed.), *Systems Biology in Drug Discovery and Development: Methods and Protocols*, Methods in Molecular Biology, vol. 662, DOI 10.1007/978-1-60761-800-3_14, © Springer Science+Business Media, LLC 2010

result of influenza (1). The influenza pandemic in 1918 caused at least 675,000 U.S. deaths and up to 50 million deaths worldwide (2). The World Health Organization (WHO) phase of pandemic alert in 2009 is six, which indicates that a global pandemic is under way (3). It is urgent to develop effective methods and tools for the prevention and treatment of influenza.

Many challenges need to be solved in order to develop better therapeutic strategies for influenza. These challenges include drug resistance, viral divergence and antigenic shifts, and the time lag in vaccine production. It has been suggested that "multidisciplinary and coordinated efforts by healthcare workers and scientists around the world" are needed to solve these problems (4).

Furthermore, as a clinician noted, "not every vaccine is equally safe or equally effective in every person." (5) Personalized and safer vaccines and drugs are needed for achieving better clinical outcomes. As the basis of personalized medicine, pharmacogenomics is an emerging area that studies variations in patient responses to therapeutics. However, many problems need to be resolved before pharmacogenomics can be applied in the clinic, especially the understanding of biomarkers, pathways, and interactions between genes and therapeutics through systems biology studies (6).

Because of highly diversified genetics and multiple phases of the infectious process, systems biology studies in influenza virus are especially complicated. In addition, the two organisms involved in the infection, the virus and the host, and host–influenza interactions add more dimensions to the complexity. The integration of such multidimensional information at the systems level is important for understanding viral infections and for analyzing the efficacies of vaccines or antiviral drugs.

As shown in Table 1, various bioinformatics sources are available for systems biology studies of influenza. These sources include the Influenza Virus Resource at National Center for Biotechnology Information (NCBI), Influenza Research Database (IRD), and Influenza Primer Design Resource. Relevant information from immunoinformatics resources such as the international ImMunoGeneTics information system (IMGT) is helpful for analyses of immune epitopes, responses, and vaccine design. IDPM is a database containing systematic and updated information on personalized prevention and treatment of influenza. The following sections will briefly discuss current studies of influenza systems biology from different aspects, as well as their applications in vaccine and drug development.

Table 1
Bioinformatics resources for influenza systems biology studies

Name	URL[a]	Content
Influenza virus resource	http://www.ncbi.nih.gov/genomes/FLU/FLU.html	Influenza sequence database and tools
Influenza Research Database (IRD)	http://www.fludb.org/brc/home.do?decorator=influenza	Database of sequences, structure, epitopes, phenotypes
Influenza Primer Design Resource (IPDR)	http://www.ipdr.mcw.edu	Influenza sequence information relevant to diagnostics
Reactome	http://www.reactome.org/cgi-bin/eventbrowser?DB=gk_current&ID=168254&ZOOM=2	Influenza infection pathways
KEGG	http://www.genome.jp/kegg/pathway.html	Pathway analysis
VirusMINT	http://mint.bio.uniroma2.it/virusmint/graph.do	Interactions between human and viral proteins
IMGT	http://imgt.cines.fr/	Immunogenetics information system
Influenza Database for Personalized Medicine (IDPM)	http://flu.pharmtao.com/	Information on personalized and systemic prevention and treatment of influenza

[a]Websites were accessed in July 2009

2. Genotype – Phenotype Correlations and Personalized Medicine

Genotype-phenotype correlations play a crucial role in the translation of pharmacogenomics into clinical personalized medicine (7). Here phenotype is defined as visible traits, such as clinical measurements. The association between virus genotypes and infectious phenotypic features such as infectivity has been studied extensively and covered by some databases such as IRD (see Table 1). The elucidation of how human genetic variants influence vaccine or drug response phenotypes can help identify patient subgroups for individualized prevention and therapy.

Influenza viruses are enveloped viruses in the *Orthomyxoviridae* family. There are three types of influenza viruses, A, B, and C. Influenza A and B viruses are major causes of epidemics in humans. The A and B viruses contain a genome of eight negative-stranded RNA segments. The eight segments in influenza A viruses encode proteins including haemagglutinin (HA), matrix protein (M), neuraminidase (NA), nucleoprotein (NP), nonstructural protein

(NS), polymerase basic 1 protein (PB1), polymerase basic 2 protein (PB2), and polymerase acidic protein (PA).

For influenza prevention and treatment, it is essential to understand how genetic variants affect immune responses, especially responses to the HA and NA transmembrane glycoproteins (8). HA and NA are the major components of inactivated influenza vaccines. NA is also the target of antiviral drugs such as zanamivir and oseltamivir (4).

For example, human leucocyte antigen (HLA) polymorphisms are important contributors to vaccine-induced immune responses. The HLA-DRB1*0701 allele was over-expressed among persons who failed to mount a neutralizing antibody response (9). Such mechanisms are critical for the identification of patient subgroups that may not be protected by current vaccination strategies or may be resistant to drugs. The differentiation of such patient subgroups based on genotype–phenotype correlations may establish the foundation of personalized therapeutic strategies.

3. Host–Influenza Interactions

At this time, the detailed mechanistic examination of host–pathogen systems is still in its infancy (10). Classical virology has mostly focused on the virus itself and somewhat ignored many complex processes from the host cells (11). However, new personalized vaccines, adjuvants, and drugs may result from the understanding of interactions of host genetic, environmental, and other factors that control immune responses (8). The systems biology approach is to simulate and model large networks of the interacting components, which can be excellent targets for antiviral therapies.

Host responses to viral infections at various systems levels such as cellular, tissue, and organ levels are important for drug target identification. For example, toll-like receptor (TLR) signaling pathways play a central role in mediating the antiviral and inflammatory responses to viral infections (12). TLR-3 is expressed in various cells and tissues, including myeloid dendritic cells, macrophages, and alveolar and bronchial epithelial cells. TLR-3 agonists have been suggested as potential drugs to protect against lethal seasonal influenza virus infections. In another example, the binding of influenza NS1 protein to host molecules may result in the inhibition of host mRNA processing and inhibition of interferon synthesis. Counteractions of such inhibitions of host's antiviral mechanisms are important for the development of therapeutics.

Host intracellular molecules can interact with influenza and play significant roles in influenza life cycle. These molecules can be essential targets for the development of antiviral therapeutics. Table 2 lists some human molecules that interact with influenza A viruses. Tables 3–5 summarize how these molecules may interact with the virus and the results of the interactions. Detailed descriptions and examples of such interactions and pathways at various levels will be given in the following sections. For instance, influenza infections can induce different molecular and cellular responses in different cell types, including airway epithelium cells and immune system cells. Potential therapeutics can be designed based on the understanding of these interactions. A database containing relevant information is available at http://flu.pharmtao.com.

3.1. Influenza NS1 Protein Binding and Inhibition of Antiviral Activities

The NS1 protein of influenza is a nuclear, dimeric protein expressed abundantly in infected cells. It is a virulence factor that can reverse cellular antiviral activities. Its N-terminal RNA-binding domain can bind double strand RNA (dsRNA) (13), which represents the signal for virus infection. Such dsRNA-binding activity plays a crucial role in influenza–host interactions (14), which may lead to a series of inhibition. These include the inhibition of host mRNA processing, inhibition of interferon synthesis, and inhibition of protein kinase R (PKR, or Eukaryotic translation initiation factor 2-alpha kinase 2) (see Table 3).

During the inhibition of host mRNA processing, NS1 binds to the host cell's cleavage and host polyadenylation specificity factor 4 (CPSF4). CPSF4 is essential for the 3′ end processing mechanism of cellular pre-mRNAs. The NS1-CPSF4 binding blocks efficient 3′-end processing, preventing the export of host cell mRNAs out of the nucleus. Such interaction is largely responsible for the posttranscriptional inhibition of the processing of the cellular antiviral pre-mRNAs (15).

The NS1 molecule can also interact with the host cell's poly(A)-binding protein II (PABII) and prevent PABII from properly extending the poly-A tail of pre-mRNA in the host cell nucleus. This binding stops pre-mRNAs from exiting the nucleus. Both CPSF4 and PABII proteins bind to nonoverlapping regions of the NS1A protein effector domain (16).

The amino-terminal region of the NS1 protein is critical in blocking the induction of beta interferon (IFN-beta) in virus-infected cells. The IFN antagonist property depends on the ability of NS1 to bind dsRNA (17). This interaction may partially stop host IFN synthesis through inhibiting the 2′–5′ oligo (A) synthetase/RNase L pathway, which would otherwise be activated by intracellular dsRNA (13).

The latent protein kinase PKR, also called EIF2AK2, is an important element of the cellular antiviral system. It is activated by binding to either dsRNA or the cellular PACT

Table 2
A sample list of human molecules that interact with influenza A viruses

Gene symbol	Full name	Chromosome location	Gene family
BCL2 (Bcl-2)	B-cell CLL/lymphoma 2	18q21.3	
CASP1 (caspase 1)	Caspase 1, apoptosis-related cysteine peptidase (interleukin 1, beta, convertase)	11q23	Cysteine-aspartic acid protease (caspase) family
CDC42	Cell division cycle 42	1p36.1	GTPase of the Rho-subfamily
CPSF4	Cleavage and polyadenylation specific factor 4	7q22.1	
DEFA1 (HNP-1)	Defensin, alpha 1	8p23.1	A family of microbicidal and cytotoxic peptides involved in host defense, abundant in neutrophils
DMBT1 (gp-340)	Deleted in malignant brain tumors 1	10q25.3-q26.1	
EIF2AK2 (PKR)	Eukaryotic translation initiation factor 2-alpha kinase 2	2p22-p21	
GSS (GSH)	Glutathione synthetase	20q11.2	
HLA-A	Major histocompatibility complex, class I, A	6p21.3	HLA class I heavy chain paralogues
HLA-B	Major histocompatibility complex, class I, B	6p21.3	HLA class I heavy chain paralogues
IFNA (IFN-alpha)	Interferon, alpha	9p22	Interferon
IFNB (IFN-beta)	Interferon, beta, fibroblast	9p21	Interferon
IKBKB (IKK)	Inhibitor of kappa light polypeptide gene enhancer in B-cells, kinase beta	8p11.2	NFKB complex
IL8	Interleukin 8	4q13-q21	CXC chemokine family
KIR2DL1	Killer cell immunoglobulin-like receptor, two domains, long cytoplasmic tail, 1	19q13.4	Killer cell Ig-like receptors, or KIRs
LILRB1	Leukocyte immunoglobulin-like receptor, subfamily B (with TM and ITIM domains), member 1	19q13.4	Leukocyte immunoglobulin-like receptor

(continued)

Table 2 (continued)

Gene symbol	Full name	Chromosome location	Gene family
MAPK8IP3 (JNK/ SAPK-associated protein-1)	Mitogen-activated protein kinase 8 interacting protein 3	16p13.3	
MIF	Macrophage migration inhibitory factor (glycosylation-inhibiting factor)	22q11.23	Lymphokine
MX1 (MxA)	Myxovirus (influenza virus) resistance 1	21q22.3	GTPases
NFKB1	Nuclear factor of kappa light polypeptide gene enhancer in B-cells	4q24	NF-kappa-B
PABII	PolyA-binding protein II	19q13.41	
SFTPD (SP-D)	Surfactant, pulmonary-associated protein D	10q22.2-q23.1	Surfactant protein
SLC25A6 (ANT3)	Solute carrier family 25 (mitochondrial carrier; adenine nucleotide translocator), member 6	Xp22.32 and Yp11.3	Solute carrier family
TLR3	Toll-like receptor 3	4q35	Toll-like receptor (TLR) family which plays a fundamental role in pathogen recognition and activation of innate immunity
TNF	Tumor necrosis factor (TNF superfamily, member 2)	6p21.3	Tumor necrosis factor (TNF) superfamily
TNFSF10 (TRAIL)	Tumor necrosis factor (ligand) superfamily, member 10	3q26	Tumor necrosis factor (TNF) ligand family
TXN (TRX)	Thioredoxin	9q31	Oxidoreductase enzyme
VDAC1	Voltage-dependent anion channel 1	5q31	Anion channel
VPS28	Vacuolar protein sorting 28 homolog	8q24.3	A component of the ESCRT-I complex
XDH	Xanthine dehydrogenase	2p23.1	Molybdenum-containing hydroxylases involved in the oxidative metabolism of purines

Table 3
Molecules and interactions involved in influenza NS1 protein binding and inhibition of antiviral activities

Interactive molecules	Interactions	Effects	References
CPSF4	NS1-CPSF4 binding blocks 3′-end processing, stops the export of host mRNAs out of the nucleus	Inhibition of the host antiviral pre-mRNAs processing	(15)
PABII	NS1-PABII binding prevents PABII from extending the poly-A tail of pre-mRNA in the nucleus, stops the export of pre-mRNAs out of the nucleus	Inhibition of the host antiviral pre-mRNAs processing	(16)
dsRNA, IFN-beta	NS1-dsRNA binding inhibits the 2′–5′ oligo (A) synthetase/RNase L pathway	Inhibition of the host IFN-beta synthesis and antiviral pathways	(13, 17)
PKR	NS1A-PKR binding stops phosphorylation of eIF2	Inhibition of PKR activation and antiviral pathways	(18)
PI3K/Akt, caspase 9, glycogen synthase-kinase 3beta	NS1 activates PI3K/Akt pathway, and influences caspase 9 and glycogen synthase-kinase 3beta	Virus replication	(19)
Caspase-1 (in primary macrophages)	Mutant viruses with altered NS1 provoke caspase-1 activation	Fast apoptosis, release of interleukins 1beta and 18	(30)

(protein activator of PKR) protein. Activated PKR in turn phosphorylates the translation initiation factor eIF2 and prevents viral and cellular protein synthesis and virus replication (18). However, the direct binding of the NS1A protein to the N-terminal 230 amino acid region of PKR may inhibit PKR activation by PACT and dsRNA (18).

Furthermore, the NS1 protein may also cause the activation of the phosphatidylinositol 3-kinase (PI3K)/Akt pathway and influence caspase 9 and glycogen synthase-kinase 3beta. Such interactions may affect viral replication (19). Through such mechanism, NS1 not only stops but also stimulates signaling pathways that enable efficient virus replication.

These NS1 interactions can inhibit the antiviral processes of the host cells (see Table 3). Understanding of such NS1 binding mechanisms is important for locating targets for the development of antiviral therapeutics. For example, RNA oligonucleotides targeting at the H5N1 avian influenza virus (AIV) NS1 gene was found to inhibit the virus reproduction (20). RNA oligonucleotides may have potentials as prophylaxis and therapy for H5N1 influenza virus infection in humans.

3.2. Host Molecules Involved in the Influenza Virus Life Cycle

The replication of influenza A virus relies on certain intracellular pathways in host cells. Host intracellular molecules may interact with influenza virus proteins and participate in the influenza virus life cycle. The influenza virion enters the host cell through endocytosis, fuses to the host cell endosome, and becomes uncoated (14). Viral ribonucleoproteins (RNP) can be transported into the host nucleus, and viral RNA can be synthesized and replicated. After exporting viral RNP from the nucleus, components of the virus are assembled, packaged, and released.

Many host molecules have significant roles in these processes and may become potential antiviral targets (see Table 4). The identification of these molecules and their roles in the host–influenza interactions may help with the discovery of better vaccines or drugs. For example, vacuolar protein sorting 28 homolog (VPS28) is a component of the endosomal sorting complex required for transport (ESCRT)-I. Cell division cycle 42 (Cdc42) is an element that belongs to the Rho family GTP-binding

Table 4
Host molecules and interactions involved in influenza life cycle and apoptosis

Interactive molecules	Interactions	Effects	References
VPS28, Cdc42	Viral M1 YRKL motif interacts with VPS28 and Cdc42	Inhibition of decrease of influenza production	(21)
GSH, Bcl-2, MAPK	Apoptosis pathways	Persistent viral infection	(22, 23)
XDH	Be changed to xanthine oxidase	Xanthinuria, adult respiratory stress syndrome, potentiate influenza infection	(24)
TGF-beta	Viral NA, M1 and M2 activate TGF-beta	Induction of apoptosis	(26, 27)
ANT3, VDAC1	Viral PB1-F2 binds ANT3, VDAC1		(28, 29)
TRAIL (in MDMs)	Cytotoxicity, enhanced sensitization to DRLs		(31)
JNK/SAPK	The JNK/SAPK cascade modifies c-Jun/AP-1 and cytokine TGF-beta		(32)
MxA	Nuclear MxA interacts with viral PB2 and NP		(33–35)
CXCL8/ IL-8 (in primary lung epithelial cells)	Influenza infection triggers the release of CXCL8/ IL-8	Induction of cell necrosis	(36)

proteins. Both of the two molecules, VPS28 and Cdc42 can interact with influenza M1 protein through the YRKL motif, the L domain motif in the influenza virus. This interaction is important in the influenza virus life cycle, as the depletion of VPS28 and Cdc42 can result in the decrease of influenza virus production (21).

Glutathione (GSH) and Bcl-2 are components of antioxidant and apoptosis pathways and may also be crucial for influenza A virus replication. Bcl-2 expression and GSH are associated with persistent viral infection, although they have effects at different stages of the viral life cycle (22). In addition, the cell type being targeted by the virus may partly decide the viral pathological effects. For example, the influence of p38 mitogen-activated protein kinases (MAPK) on the influenza life cycle and the host apoptotic response may depend on if the cells express Bcl-2 (23). Such mechanisms need to be considered during the development of antiviral therapeutics.

In another example, the enzyme xanthine dehydrogenase (XDH) is a component of the group of molybdenum-containing hydroxylases associated with the oxidative metabolism of purines (24). XDH can be changed to xanthine oxidase by reversible sulfhydryl oxidation or by irreversible proteolytic modification. Altered XDH may result in xanthinuria and lead to adult respiratory stress syndrome. Such changes may potentiate influenza virus infection through an oxygen metabolite-dependent mechanism.

3.3. Influenza and Apoptosis Pathways

After the infection, influenza A virus may induce apoptosis in host cells, such as lymphocytes and monocytes. This may be the mechanism of influenza to destroy the human immune defense and cause susceptibility to a secondary infection (25).

The process of apoptosis may start with the activation of host transforming growth factor beta (TGF-beta) through the expression of viral NA, M1 and M2 proteins (26) (see Table 4). M2 integral membrane protein may stop autophagosome maturation and affect host cell apoptosis (27). Another mechanism is the binding of viral PB1-F2 (28) with host molecules, including mitochondrial adenine nucleotide translocator 3 (ANT3) and voltage-dependent anion channel 1 (VDAC1) (29).

Different cell types may have different apoptosis pathways. For instance, in infected primary human macrophages, mutant viruses with altered NS1 may provoke caspase-1 activation. Such changes may lead to fast apoptosis and the production of high levels of interleukins 1beta and 18 (30).

In infected human monocyte-derived macrophages (MDMs), functional tumor necrosis factor (TNF)-related apoptosis-inducing ligand (TRAIL) can be produced. This molecule is associated with the cytotoxicity, and the enhanced sensitization to death receptor ligands (DRLs)-induced apoptosis upon avian influenza virus infection (31).

Influenza infection may also go through stress-activated pathways to start the c-Jun N-terminal kinase/stress-activated protein kinase (JNK/SAPK) cascade. The JNK/SAPK cascade can modify the activity of apoptosis-promoting regulatory factor c-Jun/AP-1 (activator protein 1) and cytokine TGF-beta (32).

Another pathway involves Mx proteins, which are in the dynamin superfamily of high molecular weight GTPases. Human MxA is an interferon- alpha/beta (IFN-alpha/beta)-inducible protein and has effects on cellular functions including the apoptotic pathway. The C-terminal and N-terminal regions of MxA may be involved in the promotion of cell death (33, 34). Furthermore, nuclear MxA can influence the influenza virus transcription by interacting with viral proteins PB2 and NP (35).

As mentioned above, the cell type rather than the virus determines which pathway will be followed. For instance, in monocytes and epithelial cells from origins other than the lung, influenza A virus infection may induce apoptosis. However, in primary lung epithelial cells, influenza infection can trigger the release of CXCL8/interleukin-8 (IL-8) and cause cell necrosis (36). Moreover, there is a massive release of macrophage migration inhibitory factor (MIF) from virus-infected lung cells. Increased levels of MIF may lead to the host immune response during the acute phase of influenza A virus infection in humans.

In primary human natural killer (NK) cells, influenza viruses can enter the cells via clathrin- and caveolin-dependent endocytosis through the sialic acids on cell surfaces (37). The virus infection may induce apoptosis of NK cells, escape the NK cell innate immune defense, and lead to virus pathogenesis.

3.4. Influenza Infection-Induced Host Responses in Airway Epithelia Cells

As discussed before, in different cell types, influenza infection can trigger different host responses. Elucidation of such different mechanisms can help understand human responses at the systems level. For example, in airway epithelial cells, host molecules that may participate in antiviral responses include glycoprotein-340 (gp-340), toll-like receptor 3 (TLR3), IFN-alpha, TNF-alpha, IkappaB kinase (IKK), and IL8 (see Table 5).

For instance, during influenza A virus infection, bronchiolar epithelial cells are the prime targets. The lung scavenger receptor-rich protein glycoprotein-340 (gp-340) (also called DMBT1) exists in bronchoalveolar lavage (BAL) fluids and saliva. This protein can stop the HA activity and infectivity of influenza A viruses, and agglutinate the virions (38). The underlying mechanisms of such antiviral effects are noncalcium-dependent interactions between the virus and sialic acid-bearing carbohydrates on the gp-340 protein.

Table 5
Host molecules and interactions involved in inflammation and immune responses

Interactive molecules	Interactions	Effects	References
Gp-340 (in bronchiolar epithelial cells)	Interactions between the virus and sialic acid-bearing carbohydrates on gp-340	Inhibition of HA and viral infectivity, and agglutination of the virions	(38)
TLR3, MAPK, phosphatidylinositol 3-kinase/Akt; TRIF, NF-kappaB; interferon-beta; IL-8, IL-6, RANTES, ICAM-1	Influenza-TLR3	Induction of immune responses	(39)
TRX	Influenza infection	Inhibition of the inflammatory overshoot of viral pneumonia	(40)
IFN-alpha, TNF-alpha	Influenza infection	Excessive inflammation, febrile seizures	(41, 42)
IKK (in pulmonary epithelium)	Influenza infection → IKK → activation of NF-kappaB	Inflammation	(43)
IL-10 (in T cells)	Influenza infection	Controls inflammation	(44)
HLA-A and –B (in CTL)	Influenza infection	Influences on CTL responses	(45)
NK receptors (in NK cells), KIR2DL1, LILRB1	Influenza infection	Influences on the killing of target cells	(46)
SP-D (neutrophils), HNPs; SP-A, lung glycoprotein-340, mucin	Influenza infection	Influences on viral neutralization, opsonization, virus uptake	(48–50)

The TLR3 protein contributes directly to the immune responses in respiratory epithelial cells to influenza A viruses (see Table 5). One of the mechanisms in these responses involves mitogen-activated protein kinases (MAPK) and phosphatidylinositol 3-kinase/Akt signaling (39). Another mechanism involves the TLR3-associated adaptor molecule, TIR-domain-containing adapter-inducing interferon-β (TRIF) activation of the transcription factors NF-kappaB.

Other networks, such as interferon regulatory factor/interferon-sensitive response-element pathways are also involved (39). In addition, TLR3 is associated with the secretion of the cytokines IL-8, IL-6, RANTES (regulated on activation normal T cell expressed and secreted), and interferon-beta. It is involved in the up-regulation of the major adhesion molecule ICAM-1 (39). As mentioned before, TLR3 pathways are important targets for the development of anti-influenza therapeutics.

3.5. Influenza Infection, Host Responses, and Inflammation

Inflammation is an important process in host–influenza interactions (see Table 5). For example, thioredoxin (TRX, also called TXN) is a small redox-active protein with antioxidant features and redox-regulating functions. Over-expression of TRX in TRX transgenic (Tg) mice has been found to inhibit the inflammatory overshoot of viral pneumonia caused by influenza virus infection. This may lead to the reduction of mortality without affecting the host's systemic immune responses to the infection. Therefore, TRX may be important in regulating the inflammatory process in the primary host defense against infection (40).

In influenza pathogenesis, excessive inflammation is critical. Excessive inflammation may be caused by overabundant production of proinflammatory cytokines from airway epithelial cells. Increased levels of serum IFN-alpha are correlated with febrile seizures in influenza (41). Moreover, IFN-alpha and TNF-alpha have significant roles in priming epithelial cells for higher cytokine and chemokine production in influenza A virus infection (42).

Another important factor in activating influenza-induced inflammatory reactions in pulmonary epithelium is IkappaB kinase (IKK, also called IKBKB) (see Table 5). The activation of IKK by influenza infection leads to persistent activation of nuclear factor-kappaB (NF-kappaB), a key regulator of the inflammatory response (43).

During acute virus infection, antiviral CD8+ and CD4+ effector T cells in the infected periphery have an anti-inflammatory property (44). The T cells can regulate the extent of lung inflammation and injury caused by influenza infection through producing anti-inflammatory cytokines such as interleukin-10 (IL-10).

3.6. Influenza Infection-Induced Host Responses in Immune Cells

Influenza infection can lead to a variety of responses in host immune system cells (see Table 5). Such immune responses are crucial in the design of anti-influenza therapeutics. For instance, the repertoire of human cytotoxic T-lymphocytes (CTL) in response to influenza A viruses are directed toward multiple epitopes. The magnitude and specificity of CTL responses in humans are associated with HLA-A and -B phenotypes (45).

Innate immune cells such as natural killer (NK) cells, alveolar macrophages, and dendritic cells are essential after influenza A infection (46). This line of immune defense controls viral replication directly and regulates virus-specific adaptive immune responses. For example, the cytotoxic activity of NK cells is regulated through both inhibitory and activating NK receptors. Alterations in the expression levels and in the affinity or avidity of those receptors may affect the killing of target cells.

Upon the influenza infection, the binding of NK-inhibitory receptors is increased, which involves the generation of major histocompatibility complex (MHC) class I complexes in infected cells. The increased binding may happen in both the killer cell

Ig-like receptor 2 domain long tail 1 (KIR2DL1) and leukocyte Ig-like receptor-1 (LILRB1) (47).

The collectin, surfactant protein D (SP-D), binds to a variety of pathogens through its carbohydrate recognition domain (see Table 5). SP-D plays important roles in innate host defense against influenza A virus infection, partly by modifying interactions with neutrophils. Multimerization mediated by the N-peptide may be important for viral neutralization and opsonization, while the collagen domain may affect the antiviral activity of multimerized forms of SP-D (48).

Interactions during the early phase of host defense against influenza A virus are done through a complex interplay between SP-D and human neutrophil defensins (HNPs) at sites of active inflammation (49). However, such ability of SP-D to increase neutrophil uptake of influenza A virus can be dissociated from enhancement of oxidant responses. Some innate immune proteins that bind to SP-D, such as SP-A, lung glycoprotein-340 or mucin, can significantly decrease the ability of SP-D to promote neutrophil oxidant response. Such effects may result in the increase of neutrophil uptake of influenza A virus, and the reduction of the respiratory burst response to virus (50).

With more studies done about the host–influenza interactions, more molecules and pathways involved in these processes will be elucidated, and a more systemic picture will be able to be drawn. Antiviral therapeutics developed based on the understanding of these mechanisms should provide safer and more effective treatment for the diseases.

4. Host–Influenza–Vaccine–Drug Interactions and Optimal Therapeutic Strategies

Knowledge of host–virus–vaccine–drug–environment interactions will enable predictive models for interaction networks and therapeutic responses (see Fig. 1). Such multidimensional interactions include network interrelationships among host–virus–vaccine, host–virus–drug, host–drug–drug, and host–drug–vaccine. The study of the network interactions may lead to the development of the optimal treatment strategies.

Currently, most of the circulating strains of influenza A virus are resistant to anti-influenza drugs, including the adamantanes and neuraminidase inhibitors. It is urgent to develop novel therapeutic strategies that can be rapidly utilized to address the resistance issue. Understanding of the host–influenza–drug and host–drug–drug interactions may contribute to the selection of drug combination therapies for effective treatments. Drug combination therapies may enable the efficient applications of the existing drug supplies.

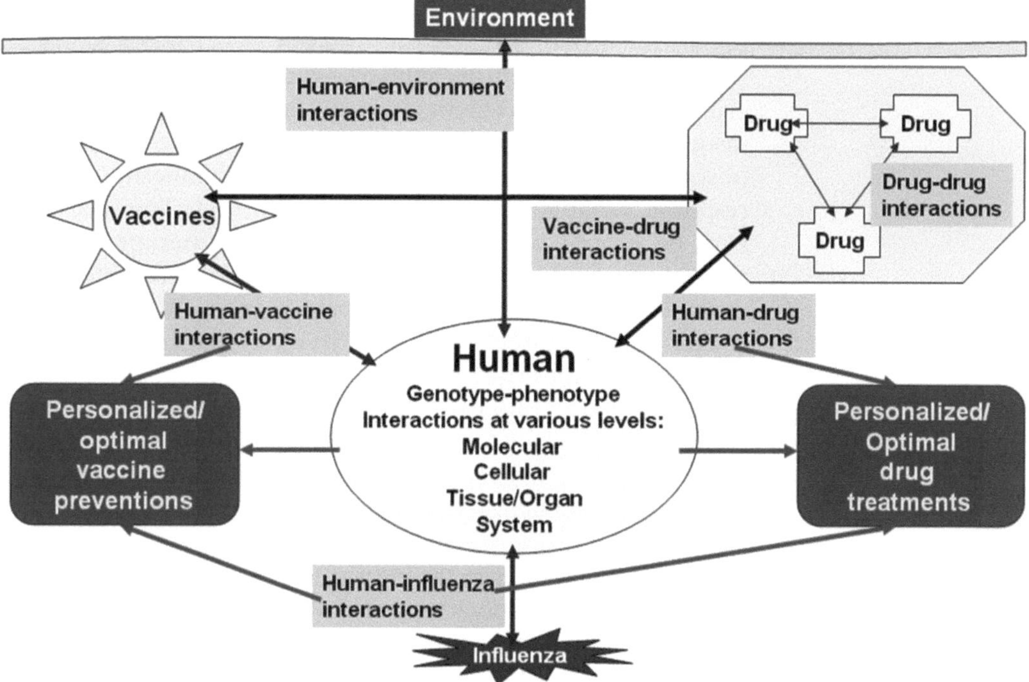

Fig. 1. The information flow in systemic interactions toward personalized medicine for influenza.

For instance, a triple combination of antivirals oseltamivir, amantadine, and ribavirin may be highly synergistic against influenza A virus (51). The synergy of the triple combination was found to be 2- to 13-fold greater than the synergy of any double combinations. Such combination therapies can be used as effective treatments for both seasonal and pandemic influenza viruses.

In another example, the NA inhibitor oseltamivir and the inhibitor of influenza virus polymerases ribavirin can be combined to have better results than being used alone (52). Different dosages and combinations of the two drugs may have different effects on different virus strains. The optimal combinations prevented the spread of H5N1 viruses beyond the respiratory tract and abrogated cytokine responses including interleukin-1α (IL-1α). Understanding of such multidimensional interactions can help decide the optimal drug dosages and combinations to achieve the best possible outcomes.

Host genetic variations may influence drug responses through host–influenza–drug interactions. Such interactions are usually meaningful for avoiding drug adverse reactions and for the selection of patient subgroups for the optimized treatment. For instance, a nonsynonymous single nucleotide polymorphism (SNP) R41Q in human cytosolic sialidase in a small Asian population may cause reduced enzyme activity. Such effects may have potential associations with severe adverse reactions to oseltamivir in this population group (53).

Studies of the host–influenza–vaccine interactions, especially the immunogenetics of vaccine response, may offer insights for the development of better vaccines. It has been observed that influenza vaccines have significant immune response variability across the population (8). This variability may be caused by the polymorphisms of immune response genes including HLA, cytokines, and cytokine receptors. For example, the –1082 allele polymorphism in the IL-10 promoter region may be associated with adverse responses induced by influenza vaccines (54).

Furthermore, information on influenza drug–vaccine interactions is important for avoiding certain adverse reactions and side effects. For instance, in some cases, warfarin therapy may interact with influenza vaccination and cause increased anticoagulation (55).

Figure 1 summarizes the information flow in the multidimensional and systemic interactions that need to be considered for the development of personalized antiviral therapeutics. Several major entities are involved in the interactions, including the human body, influenza virus, vaccine, drug, and the environment.

These entities interact with each other to form an interwoven network. The information of human–influenza–vaccine–drug interactions, together with the genotype–phenotype correlation and interactions at different systems levels will contribute to the development of personalized and systems medicine for influenza.

References

1. Center for Disease Control and Prevention (CDC). Available at: http://www.cdc.gov/flu/about/disease/index.htm. Accessed July 2009
2. http://flu.gov. Accessed July 2009
3. World Health Organization (WHO). Available at: http://www.who.int/csr/disease/avian_influenza/phase/en/index.html. Accessed July 2009
4. Memoli MJ, Morens DM, Taubenberger JK (2008) Pandemic and seasonal influenza: therapeutic challenges. Drug Discov Today 13:590–595
5. Brady MT (2006) Immunization recommendations for children with metabolic disorders: more data would help. Pediatrics 118:810–813
6. Yan Q (2008) Pharmacogenomics in drug discovery and development. Preface. Methods Mol Biol 448:v–vii
7. Yan Q (2005) Pharmacogenomics and systems biology of membrane transporters. Mol Biotechnol 29:75–88
8. Poland GA, Ovsyannikova IG, Jacobson RM (2009) Application of pharmacogenomics to vaccines. Pharmacogenomics 10:837–852
9. Lambkin R, Novelli P, Oxford J et al (2004) Human genetics and responses to influenza vaccination: clinical implications. Am J Pharmacogenomics 4:293–298
10. Forst CV (2006) Host-pathogen systems biology. Drug Discov Today 11:220–227
11. Damm EM, Pelkmans L (2006) Systems biology of virus entry in mammalian cells. Cell Microbiol 8:1219–1227
12. Wong JP, Christopher ME, Viswanathan S et al (2009) Activation of toll-like receptor signaling pathway for protection against influenza virus infection. Vaccine 27:3481–3483

13. Min JY, Krug RM (2006) The primary function of RNA binding by the influenza A virus NS1 protein in infected cells: Inhibiting the 2′–5′ oligo (A) synthetase/RNase L pathway. Proc Natl Acad Sci USA 103:7100–7105
14. KrugRM,LambRA(2001)Orthomyxoviridae: the viruses and their replication. In: Knipe DM, Howley PM (eds) Fields virology, 4th edn. Lippincott Williams & Wilkins, Philadelphia
15. Noah DL, Twu KY, Krug RM (2003) Cellular antiviral responses against influenza A virus are countered at the posttranscriptional level by the viral NS1A protein via its binding to a cellular protein required for the 3′ end processing of cellular pre-mRNAS. Virology 307:386–395
16. Chen Z, Li Y, Krug RM (1999) Influenza A virus NS1 protein targets poly(A)-binding protein II of the cellular 3′-end processing machinery. EMBO J 18:2273–2283
17. Donelan NR, Basler CF, Garcia-Sastre A (2003) A recombinant influenza A virus expressing an RNA-binding-defective NS1 protein induces high levels of beta interferon and is attenuated in mice. J Virol 77:13257–13266
18. Li S, Min JY, Krug RM et al (2006) Binding of the influenza A virus NS1 protein to PKR mediates the inhibition of its activation by either PACT or double-stranded RNA. Virology 349:13–21
19. Ehrhardt C, Wolff T, Pleschka S et al (2007) Influenza A virus NS1 protein activates the PI3K/Akt pathway to mediate antiapoptotic signaling responses. J Virol 81:3058–3067
20. Wu Y, Zhang G, Li Y et al (2008) Inhibition of highly pathogenic avian H5N1 influenza virus replication by RNA oligonucleotides targeting NS1 gene. Biochem Biophys Res Commun 365:369–374
21. Hui EK, Barman S, Tang DH, France B, Nayak DP (2006) YRKL sequence of influenza virus M1 functions as the L domain motif and interacts with VPS28 and Cdc42. J Virol 80:2291–2308
22. Nencioni L, Iuvara A, Aquilano K et al (2003) Influenza A virus replication is dependent on an antioxidant pathway that involves GSH and Bcl-2. FASEB J 17:758–760
23. Nencioni L, De Chiara G, Sgarbanti R et al (2009) Bcl-2 expression and p38MAPK activity in cells infected with influenza A virus: impact on virally induced apoptosis and viral replication. J Biol Chem 284:16004–16015
24. Akaike T, Ando M, Oda T et al (1990) Dependence on O2- generation by xanthine oxidase of pathogenesis of influenza virus infection in mice. J Clin Invest 85:739–745
25. Xie D, Bai H, Liu L et al (2009) Apoptosis of lymphocytes and monocytes infected with influenza virus might be the mechanism of combating virus and causing secondary infection by influenza. Int Immunol 21:1251–1262
26. Morris SJ, Smith H, Sweet C (2002) Exploitation of the Herpes simplex virus translocating protein VP22 to carry influenza virus proteins into cells for studies of apoptosis: direct confirmation that neuraminidase induces apoptosis and indications that other proteins may have a role. Arch Virol 147:961–979
27. Rossman JS, Lamb RA (2009) Autophagy, apoptosis, and the influenza virus M2 protein. Cell Host Microbe 6:299–300
28. Chanturiya AN, Basanez G, Schubert U et al (2004) PB1-F2, an influenza A virus-encoded proapoptotic mitochondrial protein, creates variably sized pores in planar lipid membranes. J Virol 78:6304–6312
29. Zamarin D, Garcia-Sastre A, Xiao X et al (2005) Influenza virus PB1-F2 protein induces cell death through mitochondrial ANT3 and VDAC1. PLoS Pathog 1:e4
30. Stasakova J, Ferko B, Kittel C et al (2005) Influenza A mutant viruses with altered NS1 protein function provoke caspase-1 activation in primary human macrophages, resulting in fast apoptosis and release of high levels of interleukins 1beta and 18. J Gen Virol 86:185–195
31. Zhou J, Law HK, Cheung CY et al (2006) Functional tumor necrosis factor-related apoptosis-inducing ligand production by avian influenza virus-infected macrophages. J Infect Dis 193:945–953
32. Lin C, Zimmer SG, Lu Z et al (2001) The involvement of a stress-activated pathway in equine influenza virus-mediated apoptosis. Virology 287:202–213
33. Numajiri A, Mibayashi M, Nagata K (2006) Stimulus-dependent and domain-dependent cell death acceleration by an IFN-inducible protein, human MxA. J Interferon Cytokine Res 26:214–219
34. Mibayashi M, Nakad K, Nagata K (2002) Promoted cell death of cells expressing human MxA by influenza virus infection. Microbiol Immunol 46:29–36
35. Turan K, Mibayashi M, Sugiyama K et al (2004) Nuclear MxA proteins form a complex with influenza virus NP and inhibit the transcription of the engineered influenza virus genome. Nucleic Acids Res 32:643–652

36. Arndt U, Wennemuth G, Barth P et al (2002) Release of macrophage migration inhibitory factor and CXCL8/interleukin-8 from lung epithelial cells rendered necrotic by influenza A virus infection. J Virol 76:9298–9306
37. Mao H, Tu W, Qin G et al (2009) Influenza virus directly infects human natural killer cells and induces cell apoptosis. J Virol 83:9215–9222
38. Hartshorn KL, White MR, Mogues T et al (2003) Lung and salivary scavenger receptor glycoprotein-340 contribute to the host defense against influenza A viruses. Am J Physiol Lung Cell Mol Physiol 285:L1066–L1076
39. Guillot L, Le Goffic R, Bloch S et al (2005) Involvement of toll-like receptor 3 in the immune response of lung epithelial cells to double-stranded RNA and influenza A virus. J Biol Chem 280:5571–5580
40. Nakamura H, Tamura S, Watanabe I et al (2002) Enhanced resistancy of thioredoxin-transgenic mice against influenza virus-induced pneumonia. Immunol Lett 82:165–170
41. Masuyama T, Matsuo M, Ichimaru T et al (2002) Possible contribution of interferon-alpha to febrile seizures in influenza. Pediatr Neurol 27:289–292
42. Veckman V, Osterlund P, Fagerlund R et al (2006) TNF-alpha and IFN-alpha enhance influenza-A-virus-induced chemokine gene expression in human A549 lung epithelial cells. Virology 345:96–104
43. Bernasconi D, Amici C, La Frazia S et al (2005) The IkappaB kinase is a key factor in triggering influenza A virus-induced inflammatory cytokine production in airway epithelial cells. J Biol Chem 280:24127–24134
44. Sun J, Madan R, Karp CL et al (2009) Effector T cells control lung inflammation during acute influenza virus infection by producing IL-10. Nat Med 15:277–284
45. Boon AC, de Mutsert G, Graus YM et al (2002) The magnitude and specificity of influenza A virus-specific cytotoxic T-lymphocyte responses in humans is related to HLA-A and -B phenotype. J Virol 76:582–590
46. McGill J, Heusel JW, Legge KL (2009) Innate immune control and regulation of influenza virus infections. J Leukoc Biol 86:803–812
47. Achdout H, Arnon TI, Markel G et al (2003) Enhanced recognition of human NK receptors after influenza virus infection. J Immunol 171:915–923
48. White M, Kingma P, Tecle T et al (2008) Multimerization of surfactant protein D, but not its collagen domain, is required for antiviral and opsonic activities related to influenza virus. J Immunol 181:7936–7943
49. Hartshorn KL, White MR, Tecle T et al (2006) Innate defense against influenza A virus: activity of human neutrophil defensins and interactions of defensins with surfactant protein D. J Immunol 176:6962–6972
50. White MR, Crouch E, Vesona J et al (2005) Respiratory innate immune proteins differentially modulate the neutrophil respiratory burst response to influenza A virus. Am J Physiol Lung Cell Mol Physiol 289:L606–L616
51. Nguyen JT, Hoopes JD, Smee DF et al (2009) Triple combination of oseltamivir, amantadine, and ribavirin displays synergistic activity against multiple influenza virus strains in vitro. Antimicrob Agents Chemother 53:4115–4126
52. Ilyushina NA, Hay A, Yilmaz N et al (2008) Oseltamivir-ribavirin combination therapy for highly pathogenic H5N1 influenza virus infection in mice. Antimicrob Agents Chemother 52:3889–3897
53. Li CY, Yu Q, Ye ZQ et al (2007) A nonsynonymous SNP in human cytosolic sialidase in a small Asian population results in reduced enzyme activity: potential link with severe adverse reactions to oseltamivir. Cell Res 17:357–362
54. Tang YW, Li H, Wu H et al (2007) Host single-nucleotide polymorphisms and altered responses to inactivated influenza vaccine. J Infect Dis 196:1021–1025
55. Carroll DN, Carroll DG (2009) Fatal intracranial bleed potentially due to a warfarin and influenza vaccine interaction. Ann Pharmacother 43:754–760

Chapter 15

Methods in Systems Biology of Experimental Methamphetamine Drug Abuse

Firas H. Kobeissy, Shankar Sadasivan, Melinda Buchanan, Zhiqun Zhang, Mark S. Gold, and Kevin K.W. Wang

Abstract

The use of methamphetamine (METH) as recreational drugs is a growing problem worldwide with recent concerns that it might cause long-lasting harmful effects to the human brain. METH is an illicit drug that is known to exert neurotoxic effects on both dopaminergic and serotonergic neural systems. Our laboratory has been studying the biochemical mechanisms underlying METH-induced neurotoxic effects both in vivo and in vitro. Our psychoproteomics METH abuse research focuses on the global alteration of cortical protein expression in rats treated with acute METH. In our analysis, an altered protein expression was identified using a multistep protein separation/proteomic platform. Differential changes of the selected proteins were further confirmed by quantitative immunoblotting. Our study identified 82 differentially expressed proteins, 40 of which were downregulated and 42 of which were upregulated post acute METH treatment. In this chapter, we describe the current protocols for the neuronal cell culture in vitro and the in vivo rat model of acute METH treatment (4×10 mg/kg) coupled with the description current bioinformatics analysis utilized to analyze the different implicated interaction protein/gene maps that reflected on the altered functions observed. These methods and protocols are discussed in the paradigm of the acute model of METH drug abuse and neuronal cell culture and can be applied on other models of substance abuse such as on MDMA or cocaine.

Key words: Neurotoxicity, Methamphetamine, Proteomics, Drug of abuse, Proteomics, Systems biology, Genomics

1. Introduction

Methamphetamine (METH) is rapidly growing as the drug of choice for abuse worldwide. This is probably due to the ease of its availability. METH abuse has increased radically in the last 20 years (1). METH has been defined as a psychostimulant with addictive potential, known to primarily affect the reward pathway

Qing Yan (ed.), *Systems Biology in Drug Discovery and Development: Methods and Protocols*, Methods in Molecular Biology, vol. 662, DOI 10.1007/978-1-60761-800-3_15, mechanisms in the brain. Acute overdose of METH has been demonstrated to be neurotoxic with acute abuse primarily affecting dopaminergic and serotonergic neurotransmission. The neurotoxic nature of METH abuse has been shown to affect the striatum and other functional regions in the brain, such as the frontal and the prefrontal cortex. Studies have also demonstrated structural abnormalities in neurons and defective protein generation in the hippocampus (the memory forming center of the brain) and the cerebellum, following Acute METH use, ultimately leading to neuronal degeneration (2–4).

Neuronal cell death following acute exposure to METH has been attributed to necrosis and apoptosis. Studies from our lab and others have demonstrated necrosis and apoptosis induction in both in vivo and in vitro experimental models (5–8). METH exposure has been demonstrated to cause an increase in neuronal intracellular calcium in the neurons resulting in the activation of calpain proteases causing necrosis. Furthermore, increases in neuronal intracellular calcium has also been postulated to cause mitochondrial and endoplasmic reticulum stress (ER) which further signal the activation of calpain and caspase proteases culminating in cell death (9–11). Recent reports by Larsen et al. and Kanthasamy et al. have suggested that acute METH exposure promotes oxidative stress resulting in the formation of autophagic bodies called autophagosomes within the cell bodies of dopamine neurons, an indication of cell stress (12, 13). The induction of autophagy and putative autophagic cell death may represent another pathway for neurons to die following METH exposure. Thus, there may exist a complicated cross talk between the apoptotic, autophagic, and the necrotic pathways, following acute METH exposure.

The application of computational techniques has provided immense promise to outline some of the complex cross talk occurring between pathways in a complete experimental system. Systems biology incorporates data from different experimental fields such as pharmacology, proteomics, biochemistry, and genetics to suggest interactive pathways and potential therapeutic avenues. By causing manipulations (mutation of genes, use of pharmacological agonists/antagonists) to perturb specific pathways in a biological system, complex potential interactions can be elucidated at the molecular level. In an attempt to provide a more comprehensive understanding of the high throughput data, advanced bioinformatics software are used to construct functional interaction maps. These maps aim at correlating biochemical, microarray, and proteomic data to provide a whole functional unit related to a specific brain disorder. These data are then integrated in a functional network map relating altered subsets of genes and/or proteins describing altered function relevant to the disorder in question.

The analysis of interaction maps predicts the different potential functions of identified proteins and/or genes with unknown physiological roles as well as identified proteins and/or genes that have been missed by experimental analysis. There are a number of software that can construct these interaction maps, such as Ingenuity Pathway Analysis, Pathway Studio™ and PathwayArchitect™. In the context of METH abuse, combining the data obtained from the in vitro and in vivo experiments, we can generate specific hypotheses to study potential targets of neurotoxicity in brain structures resulting from METH exposure. The use of systems biology approach to study METH-associated neurotoxicity can thus help in better understanding the effects not only at the biochemical level but also at the genetic level (14). A schematic of the sequential steps used in our multidimensional proteomic platform is shown in Fig. 1.

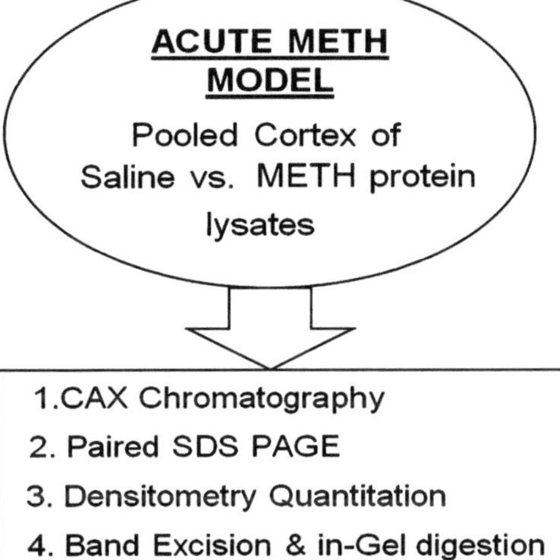

Fig. 1. A schematic illustration of the differential cation–anion exchange chromatography (CAX-PAGE) psychoproteomic platform. The schematic diagram illustrates the nine sequential steps following acute administration of METH followed by CAX chromatography and 1D-PAGE separation as the first and second dimension. After CAX-PAGE separation, selected differential protein bands are excised and in-gel digested followed by RPLC/MSMS generating a differential protein list. Selected protein subsets are then subjected for validation via immunoblotting. Subsequently, using PathwayArchitect™ software, a functional interactive map is constructed based on the psychoproteomic data.

2. Materials

2.1. Animal Model and Methamphetamine

1. Adult male Sprague–Dawley rats (Harlan, Indianapolis, IN, USA) that were aged 60 days and weighed between 250 and 275 g.
2. Pharmacologic agent (+/−) methamphetamine hydrochloride (Sigma-Aldrich, St. Louis, MO) (see Note 1).
3. 0.9% physiological saline.
4. HPLC Grade Water (HPLC grade, Burdick & Jackson, Muskegon, MI).
5. Methanol, HPLC grade (Fisher Scientific, Suwanee, GA).
6. Glacial acetic acid, HPLC grade (Fisher Scientific, Suwanee, GA).

2.2. Cortical Tissue Collection and Protein Extraction

1. Liquid Nitrogen to snap freeze the brain tissue. Dry Ice to cool mortar and pestle (Fisher Scientific, Suwanee, GA) to crush the tissue.
2. 1× SDS lysis buffer: 0.1% SDS lysis buffer containing 150 mM sodium chloride, 1 mM sodium vanadate, 1 mM dithiothreitol (DTT) (Sigma-Aldrich, St. Louis, MO). Prepare fresh and store at 4°C (see Note 2).
3. Complete Mini® protease inhibitor cocktail tablet (Roche Biochemicals, Indianapolis, IN) (see Note 3).
4. BIO-RAD DC Protein Assay (BIO-RAD Laboratories, Inc., Hercules, CA, USA).
5. Bovine serum albumin (BSA) standard: 2 mg/ml ampoules (Pierce Cat #23210).

2.3. 1D-SDS-Polyacrylamide Gel Electrophoresis (1D-SDS-PAGE)

1. BIO-RAD Molecular Weight Markers: Precision Plus® Protein All Blue Standards to determine the molecular weights of the proteins.
2. Precast 4–20% or 10–20% gradient Tris-glycine or Tricine polyacrylamide gels, 1.0 mm, 10 well (Invitrogen Life Technologies, Carlsbad, CA).
3. 10× Tris-SDS/Tricine-SDS running buffer: 100 mM Tris, pH 8.3, 100 mM Tricine, 0.1% SDS, kept at room temperature (BIO-RAD Laboratories, Inc., Hercules, CA, USA).
4. 2× Laemmli sample buffer (BIO-RAD Laboratories, Inc., Hercules, CA, USA) with 5% β mercaptoethanol (see Note 4).
5. X-Cell Sure Lock Mini Cell Apparatus (Invitrogen Life Technologies, Carlsbad, CA).
6. BIO-RAD Power PAC-3000.

2.4. Immunoblotting

1. Semi-dry transfer method in a transfer buffer (39 mM glycine, 48 mM Tris, and 5% methanol) stored at room temperature.
2. Wash buffer: 1× PBS, 0.1% Tween-20. Store at room temperature.
3. Thick pads (BIO-RAD Laboratories, Inc., Hercules, CA, USA).
4. Thin pads (Invitrogen Life Technologies, Carlsbad, CA).
5. Polyvinylidene fluoride (PVDF) membrane (Invitrogen Life Technologies, Carlsbad, CA).
6. Tris-buffered saline with Tween-20 (TBST; 20 mM Tris–HCl, 150 mM NaCl, and 0.003% Tween-20, pH 7.5; Sigma-Aldrich, St. Louis, MO).
7. Blocking buffer: 5% nonfat dry milk in TBST.
8. Primary antibodies: anti-αII-spectrin (Affiniti Research Products, Ltd., UK) anti-β actin (Sigma-Aldrich, St. Louis, MO), anti-synapsin-1 (BD Biosciences, NJ, USA), UCH-L1 (gift from Dr. Monica Oli, Banyan Biomarkers, Inc, Alachua, Florida), anti-light chain 3 (LC3) (Anti-LC3 antibody was raised in rabbits against a synthetic peptide corresponding to the N-terminal of LC3) anti-Map kinase kinase-1 (MKK-1) (Cell Signaling Technology, Beverly, MA), superoxide dismutase1 (SOD1) (gift from Dr. David Borchelt laboratory at the McKnight Brain Institute of the University of Florida, Gainesville, Fl), antiphosphatidylethanolamine-binding protein-1 (PEBP-1) (Abcam Ltd, Cambridge, UK), and anti-CRMP-2 (IBL, Japan) (see Note 5).
9. Secondary biotinylated antibodies (Amersham Biosciences, United Kingdom).
10. Streptavidin conjugated alkaline phosphatase (Amersham Biosciences, United Kingdom).
11. BIO-RAD Transblot SD Semi-Dry Transfer Cell.

2.5. Coomassie Blue Gel Staining

1. Coomassie Brilliant Blue R-250 (BIO-RAD) to stain the gels.
2. Destaining solution: 40% methanol, 50% deionized water, and 10% acetic acid.

2.6. Gel Band Visualization and Image Quantification

1. NIH ImageJ densitometry software (version 1.6, NIH, Bethesda, MD).
2. Epson Expression 8836XL high resolution flatbed scanner (Epson, Long Beach, CA).
3. SigmaStat software (Version 2.03, Systat Software Inc.).

2.7. Systems Biology

1. Ingenuity Pathway Analysis software (IPA Ingenuity Systems, Mountain View, CA).

3. Methods

3.1. In Vitro Experiments

1. Primary cerebrocortical cultures harvested from a homogenized pool of ten deeply anesthetized P1 Sprague–Dawley rat pups. Rat brain cortices are plated on culture plates coated with poly-l-lysine, similar to previously described methods (15) at a density of 3×10^6 cells/ml.
2. Cultures are maintained in high glucose-supplemented Dulbecco's modified Eagle's medium (DMEM) in a humidified incubator in an atmosphere of 10% CO_2 at 37°C.
3. Three days following culturing, the DMEM solution is replaced with high glucose DMEM containing 1% cytosine arabinoside (ARC). The ARC-DMEM solution is replaced by high glucose DMEM medium after 2 days (day 5).
4. The cells remained in culture to mature for an additional 10 days before treatment (day 15).
5. In addition to untreated controls, animals were injected with METH (1 mM and 2 mM). For pharmacological intervention, additional cultures were pretreated 1 h with either 30 mM of the calpain inhibitor SJA6017 or caspase inhibitors z-VAD-fmk or Z-D-DCB prior to METH (2 mM) challenge. Brain tissue was harvested and collected after 24 h or 48 h, posttreatment.

3.2. In Vivo Experiments

3.2.1. Animal Habituation

All procedures involving animal handling and processing were done in compliance with the Animal Welfare Act and the University of Florida Institutional Animal Care and Use Committee (IACUC) and the National Institutes of Health guidelines detailed in the Guide for the Care and use of Laboratory Animals. Adult male (280–300 g) Sprague–Dawley rats were habituated for at least 10 days prior to treatment. Animals were housed in pairs in polyethylene cages containing hardwood bedding in a temperature-controlled (approximately 22°C) room with a 12 h light: dark cycle. Animals were given access to rat chow and tap water ad libitum (see Note 6).

3.2.2. Animal Drug Injection

Following habituation, experimental groups were divided into two groups ($n = 7$), each group was injected intraperitoneally (i.p.) with either racemic METH–HCl or an equivalent volume of 0.9% saline. Rats were given 4 mg/kg intraperitoneal injections of METH at 0 h, 2 h, 4 h, and 6 h to simulate METH influence model. The saline group (vehicle group) received similar injection schedules of physiological saline. Paradigm of METH abuse is shown in Fig. 2.

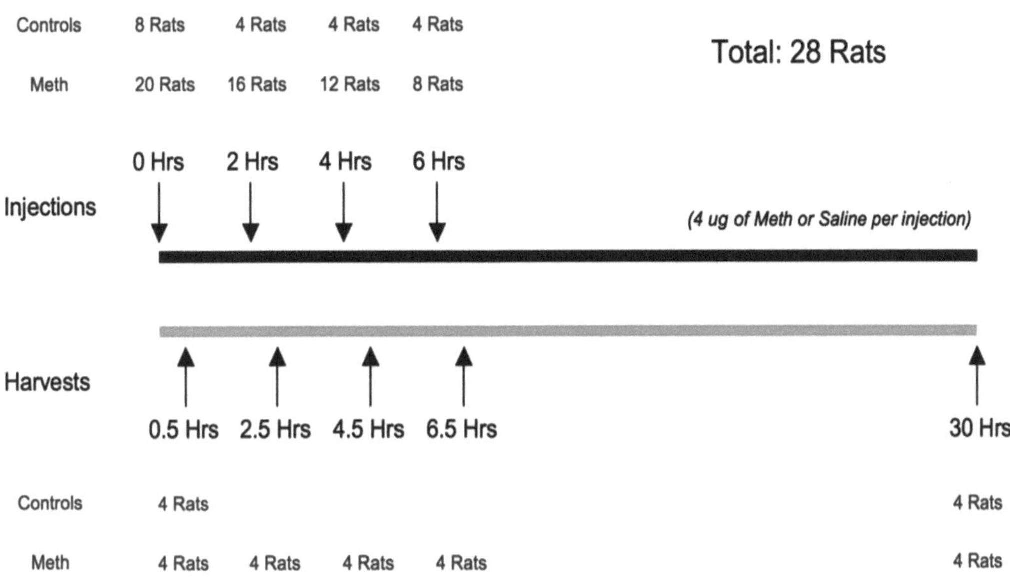

Fig. 2. Experimental design to establish acute methamphetamine model illustrating time points in which rats were injected with saline or METH and sacrificed for tissue harvesting.

3.2.3. Cortical Tissue Collection

For a successful profiling of brain psychoproteome after METH administration, special care should be taken upon harvesting the brain tissue to avoid any possible brain tissue proteolysis. At 24 h postintraperitoneal injection, treated animals were briefly anesthetized with 3–4% isoflurane and were sacrificed by decapitation. The brains are harvested from the experimental animals and snap frozen using liquid nitrogen prior to further analysis. The effective study of changes in protein profiles in the brain tissue is dependent on prompt removal and proper handling conditions.

3.3. SDS-PAGE

1. Cortical samples are adjusted to a concentration of 2 μg/μl by mixing with distilled water and then adding an equivalent volume of Novex 2× Laemmli sample buffer tricine/glycine-SDS sample buffer to achieve a concentration of 1 μg/μl.
2. Samples are heated for 2–3 min at 90°C then vortexed prior to loading.
3. Fill the BIO-RAD chamber with 1× Tris/SDS running buffer.
4. Insert the precast gels and wash the ten wells prior to sample loading (see Note 7).
5. Run the power supply unit at constant voltage (120 V) at 4°C for 2 h until the tracking dye has just migrated out from the gel (see Note 8).

3.4. Coomassie Blue Staining

1. Once the electrophoresis step is done, gels are washed three times with 100 ml distilled water (5 min each wash) (see Note 9).
2. Fixing step follows washing and this is achieved by incubating the gel for 1 h in 10% acetic acid on a shaker.
3. Staining step is achieved by incubating the gel in staining solution for 1 h on a shaker.
4. Destaining step is achieved by the periodical rinsing of the gel in destaining solution until desired background versus gel band contrast is obtained (see Note 10).
5. Gel is kept at 4°C in 10% acetic acid for Gel Band Visualization.

3.5. Gel Band Visualization and Statistical Analysis of Immunoblotting Quantification

1. For the SDS-1D PAGE visualization, protein fractions were run side-by-side on 10–20% gradient Tris–HCl gels (BIO-RAD) and were visualized with Coomassie Blue staining for differential gel bands selection.
2. NIH ImageJ densitometry software was used for lane and band detection, providing differential comparison between the saline and METH band densitometric analysis.
3. Fold increase or decrease between METH and saline samples was computed by dividing the greater value by the lesser value with a negative sign to indicate a decrease after METH treatment.
4. The densitometric quantification of the immunoblot bands was performed using an Epson Expression 8836XL high resolution flatbed scanner and NIH ImageJ software.
5. Data were acquired using integrated densitometric values and transformed to percentages of the densitometric levels.
6. Densitometry values of the four different individuals of saline and METH samples were evaluated for statistical significance using SigmaStat software using a Student's *t*-test.
7. *P*-value of <0.05 was considered to be significant for data acquired in arbitrary density units.

3.6. Immunoblotting Technique

1. For immunoblotting procedure, precast gels are opened from the two plastic covers using a scalpel; the gel is washed with distilled water.
2. PVDF membranes are soaked with 75% transfer buffer 25% methanol for 30 s.
3. Thick and thin pads are soaked in 1× transfer buffer.
4. Prepare the semi-dry transfer unit to achieve a sandwich format as follows:

5. Place the sandwich format as follows: a thin membrane followed by a thick pad, PVDF membrane, precast gel, thick pad, and finally add a thin pad.
6. Use a pencil of a pipette to push any air bubbles within the pads facilitating a uniform protein transfer onto the membrane.
7. Run the transfer unit for 2 h at 20 V at room temperature.
8. Following the transfer, the membrane is blocked in 5% nonfat dry milk in TBST (20 mM Tris–HCl, 150 mM NaCl, and 0.003% Tween-20, pH 7.5) for an hour at room temperature on a shaker platform (see Note 11).
9. After blocking, incubate the membrane overnight with the primary antibody at 4°C diluted at the proper concentrations on a shaker platform (see Note 12).
10. On the following day, the membranes were washed with excess TBST, three times, for 5 min each.
11. Membranes are probed with the secondary antibody for an hour at room temperature on a shaker platform.
12. Primary antibodies were used at a dilution of 1:1,000 in 5% milk.
13. Secondary biotinylated antibodies (Amersham Biosciences, United Kingdom) were used at a dilution of 1:3,000 in 5% milk.
14. Discard the secondary antibodies and wash the membranes with excess TBST, three times, for 5 min each.
15. Immunoreactivity was detected by using streptavidin-conjugated alkaline phosphatase.
16. Streptavidin-conjugated alkaline phosphatase were used at a dilution of 1:3,000 in 5% milk. Example of one immunoblot of identified upregulated proteins is shown in Fig. 3.

3.7. NeuroSystems Biology Analysis

1. To explore gene–gene interactions and functional modules of interest from the experimental data sets, we used the Ingenuity Pathway Analysis software (IPA Ingenuity Systems, Mountain View, CA). The IPA uses the Ingenuity Pathways Knowledge Base, which is a curated database of biological networks consisting of millions of individually modeled, peer-reviewed pathway relationships. From our data set of 1,743 significant ($p < 0.005$) probe sets, fold changes were calculated for each of the time points in relationship to the 0.5 h Saline sacrifices. This list of fold changes was uploaded to the IPA database, and bio functions known to be affected by METH use, for example, neuronal disease, degeneration, and cell death were selected for analysis as shown in Fig. 4. Genes from these bio

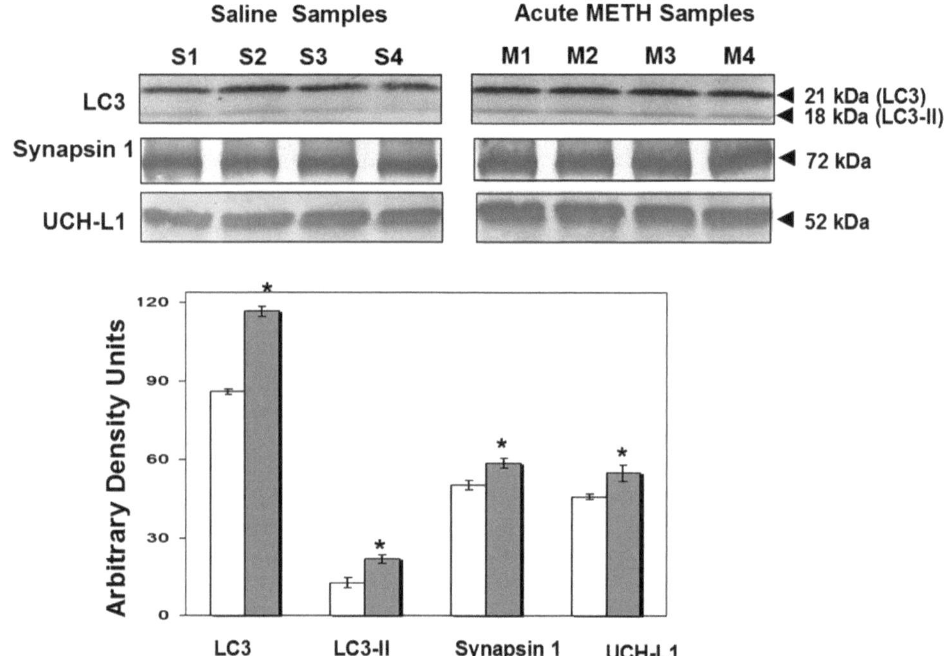

Fig. 3. Immunoblotting validation of acute METH Proteins in individual saline and METH cortex. Immunoblot analysis of intact UCHL-1 (52 kDa), synapsin-1 (72 kDa) and intact LC3 (21 kDa) and LC3-II (18 kDa) proteins from four cortical METH-treated samples and saline control samples ($n = 4$). These immunoblots show higher protein abundance in acute METH samples compared to the saline samples. A graphical densitometric analysis shows elevated proteins (UCH-L1, synapsin-1, LC3, and LC3-II) in METH-treated samples compared to the saline controls. Saline (S) samples are represented by *white bars* and acute METH (M) samples are represented by *grey bars*.

functions that were present in our data set were able to be illustrated as an interacting Network for each of the fold change time points. Blue shades represented a lower level of expression relative to the 0.5 h Saline expression level, and red shades represented a higher level of expression relative to the 0.5 h Saline expression level. Examples of functional network maps relating altered subsets of genes and/or proteins describing the altered function relevant to the disorder in question are shown in Fig. 5a, b.

4. Notes

1. (+/−) METH hydrochloride can be prepared in saline and kept at 4°C, you can make a stock and dilute it for other concentrations.
2. DTT is prepared fresh for the lysis buffer.
3. Protease inhibitor cocktail tablet should be prepared freshly every time you prepare the lysis buffer.

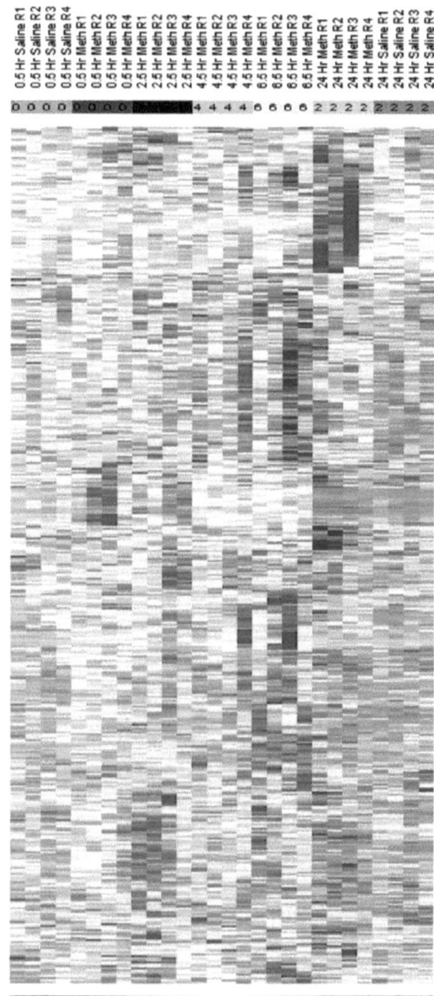

Select Bio Function Molecules from f-test: 429 Total

Cell Death: 202 Molecules
Cell Death of Neurons: 37 Molecules
Cell Death of Leukocytes: 32 Molecules
Apoptosis: 179 Molecules
Apoptosis of Neurons: 25 Molecules
Survival of Neurons: 17 Molecules
Signaling Pathway: 248 Molecules
Calcium Signaling: 21 Molecules
Circadian Rhythm Signaling: 7 Molecules
IL-6 Signaling: 12 Molecules
Endoplasmic Reticulum Stress Pathway: 5
Metabolic Disorder: 70
Autoimmune Disease: 53
Immunological Disorder: 60
Neurological Disorder: 83
Degeneration of Neurons: 13
Inflammatory Disorder: 79
Circadian Rhythm: 11
Behavior: 38
Neurological Process of Rats: 10
Activation of Neurons: 8
Immune Response: 51
Inflammatory Response: 36
Cell Movement of Leukocytes: 34
Degeneration of Cells: 16
Endoplasmic Reticulum Stress Response of Cells: 8
Anxiety of Rodents: 8
Experimentally Induced Diabetes: 21 Molecules

Fig. 4. Time series analysis of differentially expressed genes following METH or Saline injections. One thousand seven hundred and forty three probe sets whose hybridization signal intensities showed significant ($p<0.005$) differences between the METH and sham groups of rats by *f*-test using BRB array tools. The columns are labeled either saline or METH, with the time of when the rat brain was harvested: 0.5 h, 2.5 h, 4.5 h, and 6.5 h. The exceptions are the final eight columns where 24 h corresponds to a harvest time 24 h after the final injection at 6 h (total 30 h).

4. β-mercaptoethanol is added freshly always and make sure you are working under a hood.
5. For using primary antibodies, try to do trial runs to check what concentrations are optimal for your protein.
6. For animal housing, bedding was removed during the challenge phase of METH treatment because preliminary experiments have shown that high doses of METH cause rats to ingest excessive amounts of the bedding. This contributed to adverse health effects and/or mortality.

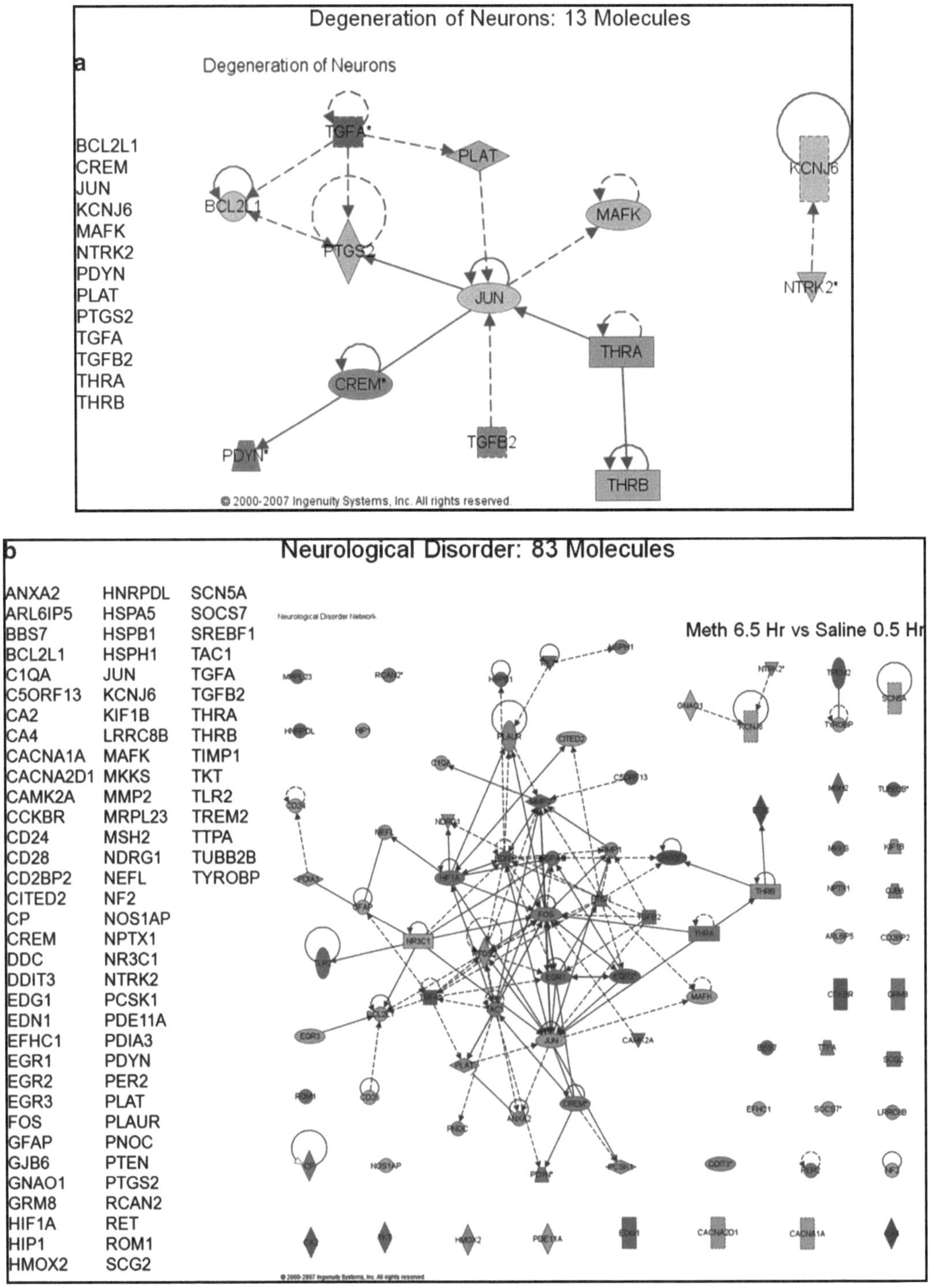

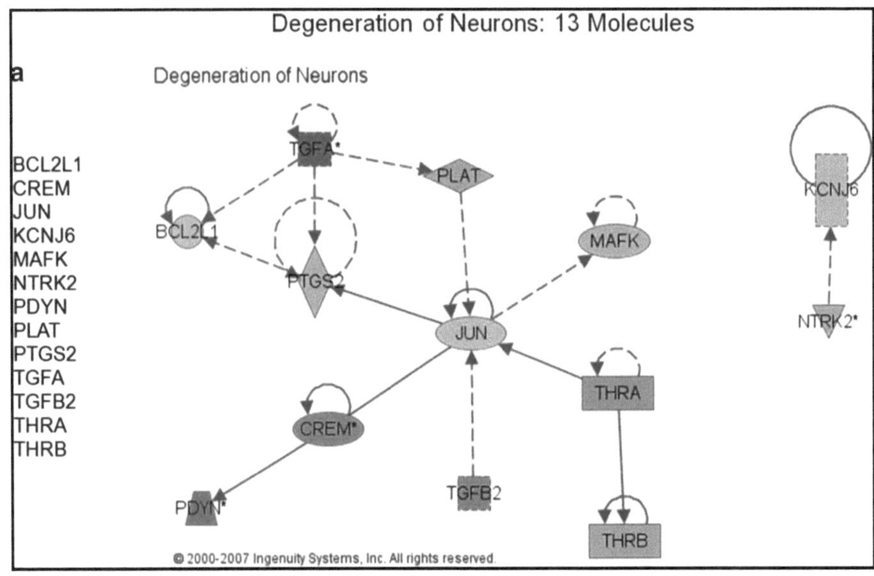

b

Neurological Disorder: 83 Molecules

Meth 6.5 Hr vs Saline 0.5 Hr

ANXA2
ARL6IP5
BBS7
BCL2L1
C1QA
C5ORF13
CA2
CA4
CACNA1A
CACNA2D1
CAMK2A
CCKBR
CD24
CD28
CD2BP2
CITED2
CP
CREM
DDC
DDIT3
EDG1
EDN1
EFHC1
EGR1
EGR2
EGR3
FOS
GFAP
GJB6
GNAO1
GRM8
HIF1A
HIP1
HMOX2
HNRPDL
HSPA5
HSPB1
HSPH1
JUN
KCNJ6
KIF1B
LRRC8B
MAFK
MKKS
MMP2
MRPL23
MSH2
NDRG1
NEFL
NF2
NOS1AP
NPTX1
NR3C1
NTRK2
PCSK1
PDE11A
PDIA3
PDYN
PER2
PLAT
PLAUR
PNOC
PTEN
PTGS2
RCAN2
RET
ROM1
SCG2
SCN5A
SOCS7
SREBF1
TAC1
TGFA
TGFB2
THRA
THRB
TIMP1
TKT
TLR2
TREM2
TTPA
TUBB2B
TYROBP

Fig. 5. Examples of functional network maps. Neurological disorder: 83 molecules (**a**) Example of functional network maps relating altered subsets of genes and/or proteins describing altered function relevant to the disorder in question. Degeneration of neurons: 13 molecules (**b**) Example of functional network maps relating altered subsets of genes and/or proteins describing altered function relevant to the disorder in question.

7. Check precast gel expiration date due to inconsistent results in expired gel.
8. It is highly recommended that you run the gel at 4°C or on ice to have a better protein migration to prevent overheating the gel.
9. Gels are washed with distilled water to remove any residual SDS left from the running buffer.
10. Destaining solution can be done by varying the methanol concentration. It can be started with 50% (v/v) methanol; 7% (v/v) acetic acid for 30 min, followed with 30% (v/v) methanol; 7% (v/v) acetic acid for 30 min, followed with 10% (v/v) methanol; 7% (v/v) acetic acid.
11. Five percent nonfat dry milk in TBST, as a blocking solution, is used to prevent nonspecific antibody binding and avoid nonspecific bands. It can be achieved at room temperature for 1 h or 24 h at 4°C.
12. Primary antibody solutions can be stored at 4°C, to be used another time; however, 0.05% of sodium azide should be added to prevent bacterial contamination.

Acknowledgments

This work was supported in part by the Donald and Irene Dizney Eminent Scholar Chair, held by Mark S. Gold, M.D. Distinguished Professor, McKnight Brain Institute and also by the Department of Defense (DOD) grant # DAMD17-03-1-0066. Supported by Florida State Center for Nano-Bio Sensors (CNBS) Center of Excellence Grant Project#2.

References

1. NSDUH-Report (2006) Methamphetamine Use, Abuse, and Dependence: 2002, 2003, and 2004, National Survey on Drug Use and Health
2. Sokolov BP, Cadet JL (2006) Methamphetamine causes alterations in the MAP kinase-related pathways in the brains of mice that display increased aggressiveness. Neuropsychopharmacology 31:956–966
3. Pu C, Broening HW, Vorhees CV (1996) Effect of methamphetamine on glutamate-positive neurons in the adult and developing rat somatosensory cortex. Synapse 23:328–334
4. Bowyer JF, Pogge AR, Delongchamp RR, O'Callaghan JP, Patel KM, Vrana KE, Freeman WM (2007) A threshold neurotoxic amphetamine exposure inhibits parietal cortex expression of synaptic plasticity-related genes. Neuroscience 144:66–76
5. Jimenez A, Jorda EG, Verdaguer E, Pubill D, Sureda FX, Canudas AM, Escubedo E, Camarasa J, Camins A, Pallas M (2004) Neurotoxicity of amphetamine derivatives is mediated by caspase pathway activation in rat cerebellar granule cells. Toxicol Appl Pharmacol 196:223–234
6. Warren MW, Zheng W, Kobeissy FH, Cheng Liu M, Hayes RL, Gold MS, Larner SF, Wang KK (2006) Calpain- and caspase-mediated alphaII-spectrin and tau proteolysis in rat cerebrocortical neuronal cultures after ecstasy or methamphetamine exposure. Int J Neuropsychopharmacol 10:1–11

7. Cadet JL, Ordonez SV, Ordonez JV (1997) Methamphetamine induces apoptosis in immortalized neural cells: protection by the proto-oncogene, bcl-2. Synapse 25:176–184
8. Cadet JL, Jayanthi S, Deng X (2003) Speed kills: cellular and molecular bases of methamphetamine-induced nerve terminal degeneration and neuronal apoptosis. FASEB J 17:1775–1788
9. Jayanthi S, Deng X, Noailles PA, Ladenheim B, Cadet JL (2004) Methamphetamine induces neuronal apoptosis via cross-talks between endoplasmic reticulum and mitochondria-dependent death cascades. FASEB J 18:238–251
10. Staszewski RD, Yamamoto BK (2006) Methamphetamine-induced spectrin proteolysis in the rat striatum. J Neurochem 96:1267–1276
11. Warren MW, Kobeissy FH, Liu MC, Hayes RL, Gold MS, Wang KK (2005) Concurrent calpain and caspase-3 mediated proteolysis of alpha II-spectrin and tau in rat brain after methamphetamine exposure: a similar profile to traumatic brain injury. Life Sci 78:301–309
12. Larsen KE, Fon EA, Hastings TG, Edwards RH, Sulzer D (2002) Methamphetamine-induced degeneration of dopaminergic neurons involves autophagy and upregulation of dopamine synthesis. J Neurosci 22:8951–8960
13. Kanthasamy A, Anantharam V, Ali SF, Kanthasamy AG (2006) Methamphetamine induces autophagy and apoptosis in a mesencephalic dopaminergic neuronal culture model: role of cathepsin-D in methamphetamine-induced apoptotic cell death. Ann N Y Acad Sci 1074:234–244
14. Kobeissy F, Sadasivan S, Liu J, Gold MS, Wang KKW (2008) Psychiatric research: psychoproteomics, degradomics and systems biology. Expert Rev Proteomics 5:293–314
15. Nath R, Probert A Jr, McGinnis KM, Wang KKW (1998) Evidence for activation of caspase-3-like protease in excitotoxin- and hypoxia/hypoglycemia-injured neurons. J Neurochem 71:186–195

Chapter 16

Systems Biology and Theranostic Approach to Drug Discovery and Development to Treat Traumatic Brain Injury

Zhiqun Zhang, Stephen F. Larner, Firas Kobeissy, Ronald L. Hayes, and Kevin K.W. Wang

Abstract

Traumatic brain injury is a significant disease affecting 1.4 to 2 million patients every year in the USA. Currently, there are no FDA-approved therapeutic remedies to treat TBI despite the fact that there have been over 200 clinical drug trials, all which have failed. These drugs used the traditional single drug-to-target approach of drug discovery and development. An alternative based upon the advances in genomics, proteomics, bioinformatic tools, and systems biology software has enabled us to use a Systems Biology-based approach to drug discovery and development for TBI. It focuses on disease-relevant converging pathways as potential therapeutic intervention points and is accompanied by downstream biomarkers that allow for the tracking of drug targeting and appears to correlate with disease mitigation. When realized, one is able to envision that a companion diagnostic will be codeveloped along the therapeutic compound. This "theranostic" approach is perfectly positioned to align with the emerging trend toward "personalized medicine".

Key words: Biomarkers, Brain injury, Theranostics, Systems biology

1. Introduction

Acute brain injury including traumatic brain injury (TBI) and stroke are diseases with unmet medical needs. The use of massive explosives, especially improvised explosive devices (IED) has become an increasingly common tactic in modern warfare and in civilian terrorism. This has resulted in a significant increase in blast injuries to the head. In addition, penetrating brain injuries, as well as other blunt impact-mediated TBI, are also significant problems in the military and civilian settings. Until now, there have been over 200 failed clinical drug trials for TBI. This has left us with no FDA-approved drugs. One could argue that a new approach is needed

Qing Yan (ed.), *Systems Biology in Drug Discovery and Development: Methods and Protocols*, Methods in Molecular Biology, vol. 662, DOI 10.1007/978-1-60761-800-3_16, © Springer Science+Business Media, LLC 2010

to mitigate the risks for future pharmaceutical development. In addition, the pharmaceutical industry is facing several significant challenges, including the rising cost of R&D, increased risk due to the increased difficulty in developing innovative drug targets, a high attrition rate due to drug toxicity or the lack of clinical efficacy, and a challenging regulatory environment.

2. Traditional Signal Therapeutic Target Approach Versus Systems Biology Approach to Drug Discovery

The adaptation of biochemical and molecular biological approaches has provided detailed information on possible single therapeutic targets linked to complex diseases such as TBI. This reductionist approach has been useful in identifying single drug targets for a disease process (Fig. 1) (1). However, it often provides neither an understanding of the interaction/interplay between molecules at cellular or systems levels nor an account for the potential compensatory mechanisms (once a drug is given) or the synergistic effects of several parallel pathways at the organism level.

The birth of genomic and proteomic studies has provided the ability to simultaneously discover multiple parts of the disease "puzzle" that were not available previously (1). However, having all the parts still does not mean you have the whole picture. In fact, in a complex disease state such as TBI, all the properties of a given disease state cannot be determined or explained by its

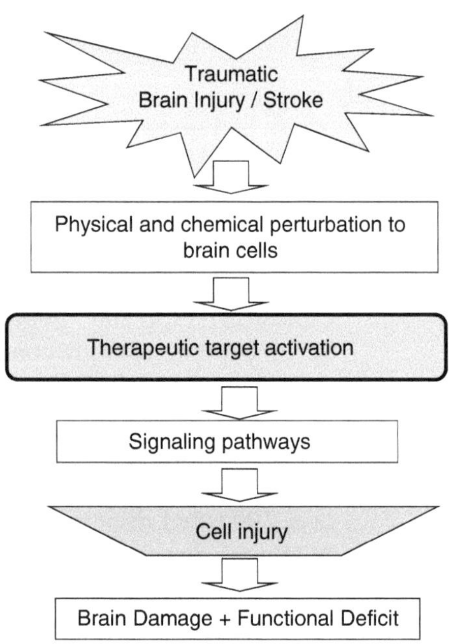

Fig. 1. Systems biology and diagnostic-assisted drug discovery and development for brain injury.

component parts alone. On the other hand, looking at the system as a whole, instead of looking at the parts in isolation, would go a long way toward enabling the determination of how the parts behave. Thus, recent discussions have suggested a "holistic" approach to the study of the disease process, including all available components parts and their interactions and interplay factored in – in other words – the Systems Biology (SB) approach. Systems biology combines experimental, basic science data sets, proteomic and genetic data sets, literature and text mining, integration with computational modeling, bioinformatics, and pathway/interaction mapping methods. When constructed properly, SB databases can provide a context or framework for understanding the biological responses within physiological networks at the organism level, rather than in isolation (2). Furthermore, it allows for hypothesis testing.

3. Systems Biology Integration: from System Components to Dynamic System Models

TBI is the outcome of complex biological systems responses, rather than of one individual gene or protein or pathway. A systems biological approach in the analysis of these networks can give important and practical clues of the underlying processes in such a way that this potential improvement of our knowledge makes therapeutic discovery possible. The system biology platform harnesses data sets that by themselves would be overwhelming, into an organized, interlinked database that can be queried to identify nonredundant brain injury pathways or convert hot spots. These can be exploited to determine their utilities as diagnostic biomarkers and/or therapeutic targets. The ultimate goals of system biology are as follows: first, by exploring the systems, it helps the biologists, pharmaceutical companies, and doctors to better understand the mechanisms underlying the disease, therefore, find suitable targets for treatment. Secondly, the systems approach enables one to be able to predict the functions and behavior of various components of the system (3).

Therefore, component identification is the essential element in systems biology. The judicious combination of the experimental and knowledge-based studies is the most practical solution to system assembly. With the rapid growth of high-throughput technology in genomic and proteomic studies, a tremendous amount of data has been generated. By using data integration and warehousing techniques, a comprehensive database of TBI-related information, including both high-throughput "omic" datasets (genomics, proteomics, metabolomics, etc.) and "targeted" pathway, pharmacology, and molecular imaging studies from the relevant scientific literature can be developed. Some general repository databases with highly relevant data include proteomic and genomic informatics data from the European Bioinformatics Institute (http://www.ebi.ac.uk/), brain proteomics data and informatics from the HUPO Brain Proteome Project (http://www.hbpp.org/40941/76179.html), 3D gene expression maps for mouse and human brains (http://www.brainmaps.org), and systems-level proteome data (http://www.hupo.org/). In addition to genomic and proteomic data, metabolomic data information is another element for the system. It reflects the dynamic metabolic response to environmental, pathophysiological, or genetic perturbations. The Human Metabolome Database (HMDB) (http://www.hmdb.ca/) is a widely used public resource in the field and includes information on metabolomics, biochemistry, clinical chemistry, biomarker discovery, medicine, nutrition, and general education. In addition to its comprehensive literature-derived data about human metabolites and metabolic enzymes, HMDB contains an extensive collection of experimental metabolite concentration data for plasma, urine, CSF (*Cerebrospinal fluid*), and/or other biofluids. It is also cross-linked to other public databases including KEGG (*Kyoto Encyclopedia of Genes and Genomes*), PubChem, PDB (protein data bank), SwissProt, and Genebank. Another web-based metabolite database, METLIN Metabolite Database, (http://metlin.scripps.edu/) has served as an important repository of current and comprehensive mass spectral metabolite data. Finally, the database PRIMe (http://prime.psc.riken.jp/) features include details on standard metabolites provided by means of multidimensional NMR spectroscopy, GC/MS, LC/MS, and CE/MS. It also provides unique tools for metabolomics, transcriptomics, along with an integrated analysis of a range of other "-omics" data.

It is the biological pathways, however, that lie at the heart of biological systems. A wealth of pathway and interaction information is available about these pathways from various web-accessible resources. For example, BIND is a system for electronically managing, finding, and/or displaying bimolecular interactions (4). This comprehensive database contains protein–protein, protein–DNA, and genetic interaction information including HPRD, MDC, HRID, CCSB, DIP, Intact and MINT. Another site with the mission to create targeted knowledge environments for molecular biomedical research so as to enable new insights into complex pathologies is the NIH's National Center for Integrative Biomedical Informatics (NCIBI) called MiMI (*Michigan Molecular Interactions*). It is based at the University of Michigan and is part of the UM Center of Computational Medicine and Biology (CCMB). It provides access to the knowledge and data merged and integrated from numerous protein interactions databases. An attraction of this particular program is that it provides a graphic interface that enables the user to extract information from many other biological sources that allows them to be integrated in a systematic way (5). Reactome (http://www.reactome.org/) is another open-source, curated resource of core pathways and reactions in human biology. It is cross-referenced to the NCBI Entrez Gene, Ensembl and UniProt databases, the UCSC and HapMap Genome Browsers, the KEGG Compound and ChEBI small molecule databases, as well as PubMed, and GO (6).

By identifying and analyzing these established networks, important and practical clues relating to biological pathways relevant to disease processes can be recognized. However, the more important underlying goal in this work is to learn what these networks have to teach us. The network is able to reveal how to predict the emergent properties of the networks, particular disease phenotypes, and to provide important clues that may suggest radically new approaches to therapeutics.

Systems modeling and simulation is now considered fundamental to the future development of effective therapies. Different model representations have been established to serve different purposes. The graphical diagrams of biological processes such as *Pathway Studio*, *Ingenuity pathway* and *Gene Go* give visual presentations of network models by incorporating genome, proteome, and metabolome data. However, different formats which incorporate quantitative data generated from or validated with directed biological studies have emerged and have found further use in system simulation and analysis. The Systems Biology Markup Language (SBML), a computer-readable format for representing models of biological processes, is applicable to simulations of metabolism, cell-signaling, and many other topics (7). SigPath is an information system designed to support quantitative studies of the signaling pathways and networks of the cell (8). CellML is currently the largest repository with more than 180 models representing cell signaling, cell cycle, electrophysiology, endocrine and metabolism, and others (9). Popular simulation tools including Celldesigner, COPASI, E-Cell 3, BIOCHAM and JDesigner and are used to bridge the existing biological knowledge with predictive model behavior. Once a pathway model is in place, experimental data can be implemented to reliably predict perturbations and to generate dynamic system models of molecular interaction networks.

Basically, rather than focusing on individual molecular components, systems biology seeks to understand the system dynamics that govern protein networks, the functional set of proteins that regulate cellular decisions related to TBI. From the perspectives of drug discovery and diagnostics, systems biology gives important and practical clues concerning the pathways relevant to TBI and the effects that drugs might have on them. Therefore, it enhances the entire biomarker and therapeutic drug discovery, development, and commercialization process.

4. Systems Biology Identification of Therapeutic Targets

Once interactive systems biology database for a disease (e.g., TBI) is constructed, one can query the system to identify specific, nonredundant disease-relevant pathways or molecular hot spots that can serve as new points of therapeutic intervention (Table 1). Candidate targets that have been identified can be further studied along with their disease linkages and specificity can be confirmed either experimentally or computationally. Another important feature is that once therapeutic pathways have been identified, potential down-stream biomarkers can be identified that would be indicative of target inhibition, predictive of drug efficacy or adverse side effects due to target inhibition (toxicology/safety screening) (Fig. 2). For example, calpain and caspase proteases have been identified taking part in two destructive proteolytic pathways that not only contribute to key forms of cell death (necrosis and apoptosis), but also in the destruction of important structural components of the axons (alphaII-spectrin breakdown products (SBDPs) and tau), dendrites (MAP2) and myelin (MBP) (Fig. 2). Interestingly, two different forms of SBDPs reflect either neuronal necrosis (SBDP150 and SBDP145 cleaved by calpain) or neuronal apoptosis (SBDP120 cleaved by caspase-3) (10). These SBDPs and other similar neural protein breakdown products can serve as target pathway specific biomarkers. Indeed there is an emerging trend to use "companion diagnostics" in drug development and testing, sometimes termed "theranostics," a word coined by the combination of *thera*peutic and diag*nostics* (11).

Table 1
Contrasting two different approaches to drug discovery and development

Traditional drug target approach	Systems biology + theranostic appraoch
Reductionist approach	Holistic approach
Inhibition of single drug target	Inhibition of one or more key targets at converging point in disease pathway
Nonaccount for organism's compensatory mechanism	Account for organism's compensatory mechanism
High risk in animal model to clinical translation	Animal model to clinical translation guided by biomarker(s)
High risk in clinical study relying solely on efficacy endpoints	Mitigated risk in clinical study as biomarker reduction used as early decision endpoint
Nonconforming to personalized medicine	Facilitating personalized medicine

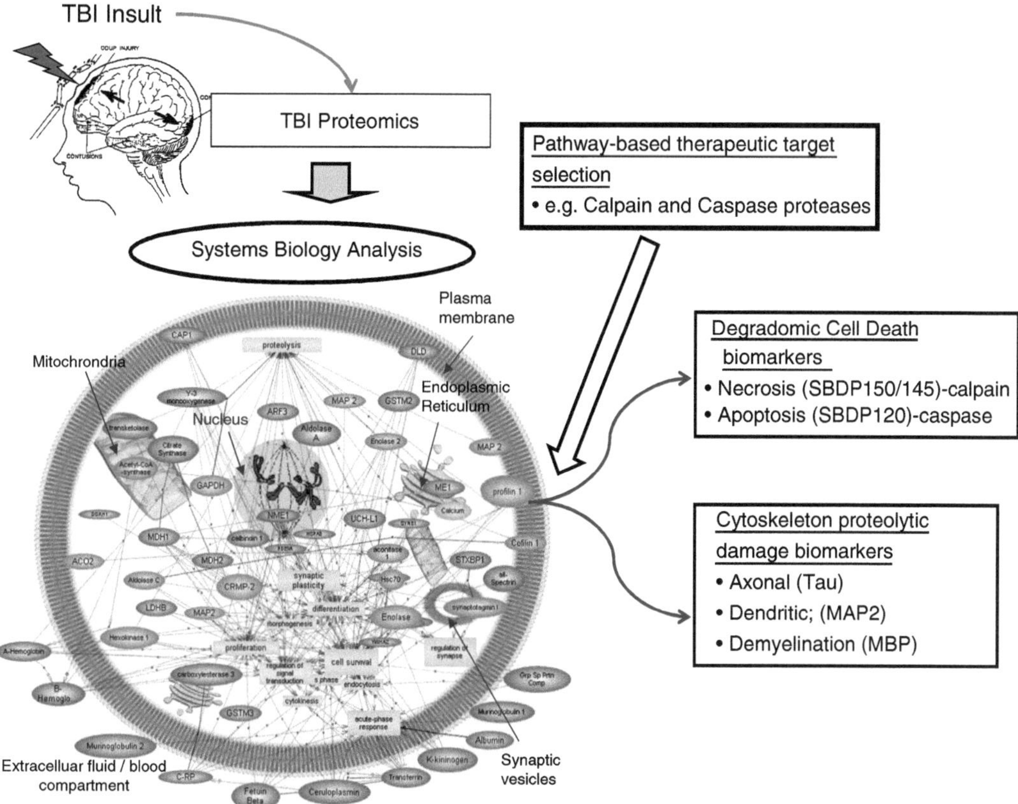

Fig. 2. Traditional single therapeutic target approach to treat brain injury.

5. Theranostic Approach to Drug Development

Theranostics represents the convergence between therapeutics and diagnostics. "Theranostics is the term used to describe the proposed process of diagnostic therapy for individual patients – to test them for possible reaction to taking a new medication and to tailor a treatment for them based on the test results" (11). Theranostics encompasses the possible utilization of a wide range of procedures including predictive medicine, personalized medicine, integrated medicine, pharmaco-diagnostics, and Diagnostics (Dx)/Prescriptions (Rx) partnering. It has been viewed as the parallel use of new therapy and diagnostic tests for a human disease or disorder so as to facilitate drug development and clinical trials and to achieve optimal clinical outcomes in a population of patients.

New therapeutic development traditionally has an extremely high triage rate. More than 90% of drugs that make it to clinical trial fail. Some argue that such extreme loss can be overcome by guiding all new therapeutic development and clinical trials with a disease-relevant diagnostic test.

Major pharmaceutical and biotech companies have been trying for years to tackle acute brain injury (TBI and ischemic stroke) without success with traditional nonbiomarker-accompanied drug development approach. It has been argued the discovery of translational biomarkers (from animal studies to clinical trials) might help to finally deliver the long sought-after clinical trial success (Table 1). Growth in molecular diagnostic technologies a driving factor leading major pharmaceutical companies to incorporate this new category in medicine. Novel, protein biomarker-based diagnostics are enabled by recent technological advances in proteomics (12–14). For example, we were able to identify therapeutic pathway-specific brain injury biomarkers that showed elevated levels in biofluids such as cerebrospinal fluid or blood after acute brain injury (Fig. 3) (15). Figure 16.3 outlines the path it takes to go from the generation of TBI biomarkers from brain tissue to their detection in CSF and blood. During brain injury, neural proteins or their breakdown products generated by proteolytic pathways (such as calpain and caspase) are released into the extracellular environment and eventually reach the CSF in relatively high concentration (16). In due time, the proteins reach the blood stream either via the compromised blood brain barrier (BBB) or via the filtration of the CSF. The clearance and half life of the biomarkers contribute to the final concentration that can be measured in the blood. The CSF volume of an adult human (CSF 125–150 ml) is about 30–40-fold less than the blood volume (4.5–5 L) which explains why the brain biomarker concentration is significantly higher in the CSF samples versus blood samples and makes the former valuable for drug development.

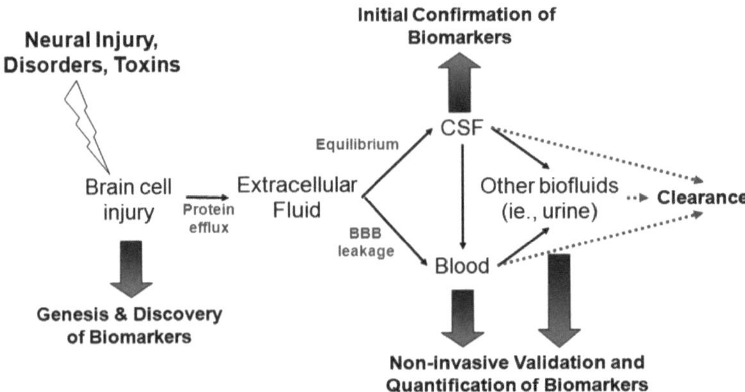

Fig. 3. Systems biology-based therapeutic target identification and target-specific biomarker selection. Systems biology-based selection of candidate brain injury therapeutic targets and target-specific biomarker selection. Calpian and caspase proteases are used here as examples of therapeutic targets with proteolytic brain biomarkers representing nonredundant pathways relevant to the pathobiology of these therapeutic targets and the disease itself.

To date, the most common biomarker detection method relies heavily on antibody-based sandwich ELISA assays (swELISA) – the formation of an immobilized capture-antibody to protein antigen to detection antibody complex. As a biomarker-based diagnostic, swELISA is indeed the method of choice since it provides antigen enrichment that significantly improves signal fidelity. Sensitive and selective sandwich ELISAs for such biomarkers have not only allowed us to quantify the extent of brain injury, but also provided the means to monitor the neuroprotective effects of new drugs in vivo (1).

6. Translation of Preclinical Models to Clinical Applications

The primary challenge facing the translation of animal model testing of drug candidates to clinical study is the uncertainties in a cross-species translation (17). Ideal theranostic biomarkers for TBI that have been identified and validated with preclinical animal models need to be validated in clinical studies as well. They should be tested in terms of their ability to detect injury magnitude as well as drug-based biomarker level reduction. A direct comparison of biomarker occurrence between preclinical models and biomarker data from human clinical studies would allow investigators to gain considerable insight into the validity (or challenges to the validity) of the employed preclinical animal models. For example, we have shown similar profiles of αII-spectrin degradation by calpain in our animal model of TBI (controlled cortical impact) and in severe human TBI (Fig. 4) (18–20).

7. Utilities of Companion Diagnostic in Clinical Trials

Clearly, a diagnostic biomarker relevant in a preclinical animal model would be very useful in the drug screening and development process (Fig. 5). The usefulness of clinically relevant biomarkers can facilitate early decision making such as Early Go/No-Go decisions, which would translate to cost savings. And with the guidance of target-relevant biomarkers, one can readily attain drug dose optimization. Patient population stratification (responders vs. nonresponders) also becomes possible. Moreover, target-based biomarkers have the potential of identifying, minimizing, or avoiding organ toxic effects due to target inhibition. All of these should increasingly translate to clinical trial success rates.

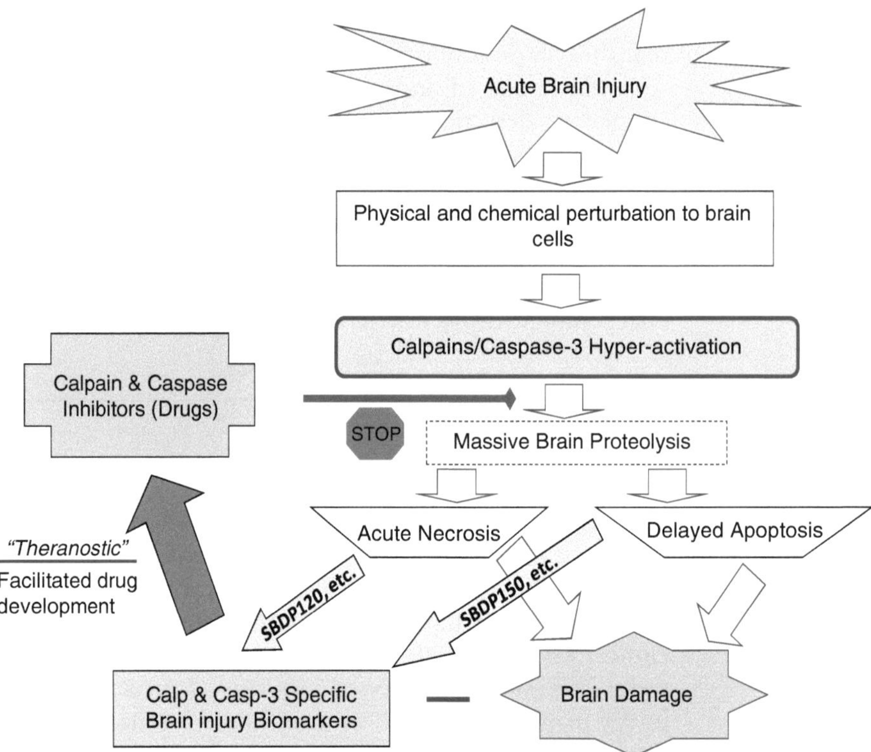

Fig. 4. Genesis and detection of brain injury biomarkers. Neural proteins or their breakdown products are released during the traumatic injury of the brain (genesis and discovery). Through exchange with the extracellular fluid in contact, these biomarkers occur in high concentration in the CSF (initial confirmation of biomarkers). Eventually, the proteins reach the blood stream either via compromised BBB or filtration of CSF (non invasive validation and quantification of biomarkers). Clearance of the biomarkers contributes to the final concentration and half life in the blood (clearance).

8. FDA Approval of a Theranostics or Tx-with Companion Diagnostics

In recognizing the emerging role of the theranostic approach, the FDA has drafted a "Drug-Diagnostic Co-Development Concept Paper" (21) with the goal of setting guidelines for the prospective codevelopment of a drug or biological therapy as a device to test the drug in a scientifically robust and efficient way. It is now possible to rely on existing FDA-approved diagnostic (Dx) biomarkers to support therapeutic (Tx) development (22). Regardless, theranostics entails the utilization of a diagnostic to classify disease subtype. The term theranostics has even been expanded to describe the use of diagnostic testing to diagnose the disease, choose the correct treatment regime and monitor the patient response (Fig. 4). Bottom line is that the theranostic approach requires a different mindset in drug development, as well as new skill-sets. The market for theranostics, while still in its infancy, is expected to grow rapidly.

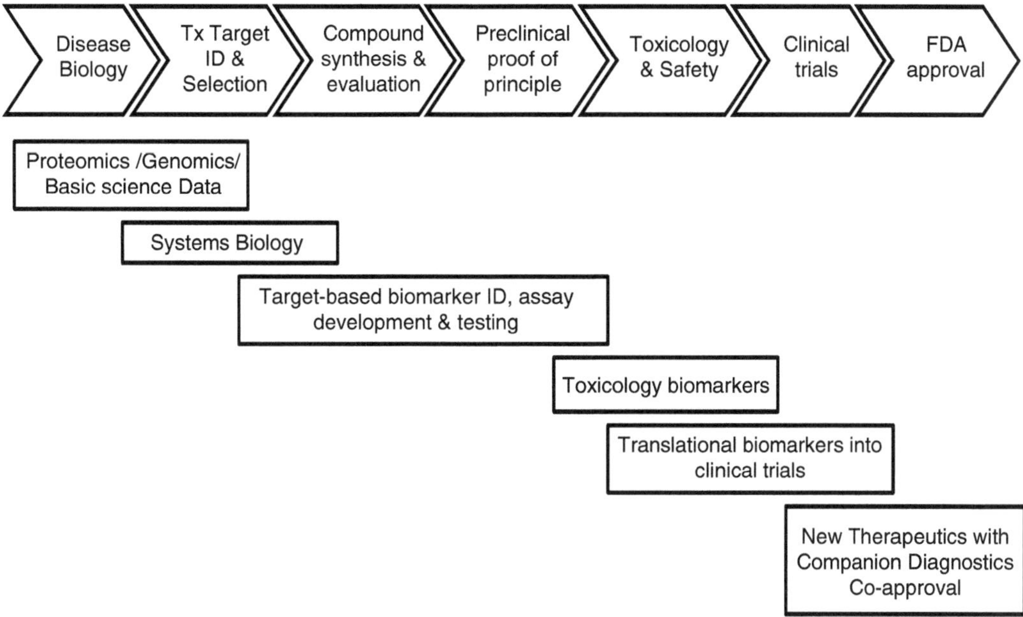

Fig. 5. Overall concept of the theranostic approach to drug development.

9. Postmarket Enhancement

Lastly, another advantage of the theranostic approach is its built-in ability to achieve a postmarketing personalized medicine paradigm (i.e., drug treatment could be tailored according to a patient's diagnostic biomarker profile over time). The availability of disease-specific diagnostic biomarkers can enhance physician education/patient-Rx management or drug selection. Blood cholesterols and triglycerol screening is an excellent example. Codevelopment of theranostic biomarkers can provide Rx-supported-diagnostic patent claims. In turn, the diagnostic industry also benefits from the theranostic approach and this will enhance the utility of diagnostic tests.

10. Summary

In summary, the Systems Biology-based approach to drug discovery and development offers a number of significant advantages over the traditional single drug targeted nondiagnostic accompanied approach (Fig. 5, Table 1). It is a natural lead-in to biomarker-assisted theranostic approach to drug development and clinical trials. It would increase the likelihood and shorten the time until the first FDA-approved therapeutic treatment for TBI occurs. When that breakthrough occurs, the drug-diagnostic codevelopment and the potential of joint postmarket education will only enhance the collaboration between the diagnostic and Rx enterprises.

References

1. Wang KK, Larner SF, Robinson G, Hayes RL (2006) Neuroprotection targets after traumatic brain injury. Curr Opin Neurol 19:514–519
2. Chen SS, Haskins WE, Ottens AK, Hayes RL, Denslow ND, Wang KKW (2007) Bioinformatics for traumatic brain injury: proteimic data mining. In: Pardalos PM, Boginski VL, Vazacopoulos A (eds) Data mining in biomedicine. Springer, New York, pp 1–26
3. Beltrao P, Kiel C, Serrano L (2007) Structures in systems biology. Curr Opin Struct Biol 17:378–384
4. Bader GD, Betel D, Hogue CW (2003) BIND: the Biomolecular Interaction Network Database. Nucleic Acids Res 31:248–250
5. Jayapandian M, Chapman A, Tarcea VG, Yu C, Elkiss A, Ianni A, Liu B, Nandi A, Santos C, Andrews P, Athey B, States D, Jagadish HV (2007) Michigan Molecular Interactions (MiMI): putting the jigsaw puzzle together. Nucleic Acids Res 35:D566–D571
6. Vastrik I, D'Eustachio P, Schmidt E, Gopinath G, Croft D, de Bono B, Gillespie M, Jassal B, Lewis S, Matthews L, Wu G, Birney E, Stein L (2007) Reactome: a knowledge base of biologic pathways and processes. Genome Biol 8:R39
7. Hucka M, Finney A, Sauro HM, Bolouri H, Doyle JC, Kitano H, Arkin AP, Bornstein BJ, Bray D, Cornish-Bowden A, Cuellar AA, Dronov S, Gilles ED, Ginkel M, Gor V, Goryanin II, Hedley WJ, Hodgman TC, Hofmeyr JH, Hunter PJ, Juty NS, Kasberger JL, Kremling A, Kummer U, Le Novère N, Loew LM, Lucio D, Mendes P, Minch E, Mjolsness ED, Nakayama Y, Nelson MR, Nielsen PF, Sakurada T, Schaff JC, Shapiro BE, Shimizu TS, Spence HD, Stelling J, Takahashi K, Tomita M, Wagner J, Wang J, SBML Forum (2003) The systems biology markup language (SBML): a medium for representation and exchange of biochemical network models. Bioinformatics 19:524–531
8. Campagne F, Neves S, Chang CW, Skrabanek L, Ram PT, Iyengar R, Weinstein H (2004) Quantitative information management for the biochemical computation of cellular networks. Sci STKE 2004:pl11
9. Nickerson D, Hunter P (2005) Using CellML in computational models of multiscale physiology. Conf Proc IEEE Eng Med Biol Soc 6:6096–6099
10. Wang KK, Ottens AK, Liu MC, Lewis SB, Meegan C, Oli MW, Tortella FC, Hayes RL (2005) Proteomic identification of biomarkers of traumatic brain injury. Expert Rev Proteomics 2:603–614
11. Warner S (2004) Diagnostics + therapy = theranostics. The Scientist 18:38
12. Vitzthum F, Behrens F, Anderson NL, Shaw JH (2005) Proteomics: from basic research to diagnostic application. A review of requirements & needs. J Proteome Res 4:1086–1097
13. Kobeissy FH, Ottens AK, Zhang Z, Liu MC, Denslow ND, Dave JR, Tortella FC, Hayes RL, Wang KK (2006) Novel differential neuroproteomics analysis of traumatic brain injury in rats. Mol Cell Proteomics 5:1887–1898
14. Liu MC, Akle V, Zheng W, Dave JR, Tortella FC, Hayes RL, Wang KK (2006) Comparing calpain- and caspase-3-mediated degradation patterns in traumatic brain injury by differential proteome analysis. Biochem J 394:715–725
15. Kobeissy FH, Sadasivan S, Liu J, Gold MS, Wang KK (2008) Psychiatric research: psychoproteomics, degradomics and systems biology. Expert Rev Proteomics 5:293–314
16. Romeo MJ, Espina V, Lowenthal M, Espina BH, Petricoin EF III, Liotta LA (2005) CSF proteome: a protein repository for potential biomarker identification. Expert Rev Proteomics 2:57–70
17. Rifai N, Gillette MA, Carr SA (2006) Protein biomarker discovery and validation: the long and uncertain path to clinical utility. Nat Biotechnol 24:971–983
18. Pike BR, Flint J, Dutta S, Johnson E, Wang KK, Hayes RL (2001) Accumulation of non-erythroid alpha II-spectrin and calpain-cleaved alpha II-spectrin breakdown products in cerebrospinal fluid after traumatic brain injury in rats. J Neurochem 78:1297–1306
19. Ringger NC, O'Steen BE, Brabham JG, Silver X, Pineda J, Wang KK, Hayes RL, Papa L (2004) A novel marker for traumatic brain

injury: CSF alphaII-spectrin breakdown product levels. J Neurotrauma 21:1443–1456

20. Pineda JA, Lewis SB, Valadka AB, Papa L, Hannay HJ, Heaton SC, Demery JA, Liu MC, Aikman JM, Akle V, Brophy GM, Tepas JJ, Wang KK, Robertson CS, Hayes RL (2007) Clinical significance of alphaII-spectrin breakdown products in cerebrospinal fluid after severe traumatic brain injury. J Neurotrauma 24:354–366
21. Hinman LM, Huang SM, Hackett J, Koch WH, Love PY, Pennello G, Torres-Cabassa A, Webster C (2006) The drug diagnostic co-development concept paper: commentary from the 3rd FDA-DIA-PWG-PhRMA-BIO Pharmacogenomics Workshop. Pharmacogenomics J 6:375–380
22. Katz R (2004) Biomarkers and surrogate markers: an FDA perspective. NeuroRx 1:189–195

INDEX

Qing Yan (ed.), *Systems Biology in Drug Discovery and Development: Methods and Protocols*, Methods in Molecular Biology, vol. 662, DOI 10.1007/978-1-60761-800-3, © Springer Science+Business Media, LLC 2010

D

E

O

P

T

U

V

W

X

Y

MIX
Papier aus verantwortungsvollen Quellen
Paper from responsible sources
FSC® C105338

If you have any concerns about our products,
you can contact us on
ProductSafety@springernature.com

In case Publisher is established outside the EU,
the EU authorized representative is:
Springer Nature Customer Service Center GmbH
Europaplatz 3, 69115 Heidelberg, Germany

Printed by Libri Plureos GmbH
in Hamburg, Germany